Advances in Thermal and Mechanical Properties of Polymeric Materials

Advances in Thermal and Mechanical Properties of Polymeric Materials

Editors

Agnieszka Kijo-Kleczkowska
Adam Gnatowski

MDPI

Basel • Beijing • Wuhan • Barcelona • Belgrade • Novi Sad • Cluj • Manchester

Editors
Agnieszka Kijo-Kleczkowska
Department of Thermal
Machinery; Faculty of
Mechanical Engineering and
Computer Science
Czestochowa University of
Technology
Czestochowa
Poland

Adam Gnatowski
Department of Technology
and Automation; Faculty of
Mechanical Engineering and
Computer Science
Czestochowa University of
Technology
Czestochowa
Poland

Editorial Office
MDPI
St. Alban-Anlage 66
4052 Basel, Switzerland

This is a reprint of articles from the Special Issue published online in the open access journal *Materials* (ISSN 1996-1944) (available at: https://www.mdpi.com/journal/materials/special_issues/Polymeric_Mat).

For citation purposes, cite each article independently as indicated on the article page online and as indicated below:

Lastname, A.A.; Lastname, B.B. Article Title. *Journal Name* **Year**, *Volume Number*, Page Range.

ISBN 978-3-7258-0288-3 (Hbk)
ISBN 978-3-7258-0287-6 (PDF)
doi.org/10.3390/books978-3-7258-0287-6

Contents

Preface

As Guest Editors of the Special Issue "Advances in Thermal and Mechanical Properties of Polymeric Materials", we congratulate all the authors of the published works and thank the reviewers for their time and invaluable comments and suggestions that raised the rank and substantive value of this reprint. We also thank the Editors-in-Chief of *Materials* for the opportunity to collaborate with the journal, as well as the Section Managing Editor, Ms. Fay Liu, for her kind support and cooperation.

We work as associate professors at the Faculty of Mechanical Engineering and Computer Science at Czestochowa University of Technology. Our scientific interests include polymeric materials and their composites, in terms of thermomechanical tests. Therefore, we decided to open this Special Issue, which turned out to be a huge success. Its goal was to publish innovative papers dealing with the issues of the thermomechanical properties of polymers and composites, structure of polymeric materials, combustion and co-combustion of polymeric materials with fuels and waste, thermal analysis of materials, emission of pollutants during thermal processes, modeling and computer simulation of polymeric material property changes, modeling and computer simulation of exploitation and thermal processes, and recycling of polymeric materials. In this reprint, we emphasized the essence of these issues.

We encourage you to read this reprint.

Agnieszka Kijo-Kleczkowska and Adam Gnatowski
Editors

materials MDPI

Editorial

Special Issue "Advances in Thermal and Mechanical Properties of Polymeric Materials"

Agnieszka Kijo-Kleczkowska * and Adam Gnatowski *

Faculty of Mechanical Engineering and Computer Science, Czestochowa University of Technology, Dabrowskiego 69, 42-201 Czestochowa, Poland
* Correspondence: a.kijo-kleczkowska@pcz.pl (A.K.-K.); adam.gnatowski@pcz.pl (A.G.)

Citation: Kijo-Kleczkowska, A.; Gnatowski, A. Special Issue "Advances in Thermal and Mechanical Properties of Polymeric Materials". *Materials* **2024**, *17*, 79. https://doi.org/10.3390/ma17010079

Received: 1 December 2023
Accepted: 14 December 2023
Published: 23 December 2023

The Special Issue "Advances in Thermal and Mechanical Properties of Polymeric Materials" aimed to publish papers that deal with the thermomechanical and electrical properties of polymers and their composites with other materials. It is important to recognize the applicability of various fillers in polymer composites in order to create new composites and modify existing composites. Advancements in the engineering of polymeric materials, including the search for innovative polymer composites with specific properties, resulted in the expansion of their application, especially in automotive, construction, energy, packaging, and medical industries. Practical application of new polymeric materials requires knowledge of their mechanical, electrical, and thermal properties, as well as the recognition of changes in these properties during the operation and destruction of polymers. The environmental aspect of research is important, including the combustion/co-combustion of polymers, the thermal use of polymer waste with energy recovery, as well as other uses of recycled polymer materials. It is important to conduct model studies on changes in the properties of polymeric materials and the computer simulation of the exploitation and thermal processes of polymers.

Starting 20 December 2021, this Special Issue has attracted the interest of respected researchers from all over the world. The success of this issue is evidenced by the publication of 12 papers that underwent a rigorous review process conducted by experts in the field. Guest Editors congratulate all the authors of the published works and thank the reviewers for their time, very valuable comments and suggestions that raised the rank and substantive value of our Special Issue. The success of this Special Issue would not be possible without the constant contact and kind support of the Section Managing Editor, Ms. Fay Liu, who should be gratefully thanked for her dedication and commitment. Guest Editors also thank the Editors-in-Chief of *Materials* for the opportunity to collaborate on the journal and congratulate them on their stewardship of a globally respected journal.

The diligence, creativity, friendly and dynamic cooperation of all those mentioned above contributed to the success of this Special Issue.

Below, a brief review of the papers published in the collection of this Special Issue is presented.

In paper [1], the authors presented deals with the aspect of the utilization of waste polyethylene (HDPE) as a matrix in composites with filler in the form of cement at 5 and 10%. Comparative thermomechanical (DSC, tensile strength, DMTA), microstructure and flammability results were presented for HDPE samples and their composites with cement. It was found that the addition of cement as a filler to polyethylene made it possible to obtain composites with good thermomechanical properties.

Bragov et al. [2] emphasized that comfort is an important quality criterion, especially for sportswear. The authors offered a statistically significant model of multiple linear regression equations to predict the thermal comfort of knitted fabric. They showed, among other aspects, that yarn's linear density, yarn short fiber hairiness, and mass per unit area of knitted fabric has the greatest impact on heat resistance.

The authors of paper [3] stated that water temperature affects the peak pressure damping of transient flows in viscoelastic pipes. In the paper, the Kelvin–Voigt model with both a quasi-steady friction model and a modified Brunone model was employed. Based on experimental data, the accuracy of simulated peak pressure damping was verified at four different water temperatures. Numerical results indicated that the viscoelastic properties of pipes have a greater impact on peak pressure damping than their frictional properties at 25, 31, and 38.5 °C.

Article [4] discussed the influence of pressure and temperature on material density and thermal conductivity of biomass compacted into briquette samples. The pressure of 200 bar was concluded to be the most economically viable in briquetting technology in the tests conducted. The average thermal conductivity for the compacted biomass was determined at a value of 0.048 ± 0.001 W/(K·m). The methodology described in the paper for thermal conductivity determination was found to be a reliable tool; therefore, it can be proposed for other applications.

In paper [5], a series of novel luminescent hybrid materials was presented, designed based on the top–down procedure. The resulting materials were characterized using Fourier transform infrared spectroscopy and thermal analysis methods, as well as luminescence spectroscopy, and the homogeneity of the resulting materials was investigated by means of optical microscopy. The authors stated, among other aspects, that all obtained materials exhibited good thermal stability in both oxidizing and inert atmospheres.

The aim of paper [6] was the study of the behavior against water of four new plaster-based composite materials using additions of two types of super-absorbent polymers (sodium polyacrylate and potassium polyacrylate) and a lightening material (vermiculite) in their manufacturing process. The transmission of water vapor through the samples and the water absorption capacity of the samples were studied, too. The results of this study showed that composites with the addition of super-absorbent polymers as well as vermiculite significantly improve their water performance compared to traditional materials up to a 7.3% water absorption with a minimal (13%) reduction in mechanical strength compared to current materials with similar additions.

Paper [7] undertook a study of the effect of storage at various relative humidities on the structure, molecular mobility, and mechanical properties of thermoplastic starch (TPS) and its nanocomposite. The reinforcing effect of TPS-montmorillonite (MMT) nanoparticles manifested itself by improving the mechanical properties of nanocomposite samples when compared to TPS samples and led to lower molecular mobility of TPS components. From the point of view of final material application in practice, the authors concluded that the best performance of the studied TPS-based nanocomposite is reached when stored at relative humidities of 55%.

In paper [8], the authors presented the impact of removing contaminants from post-consumer recycled polypropylene on its degradation and properties by implementing a systematic approach for decontamination by washing. Four lots of recycled plastics with different degrees of contamination were evaluated via Fourier transform infrared, melt flow indexer, and differential scanning calorimetry and tested for tensile strength. The authors concluded that the most promising washing procedure for improving the recycled polypropylene quality is cold washing, as it showed very similar results to washing with hot water and hot washing with cleaning agent methods, with the advantages of energy savings and environmental safety.

Paper [9] introduced novel research into specific mechanical properties of composites produced by 3D printing using Continuous Fiber Fabrication. Nylon was used as the composite base material, while carbon constituted the reinforcement element. The authors found, among other things, that the addition of carbon fiber to the base material was proven to increase the volume of defects in the sample. They observed an increase in the tensile strength of up to 12 times higher compared to the sample without reinforcement and stated that to affect the magnitude of work, the fiber reinforcement needed to break the specimens.

In paper [10], the authors presented the combustion results of selected fuels and waste: hard coal, coal sludge, coke waste, sewage sludge, paper waste, biomass waste and polymer waste. An interesting element of paper was the chemical analysis of the XRF of the materials. The authors provided a comparative analysis of pollutant emissions, especially mercury emission, during combustion, and stated that coke waste and sewage sludge are distinguished by their high mercury content. The value of Hg emission during combustion depends on the initial mercury content in the waste. The results of the conducted tests show the adequacy of mercury release during the considered materials' combustion in terms of emissions of the other compounds considered (CO, CO_2, SO_2, H_2S and NO_x). The addition of a 10% considered polymer to coal fuels leads to a reduction in mercury emissions in exhaust gases compared to coal fuels alone. This reduction is 53.72% in the case of hard coal (90%) + polymer waste (10%) and 26.36% in the case of coal sludge (90%) + polymer waste (10%).

The authors of paper [11] stated that multi-material components are difficult or even impossible to recycle. This article assessed the potential of polypropylene injection molding grades to produce parts with varied transparency (haze) in order to obtain, in the future, mono-material products with a property gradient. Several polypropylene grades of different types were identified, and the haze of parts injected under various thermal processing conditions was characterized. The results showed a significant dependence of haze on the tested materials, thermal processing conditions and thicknesses of the samples.

Rodriguez et al. [12] presented the thermal conductivity in different filaments that fused deposition modeling technology uses because there are multiple applications for these additive manufacturing products in the field of thermal insulation. The tests were carried out on a set of seven different materials with 100% fabrication density, from base materials to materials with metal inclusions. This study showed that the parts manufactured with aluminum inclusions have higher thermal conductivity, at 0.40 ± 0.05 W/m·K, compared to other materials with high mechanical and thermal resistance, such as thermoplastic polyurethane, with a conductivity of 0.26 ± 0.05 W/m·K.

Guest Editors congratulate the authors for interesting results presented in their articles and invite all scientists in this field to publish innovative articles in the new Special Issue, "Advances in Thermal and Mechanical Properties of Polymeric Materials (2nd Edition)".

Author Contributions: Conceptualization, A.K.-K. and A.G.; writing—original draft preparation, A.K.-K. and A.G.; writing—review and editing, A.K.-K. and A.G. All authors have read and agreed to the published version of the manuscript.

Funding: This research received no external funding.

Conflicts of Interest: The authors declare no conflict of interest.

References

1. Gnatowski, A.; Kijo-Kleczkowska, A.; Suchecki, Ł.; Palutkiewicz, P.; Krzywański, J. Analysis of Thermomechanical Properties of Polyethylene with Cement Addition. *Materials* **2022**, *15*, 1587. [CrossRef] [PubMed]
2. Salopek Čubrić, I.; Potočić Matković, V.M.; Pavlović, Ž.; Pavko Čuden, A. Material and Structural Functionalization of Knitted Fabrics for Sportswear. *Materials* **2022**, *15*, 3306. [CrossRef] [PubMed]
3. Sun, Q.; Zhang, Z.; Wu, Y.; Xu, Y.; Liang, H. Numerical Analysis of Transient Pressure Damping in Viscoelastic Pipes at Different Water Temperatures. *Materials* **2022**, *15*, 4904. [CrossRef] [PubMed]
4. Kurtyka, M.; Szwaja, M.; Piotrowski, A.; Tora, B.; Szwaja, S. Thermal and Stress Properties of Briquettes from Virginia Mallow Energetic Crops. *Materials* **2022**, *15*, 8458. [CrossRef] [PubMed]
5. Łyszczek, R.; Vlasyuk, D.; Podkościelna, B.; Głuchowska, H.; Piramidowicz, R.; Jusza, A. A Top-Down Approach and Thermal Characterization of Luminescent Hybrid BPA.DA-MMA@Ln_2L_3 Materials Based on Lanthanide(III) 1H-Pyrazole-3,5-Dicarboxylates. *Materials* **2022**, *15*, 8826. [CrossRef] [PubMed]
6. Álvarez, M.; Ferrández, D.; Guijarro-Miragaya, P.; Morón, C. Characterization and under Water Action Behaviour of a New Plaster-Based Lightened Composites for Precast. *Materials* **2023**, *16*, 872. [CrossRef] [PubMed]
7. Šmídová, N.; Peidayesh, H.; Baran, A.; Fričová, O.; Kovaľaková, M.; Králiková, R.; Chodák, I. Influence of Air Humidity Level on the Structure and Mechanical Properties of Thermoplastic Starch-Montmorillonite Nanocomposite during Storage. *Materials* **2023**, *16*, 900. [CrossRef] [PubMed]

8. Prior, L.; Oliveira, M.S.A.; Zhiltsova, T. Assessment of the Impact of Superficial Contamination and Thermo-Oxidative Degradation on the Properties of Post-Consumer Recycled Polypropylene. *Materials* **2023**, *16*, 1198. [CrossRef] [PubMed]
9. Pokorný, P.; Delgado Sobrino, D.R.; Václav, Š.; Petru, J.; Gołębski, R. Research into Specific Mechanical Properties of Composites Produced by 3D-Printing Additive Continuous-Fiber Fabrication Technology. *Materials* **2023**, *16*, 1459. [CrossRef] [PubMed]
10. Kijo-Kleczkowska, A.; Gnatowski, A.; Tora, B.; Kogut, K.; Bytnar, K.; Krzywanski, J.; Makowska, D. Research on Waste Combustion in the Aspect of Mercury Emissions. *Materials* **2023**, *16*, 3213. [CrossRef] [PubMed]
11. Pedroso, L.D.; Pontes, A.J.; Alves, A.; Duarte, F.M.; Carneiro, O.S. From Transparent to Opaque: A Route towards Multifunctional Parts Injected with a Single Material. *Materials* **2023**, *16*, 6219. [CrossRef] [PubMed]
12. Rodriguez, A.; Fuertes, J.P.; Oval, A.; Perez-Artieda, G. Experimental Measurement of the Thermal Conductivity of Fused Deposition Modeling Materials with a DTC-25 Conductivity Meter. *Materials* **2023**, *16*, 7384. [CrossRef] [PubMed]

materials

MDPI

Article

Experimental Measurement of the Thermal Conductivity of Fused Deposition Modeling Materials with a DTC-25 Conductivity Meter

Antonio Rodriguez [1], Juan Pablo Fuertes [1,*], Añaterve Oval [1,2] and Gurutze Perez-Artieda [1]

1 Engineering Department, Campus de Arrosadía S/N, Public University of Navarre (UPNA), 31006 Pamplona, Spain; antonio.rodriguez@unavarra.es (A.R.); aoval@arquimea.com (A.O.); gurutze.perez@unavarra.es (G.P.-A.)

2 Arquimea Research Center, Camino de Las Mantecas S/N, Parque Urbano de Las Mantecas, Edificio NANOtec, 38320 San Cristóbal de La Laguna, Spain

* Correspondence: juanpablo.fuertes@unavarra.es; Tel.: +34-948-168481; Fax: +34-948-169099

Abstract: The expansion and low cost of additive manufacturing technologies have led to a revolution in the development of materials used by these technologies. There are several varieties of materials that can be used in additive manufacturing by fused deposition modeling (FDM). However, some of the properties of these materials are unknown or confusing. This article addresses the need to know the thermal conductivity in different filaments that this FDM technology uses, because there are multiple applications for these additive manufacturing products in the field of thermal insulation. For the study of thermal conductivity, the DTC-25 commercial conductivity measurement bench was used, where the tests were carried out on a set of seven different materials with 100% fabrication density—from base materials such as acrylonitrile butadiene styrene (ABS) or polylactic acid (PLA), to materials with high mechanical and thermal resistance such as thermoplastic polyurethane (TPU), polyether ether ketone (PEEK), and high-performance polyetherimide thermoplastic (ULTEM), to materials with metal inclusions (aluminum 6061) that would later be subjected to thermal after-treatments. This study shows how the parts manufactured with aluminum inclusions have a higher thermal conductivity, at 0.40 ± 0.05 W/m·K, compared to other materials with high mechanical and thermal resistance, such as TPU, with a conductivity of 0.26 ± 0.05 W/m·K.

Keywords: additive manufacturing; fused deposition modeling; filament; thermal conductivity; 3D printing; DTC-25

Citation: Rodriguez, A.; Fuertes, J.P.; Oval, A.; Perez-Artieda, G. Experimental Measurement of the Thermal Conductivity of Fused Deposition Modeling Materials with a DTC-25 Conductivity Meter. *Materials* **2023**, *16*, 7384. https://doi.org/10.3390/ma16237384

Academic Editors: Agnieszka Kijo-Kleczkowska and Adam Gnatowski

Received: 21 October 2023
Revised: 11 November 2023
Accepted: 21 November 2023
Published: 27 November 2023

1. Introduction

The great advance of 3D printing systems using fused deposition modeling (FDM) technology has allowed their use in a variety of areas thanks to their ability to produce parts of great geometric complexity in a relatively fast, easy, and economical way, reducing the number of processing stages. In addition, FDM allows for the use of different thermoplastic materials, such as polylactic acid (PLA), acrylonitrile butadiene styrene (ABS), or polyether ether ketone (PEEK), with minimal changes in the components and configuration.

To allow the use of this technology in certain applications with thermal requirements, a critical property is the thermal conductivity of the model, which may vary depending on the amount of air inside the part [1], the material used [1–4], the direction of heat flow with respect to the printing direction [5–7], and sintering or post-treatment in filaments including metallic particles [8].

The addition of copper particles can significantly increase thermal conductivity [2], especially with contents greater than 20% by mass, going from a conductivity of 0.12 W m^{-1} K^{-1} with a pure PLA sample to 0.35 W m^{-1} K^{-1} with a copper particle content of 40% by mass. In addition, it has been verified that adding 20% Polymethyl methacrylate (PMMA) particles by mass enhances this increase in conductivity, increasing by 43% with respect to the same material without PMMA, and reaching a conductivity of 0.49 W m^{-1} K^{-1}.

Laureto et al. [1], in turn, analyzed the effects of the addition of particles of different metals and compared them with the Lichtenecker equation, which predicts the conductivity of the material as a function of the conductivity of the polymer matrix and the concentration in volume and conductivity of the metal particles. Likewise, it studies the effect of the porosity of the material, concluding that it is necessary to minimize the air content and maximize the amount of metal particles to increase thermal conductivity.

In addition, referring to additive manufacturing using fused deposition technology, articles can be found in which the inclusion of carbon fibers is used to improve thermal conductivity. Ibrahim et al. [4] analyzed the variations in the conductivity of a nylon matrix sample, with different layer configurations and fiber directions, obtaining the maximum conductivity with fibers in the direction of heat flow and reaching a conductivity 11 times higher than that of the base material. Likewise, lattice structures have been proven to be one of the best choices ever since their inception, for various structural and other commercial applications, due to their enhanced mechanical properties, especially in the case of vibration isolation, where band gaps play a vital role [9].

In the case of post-processing like the sintering of materials with copper particles, the thermal conductivity can be strongly increased, as noted by Ebrahimi et al. [8]. Starting with a material whose average volume content is 39.3% and whose conductivity is 1.5 W/K·m, it increases the copper concentration to 42.3% and reduces the porosity by sintering, resulting in a conductivity of 25.5 W/K·m.

Numerous studies have identified considerable challenges in the properties of FDM-printed components that cannot be addressed solely through optimal printing conditions [10]. On the one hand, residual stresses caused by non-uniform heating and cooling cycles during printing have been detected, which are connected to issues with surface roughness, mechanical strength, and dimensional accuracy. These cycles produce uneven temperature gradients, leading to deformations and defects, including shrinkage, warping, and twisting [11]. Another challenge is the presence of interlayer voids, which weaken the parts and contribute to mechanical failures. Partial neck growth voids are particularly significant in causing voids in FDM, resulting from incomplete neck growth between adjacent chords during the sintering process [12]. Problems that arise with solidification prior to full coalescence can stem from inherent features such as incomplete filling and incoherent material flow [13]. Additionally, addressing the staircase effect phenomenon that leads to surface roughness presents another challenge [10]. Various post-processing strategies, including mechanical and chemical techniques, have been studied to optimize the surface finish of FDM-manufactured parts [14–16]. The ultimate challenge lies in achieving automation of these processes for efficient large-scale production [17,18].

In this work, the thermal conductivity of AA6061 and several polymers, including ABS, TPU, PLA, PEEK, and PEI (ULTEM 1010 and ULTEM 9085), was studied. Acrylonitrile butadiene styrene (ABS) was among the earliest materials used in 3D printing due to its chemical and abrasion resistance, along with its remarkable impact resistance. The material's affordability further cements its place as one of the most sought-after 3D printing materials. It is crucial to consider that ABS plastic boasts a low melting point, meaning that it is unsuitable for extreme-temperature applications [18]. TPU filament is resistant to abrasion, oil, chemicals, and wear. TPU-printed parts display similar resistance to low temperatures, making them less prone to becoming brittle and challenging to handle. This flexible filament features excellent adhesion between layers and does not curl or delaminate while being 3D printed. Additionally, it is capable of withstanding significantly higher compressive and tensile forces than other more common materials, such as PLA and ABS [19].

Since 1990, polylactic acid (PLA) has been marketed as a biodegradable polymer. It is extensively researched and widely used due to its versatility in the market. PLA is also known for its easiness to print with compared to other plastics (210 °C). Additionally, it is an FDA-approved material with applications in the food and biomedicine industries [18].

Polyetheretherketone (PEEK) and polyetherimide (PEI) are examples of special engineering plastics with exceptional mechanical properties and high heat resistance [20]. PEEK's biocompatibility and superior mechanical characteristics make it a potential biomaterial that can substitute metallic or ceramic components in fields such as biomedicine or aerospace, thereby opening up interesting prospects [21]. The use of FDM with PEEK is challenging due to its high melting temperature and viscosity. These obstacles must be overcome to fully achieve the potential of PEEK in advanced engineering applications. The PEI material (ULTEM 1010 and ULTEM 9085) exhibits exceptional thermal resistance when compared to other studied materials. It is able to maintain a constant maximum working temperature of around 200 °C while displaying minimal variation in its mechanical properties. These properties make it an ideal candidate for use in creating molds—such as those subjected to high pressure and Autoclave temperature values—including short-cycle injection molding tools and carbon fiber laminating tools. In all these applications, an understanding of thermal conductivity is a crucial factor in producing molds that are faster, simpler, and less expensive than current steel or aluminum molds [22].

In this context, different base polymer materials have been studied to compare their thermal conductivities and understand how 3D printing can influence their thermal capability.

An innovative filament comprising over 65% metal and the remainder PLA has been studied alongside various other base polymer materials. For thermal conductivity measurements, a commercial bench used for testing of metal components was adapted to enable high-confidence measurements on polymeric material.

2. Materials and Methods

2.1. Test Bench

The test bench employed is a commercial system acquired by the Public University of Navarre that consists of a thermal conductivity meter, DTC-25, as shown in Figure 1. The DTC-25 thermal conductivity meter is a test instrument used for determination of the thermal conductivity of solid materials using the guarded heat flow method. Because of its simplicity to handle, small sample size, and short cycle time, it is ideally suited for quality control and the study of materials. Metals, ceramics, polymers, composites, glass, and rubber can all be tested accurately [23]. The main characteristics of the DTC-25 are shown in Table 1.

Table 1. DTC-25 main characteristics [24].

Method	Guarded Heat Flow Meter
Standard Test Method	ASTM E1530
Sample Compatibility	Solids, pastes, liquids, thin films
Temperature Range	Near ambient
Thermal Conductivity Range	0.1 to 20 W/m·K
Thermal Resistance Range	0.0004 to 0.012 m^2 K/W
Accuracy	±3%
Reproducibility	±2%

This instrument is factory-calibrated using specimens of known thermal resistance spanning the whole range of the instrument. Calibration reference sets are also available and an optional chiller to maintain fixed coolant temperature is recommended for optimal performance.

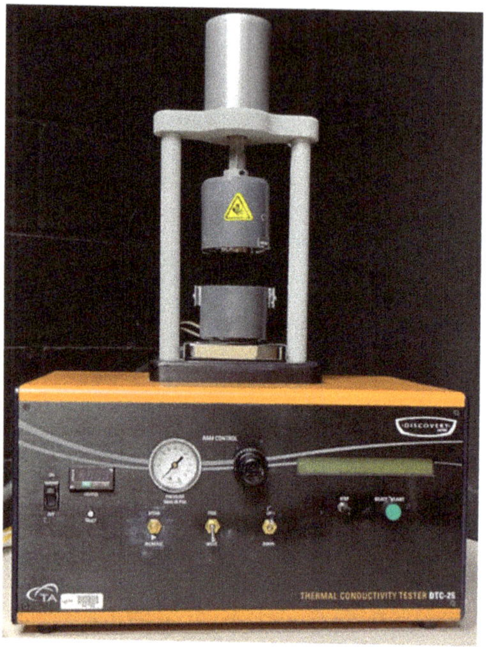

Figure 1. Test bench DTC-25 [24].

The measurement method, according to the ASTM E1530 specification [25], consists of placing the study sample under pressure between two polished metal surfaces. The upper one is heat controlled whilst the lower surface is part of a calibrated heat flux transducer and is connected to a liquid-cooled heat sink. As heat is transferred from the upper surface of the sample to its lower surface, a temperature gradient forms in the stack's axial direction. A reproducible, pneumatic load is applied to the test stack to ensure a positive thermal contact. By measuring the temperature difference across the sample along with output from the heat flux transducer, it becomes possible to determine the thermal conductivity of the sample, given that its thickness is known, as shown in Figure 2.

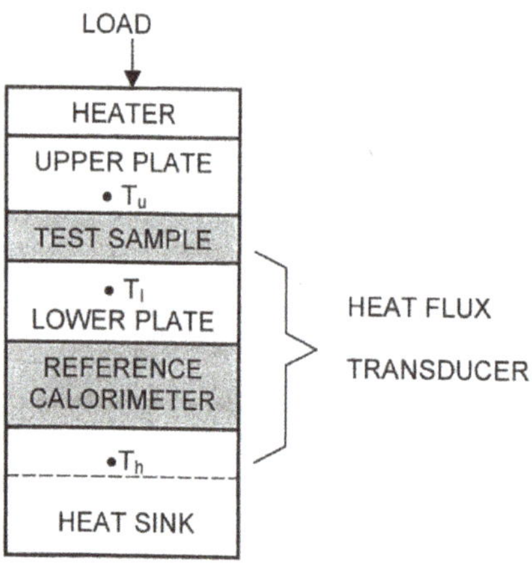

Figure 2. Guarded heat flow test method [25].

To obtain values with less uncertainty in the DTC-25, the hot focus temperature source is set to 55 °C and the cold counterpart is set to 2 °C. The hot source is adjusted by means of an electrical resistance, while the cold temperature source is controlled by means of a thermostatic bath. Moreover, a pressurized nitrogen cylinder supplies the necessary pressure to move the actuator of the DTC-25 and exert pressure on the sample under study. A maximum pressure of 45 psi, 0.3 MPa, is exerted on the samples.

Once the test has been stabilized, the values shown on the bench display are processed by the manufacturer's own software, showing the thermal conductivity value of the sample being tested.

2.2. Materials

The following subsection presents the materials used in this research work. The focus was set on affordable, common materials for 3D printing, including both polymers and metals. Specifically, we utilized aluminum 6061 as a metal and chose PLA, a biopolymer, and ULTEM, a polymer with superior mechanical properties, to be our polymeric materials. We conducted an analysis of thermal conductivity in different materials with varying properties to cover a wide range of applications. Table 2 shows the main characteristics and applications of the materials studied in this work.

Table 2. Material key properties and applications.

Material	Key Properties	Key Applications
Aluminum (≈65%)	Excellent joining characteristics, good acceptance of applied coatings. Combines relatively high strength, good workability, and high resistance to corrosion. Widely available.	Aircraft fittings, camera lens mounts, electrical fittings and connectors, hinge pins, magneto parts, brake and hydraulic pistons, appliance fittings, valve parts.
PEEK	High-performance engineering thermoplastic that belongs to the family of polyketones. Exceptional mechanical, thermal, and chemical properties	Automotive, aerospace, medical and healthcare, electrical and electronic.
TPU	It has the characteristics of both plastic and rubber. Exhibits durability, excellent tensile strength, high elongation at break, and good load-bearing capacity.	Agriculture, automotive, seals and gaskets, textile coatings, sports and leisure, tubes and hoses.
ABS	Impact-resistant engineering thermoplastic made of three monomers: acrylonitrile, butadiene, and styrene. It is the preferred choice for structural applications due to its physical properties: high rigidity, resistance to impact, abrasion, and strain.	Automotive parts, electrical and electronic, household products, pipe fittings, sports and leisure.
PLA	Rapidly growing concerns related to environmental health and safety, limiting dependence on petrochemical raw materials, and reducing carbon footprint.	Food contact packaging, healthcare and medical industry, high end structural applications, fiber and textile industries.
ULTEM	Combination of outstanding thermal (high temperature resistance, thermo-oxidative stability), mechanical (high strength-to-weight ratio), and electrical properties. ULTEM polyetherimide has found its place in high-performance applications.	Automotive, aerospace, electrical and electronic, metal replacement for industrial applications, disposable and re-usable medical applications.

2.3. Samples

As described in the previous subsection, seven different materials were studied in order to be classified according to the thermal conductivity, which was obtained from tests carried out in the DTC-25.

These are seven materials with distinct mechanical properties. To identify and exclude utility based on their thermal properties, it is necessary to determine their thermal conductivity. One of the limitations of this testing equipment is the pressure required to carry out the tests; on the DTC-25 test bench, it is 0.3 MPa. Therefore, we verified that all materials had higher maximum pressure limits; in the case of PEEK and PLA, the maximum pressure limit is 0.45 MPa, and it is up to 67 MPa in the case of thermoplastic polyurethane (TPU). It was also necessary to control the temperature of the materials. As shown in Table 3, the samples made with Filamet™ Aluminum 6061 and with a base material, PLA, had their maximum temperature limited to 55 °C. These specimens could not exceed the maximum temperature of the PLA, as the sample would lose its cohesion/integrity. In all the tests, we worked with temperatures below these limits.

Table 3. Material properties.

Material	Max Test Temp. (°C)
Aluminum (≈65%) [26]	55
PEEK [27]	140
TPU [28]	164
ABS [29]	81
PLA3080 [30]	55
ULTEM1010 [31]	213
ULTEM9085 [32]	153

Geometry

Circular samples, shown in Figure 3, were manufactured using FDM printing with a standard diameter of 50 mm and two different thicknesses, a first thickness of 5 mm and a second of 10 mm. Both diameter and thickness were accurately measured, as can be seen in Table 4.

Table 4. Study samples' geometry.

Sample	Sample	Size	
		Thickness (mm)	Diameter (mm)
Aluminum (≈65%)	1	49.75	4.60
	2	50.00	9.55
PEEK	1	50.35	5.25
	2	49.90	10.00
TPU	1	50.35	5.05
	2	50.10	10.00
ABS	1	50.00	5.20
	2	50.00	10.10
PLA3080	1	50.40	5.00
	2	50.35	10.00
ULTEM1010	1	50.00	5.30
	2	50.00	10.30
ULTEM9085	1	50.00	5.10
	2	49.80	10.20

Different densities and fill patterns were produced using the different materials to achieve suitable mechanical [33] and thermal capabilities. However, finally, a density of 100% and rectilinear fill pattern with a 45° angle offset were selected as the lowest porosity and the best mechanical behavior. The remaining main parameters for printing were 0.8 nozzle diameter, 3.000 mm/min printing speed. 0.3 mm layer height, 1.75 mm filament diameter, and between 210 and 270 °C printing temperature depending on the material. In addition, the surface finish of the samples was established as between 1 and 10 microns depending on the material. This was necessary because together with the use of thermal contact paste, it allows the samples to be within the uncertainty of the bench, minimizing the influence of the thermal contact resistance between sample and equipment [34].

Figure 3. Study samples. FDM. From left to right: (**a**) aluminum, (**b**) PEEK, (**c**) TPU, (**d**) ULTEM 1010, (**e**) ULTEM 9085, (**f**) PLA 3080, (**g**) ABS.

3. Results and Discussion

3.1. Hardness Tests

An approximation of the mechanical properties of the materials was arrived at through hardness tests [35]. A universal hardness test was used to compare all the samples. A maximum load of 1 N was applied. At least ten upload and download curves were studied for each material considering plastic and elastic deformation in the samples. Figure 4 shows an average of universal hardness at the maximum load.

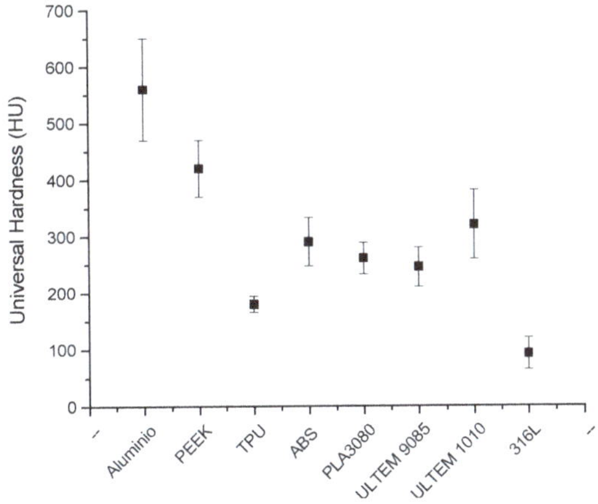

Figure 4. Hardness measurements of studied samples.

The mechanical properties evaluated through universal hardness measurements showed values between 200 and 300 HU (Universal Hardness). Only PEEK was around 400 HU, and aluminum showed values up to 550 HU.

3.2. Data Acquisition

To obtain accurate results, the system must be calibrated (Figure 5). To do this, the five standards provided by the manufacturer are tested, and the calibration curve is generated using the system software with the potential measurements taken for each standard after its study once stabilized. The temperature values measured with the respective probes are shown in voltage values. These values are the ones that will be entered into the system software for the calculation of the thermal conductivity of the sample.

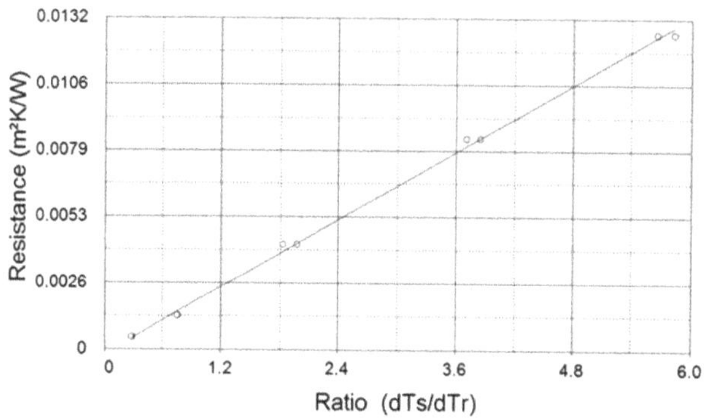

Figure 5. Calibration curve. DTC-25. Two measurements per standard.

In order to obtain more points on the calibration curve, two studies are carried out on each pattern. This way, different calibration measurements are obtained because, despite being the same piece under testing, the results may slightly differ, affected by the system itself, by the thermal paste used for the contact, or the environment (measurement deviation).

3.2.1. Stability Criterion

The stabilization in the DTC-25 conductivity meter is carried out according to experience; it is related to the thickness of the study specimen, because for greater thicknesses, since they are materials of high thermal resistance, a longer study time will be required (Table 5).

Table 5. Stability criterion. DTC-25.

Thickness (mm)	Stabilization Time (h)
5	3
10	3
15	4
20	5
25	6

3.2.2. Uncertainty

The measurements were made within the limits set by the manufacturer, so that they lay within the reproducibility and uncertainty values provided. For this purpose, work was carried out within the ranges of thickness, size, and minimum conductivity of the sample indicated on the technical data sheet. DTC-25: 50 mm diameter, sample thickness greater than 0.1 mm, and theoretical conductivity between 0.1 and 20 W/m K.

To obtain an uncertainty value, we considered the randomness in the thermal conductivity measurements, performing N = 6 tests for each material studied. The mean value $\overline{k}$ of these samples was obtained by using Equation (1).

$$\overline{k} = \frac{1}{N}\sum_{i=1}^{N} k_i \tag{1}$$

The standard uncertainty u(k) of the thermal conductivity k shown in Equation (2) was calculated in a similar way to the one shown in [35].

$$u(k) = \sqrt{\frac{\sum_{i=1}^{N}\left(k_i - \overline{k}\right)^2}{N(N-1)}} \tag{2}$$

3.3. Thermal Conductivity Tests

The thermal conductivity values obtained for each thickness and material are shown in Figure 6, based on the results calculated with the DTC-25 bench software after entering the voltage readings (V) of each test.

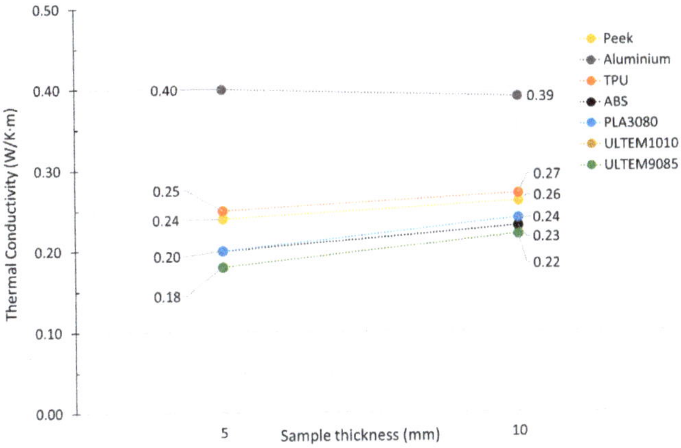

Figure 6. Average thermal conductivity.

In Table 6, we can see the results obtained for the average thermal conductivity with its measurement uncertainty. The material with the highest value is the Filamet™ of Aluminum 6061 of The Virtual Foundry (TVF), an innovative filament composed of more than 65% metal and the rest PLA, with a conductivity of 0.40 W/m·K. It is followed by the TPU, a thermoplastic polyurethane that combines hardness, elasticity, and mechanical resistance, so it maintains all the advantages of this elastomer, therefore being able to manufacture completely rigid parts, with a conductivity of 0.26 W/m·K.

Table 6. Thermal conductivity of each material.

Material	k (W/K·m)
Aluminum	0.40 ± 0.05
PEEK	0.25 ± 0.05
TPU	0.26 ± 0.05
ABS	0.22 ± 0.06
PLA3080	0.22 ± 0.06
ULTEM1010	0.20 ± 0.06
ULTEM9085	0.20 ± 0.06

This aluminum alloy, whose conductivity stands out among the rest of the materials used in FDM, arouses interest, above all, given what may happen with its conductivity after sintering (elimination of PLA from the alloy). The rest of the materials are within the thermal conductivity standard of thermoplastics, as shown in Table 6.

3.4. Discussion of Results

By comparing the results obtained in this study with the research conducted by M. C. Vu et al. [2], which examined the thermal conductivity of polylactic acid (PLA) composites with different copper (Cu) concentrations, we can highlight several findings regarding the impact of metal inclusions on thermal performance. The study by Vu et al. [2] demonstrated that for materials containing 100% PLA, the thermal conductivity was 0.12 W/m·K. As the mass percentage of Cu surpassed 20%, a notable enhancement in thermal conductivity to 0.35 W/m·K was achieved. This enhancement is ascribed to copper's intrinsically high thermal conductivity. The outcomes of this research align with those of Vu et al. [2], wherein the same effect was observed when materials comprising more than 65% Al were examined in combination with PLA. A noteworthy advancement in thermal conductivity was achieved, exhibiting a value of 0.40 W/m·K. Notably, aluminum is acknowledged to have a slightly inferior thermal conductivity in comparison to copper despite being an excellent heat conductor. Hence, the observation that metallic composites with a considerable mass fraction of aluminum exceeded those with higher copper proportion accentuates the significance of optimizing the metallic inclusions' mass percentages.

It is important to consider that thermal conductivity is influenced by factors beyond the presence of metallic inclusions, including the material's porosity. This point was emphasized in the study by Laureto et al. [1], which highlights the importance of reducing air content and increasing metal particle content to enhance thermal conductivity. The significance of design and precise material control during the FDM process is further emphasized to prevent porosity from adversely impacting the final product's heat transfer capabilities.

The comparison of Filamet™ Aluminum 6061 filament with other materials, including TPU, ABS, PEEK, PLA3080, ULTEM1010, and ULTEM9085, provides valuable insights into these materials' thermal properties. Filamet™ Aluminum 6061 demonstrates remarkable thermal conductivity of 0.40 ± 0.05 W/m.K. This innovative filament, which comprises more than 65% metal and PLA, exhibits excellent potential for achieving superior thermal performance using high-metal materials, as previously mentioned.

Conversely, TPU, a thermoplastic polyurethane, exhibits an acceptable thermal conductivity of 0.26 ± 0.05 W/m·K. This thermoplastic material exemplifies its aptness in situations calling for a combination of mechanical strength and thermal properties due to its elasticity and mechanical robustness.

The thermal conductivities of the remaining materials, ABS, PEEK, PLA3080, ULTEM1010, and ULTEM9085, were all within the expected range for thermoplastics. These materials can now be used as a benchmark for the thermal behavior of FDM materials.

3.5. Results' Limitations

The geometry chosen for this experiment relates to the limitations of the measuring equipment used. To compare the thermal conductivities of the various materials, the same conditions were maintained for all the samples in terms of geometry and test conditions. Given that changes in geometry can lead to very different outcomes. Future studies with other measuring equipment will be considered to further investigate this matter. To reduce porosity, a factor that affects the thermal conductivity of materials obtained through additive manufacturing, a rectilinear filling pattern with an angular displacement of 45° was selected. This same system was applied to all samples. Sintering or compaction could potentially enhance the porosity inherent to these processes, and a potential area for future research could be the application of post-additive manufacturing treatment techniques. Metal filaments or particles enhance thermal conductivity but also pose challenges for fabrication. While future research could investigate their potential for analysis, this study focuses on measuring commercially available and easy-to-fabricate materials.

This study's findings are of great importance to research on additive manufacturing, particularly to that involved with FDM-based 3D printing. This research significantly contributes to our understanding of the impact of various factors such as metal inclusions and material choices on the thermal conductivity of FDM printed components.

This study could provide a valuable foundation for further research. Specifically, it would be worthwhile to investigate the impact of sintering, as explored by Ebrahimi et al. [8], on materials that contain metallic particles, like Filamet™ 6061 aluminum. This could offer insights into how post-processing techniques may enhance the thermal properties of FDM printed materials.

3.6. Potential Applications

Furthermore, this research indicates that high-thermal-conductivity composites hold promising potential for improving thermal insulation in construction, aerospace, and other industries. This can enhance energy efficiency, ultimately reducing both heating and cooling costs. In addition, these materials can be utilized in electronic devices, where temperature control is vital, to enhance heat dissipation, thus pointing towards another potential research direction. Custom 3D printing can benefit from materials with high thermal conductivity, allowing us to produce personalized components with specific thermal requirements. Future research could expand upon this study to include a wider array of polymers and metals for a more comprehensive understanding. Moreover, further research could explore the effects of post-treatments, like sintering and other methods, on improving the thermal conductivity of these materials, expanding knowledge in this field of study.

4. Conclusions

The rapid expansion and cost-effectiveness of additive manufacturing technologies, particularly fused deposition modeling (FDM), have significantly altered materials' development for various applications, including the automotive industry. FDM enables the production of intricate parts with fewer processing steps, providing a versatile and economical solution for numerous industries. The thermal conductivity of materials used in additive manufacturing is critical for applications with thermal requirements. Several factors have been documented to influence thermal conductivity, including air presence within parts, material type, heat flow direction, and post-treatments that impact metal particles. Controlling thermal conductivity is crucial in optimizing the performance of these materials.

In this study, the thermal conductivity of seven materials with varying mechanical and thermal properties was analyzed. Highly precise measurements were obtained by employing a commercially available DTC-25 thermal conductivity measurement bench, adapted particularly for polymeric materials. The system's stability, uncertainty, and calibration procedures were meticulously considered, ensuring the reliability and reproducibility of the results.

The results obtained allow us to conclude that the inclusion of metallic particles in thermoplastic materials leads to a significant improvement in their thermal conductivity. Our research findings demonstrate that materials like Filamet™ 6061 Aluminum, which contain over 65% metal and the rest PLA, exhibit high thermal conductivity, boasting a value of 0.40 ± 0.05 W/m·K. Compared to other materials, such as TPU, which is renowned for its mechanical and thermal resistance, with a thermal conductivity of 0.26 ± 0.05 W/m·K, Filamet™ 6061 Aluminum proves to be a more thermally efficient option. Consequently, the outcomes of this investigation highlight the potential benefits of integrating metallic inclusions to enhance thermal performance. It is important to note that the type and concentration of metallic particles have a substantial impact on thermal conductivity.

This research work aligns with previous studies, highlighting that elevated concentrations of metallic particles result in increased thermal conductivity. For instance, it has been demonstrated that copper particles with a concentration exceeding 20% result in a significant enhancement of thermal conductivity. It has been observed that reducing the air content within the material also enhances thermal conductivity. To minimize porosity, a sintering processes or post-processing can be executed, similar to the copper particle-containing materials studied.

Additionally, the inclusion of carbon fibers can further enhance thermal conductivity, although it is crucial to consider the fibers' orientation and configuration. It has been determined that orientating fibers in the direction of heat flow leads to a noteworthy enhancement in thermal performance.

Finally, this research demonstrates the significance of thermal conductivity in additive manufacturing. It provides valuable insights into the impact of metal inclusions, particle types, and concentrations on thermal performance.

Author Contributions: Methodology, A.R. and G.P.-A.; validation, J.P.F.; formal analysis, G.P.-A.; data curation, A.O.; writing—original draft preparation, A.O.; writing—review and editing, J.P.F.; supervision, A.R.; project administration, A.R. All authors have read and agreed to the published version of the manuscript.

Funding: This research received external funding from the Department of Economic and Business Development of Navarre Government through R&D projects in 2020 under the research project acronym FORM3D (file codes: 0011-1365-2020-000084 and 0011-1365-2020-000127).

Data Availability Statement: Data are contained within the article.

Acknowledgments: The authors are indebted to the Department of Economic and Business Development of Navarre Government for the economic support of this work.

Conflicts of Interest: The authors declare no conflict of interest.

References

1. Laureto, J.; Tomasi, J.; King, J.A.; Pearce, J.M. Thermal properties of 3-D printed polylactic acid-metal composites. *Prog. Addit. Manuf.* **2017**, *2*, 57–71. [CrossRef]
2. Vu, M.C.; Jeong, T.H.; Kim, J.B.; Choi, W.K.; Kim, D.H.; Kim, S.R. 3D printing of copper particles and poly (methyl methacrylate) beads containing poly (lactic acid) composites for enhancing thermomechanical properties. *J. Appl. Polym. Sci.* **2021**, *138*, 49776. [CrossRef]
3. Zmeskal, O.; Marackova, L.; Lapcikova, T.; Mencik, P.; Prikryl, R. Thermal properties of samples prepared from polylactic acid by 3D printing. In *AIP Conference Proceedings*; AIP Publishing: College Park, ML, USA, 2020; Volume 2305.
4. Ibrahim, Y.; Elkholy, A.; Schofield, J.S.; Melenka, G.W.; Kempers, R. Effective thermal conductivity of 3D-printed continuous fiber polymer composites. *Adv. Manuf. Polym. Compos. Sci.* **2020**, *6*, 17–28. [CrossRef]
5. Shemelya, C.; De La Rosa, A.; Torrado, A.R.; Yu, K.; Domanowski, J.; Bonacuse, P.J.; Martin, R.E.; Juhasz, M.; Hurwitz, F.; Wicker, R.B.; et al. Anisotropy of thermal conductivity in 3D printed polymer matrix composites for space based cube satellites. *Addit. Manuf.* **2017**, *16*, 186–196. [CrossRef]
6. Smith, M.; Kim, S.; Lambert, A.; Walde, M.; Bougher, T.; Hasser, A.A.; Lindahl, J.; Mungale, K.; Kunc, V. Maximizing the Performance of a 3D Printed Heat Sink by Accounting for Anisotropic Thermal Conductivity During Filament Deposition. In Proceedings of the 18th IEEE Intersociety Conference on Thermal and Thermomechanical Phenomena in Electronic Systems (ITherm), Las Vegas, NV, USA, 28–31 May 2019.
7. Prajapati, H.; Ravoori, D.; Woods, R.L.; Jain, A. Measurement of anisotropic thermal conductivity and inter-layer thermal contact resistance in polymer fused deposition modeling (FDM). *Addit. Manuf.* **2018**, *21*, 84–90. [CrossRef]
8. Ebrahimi, N.D.; Ju, Y.S. Thermal conductivity of sintered copper samples prepared using 3D printing-compatible polymer composite filaments. *Addit. Manuf.* **2018**, *24*, 479–485.
9. Singh, G.; Singh, J.; Singh, N.; Farina, I.; Colangelo, F.; Pandey, P.M. Effect of unit cell shape and structure volume fraction on the mechanical and vibration properties of 3D printed lattice structures. *J. Thermoplast. Compos. Mater.* **2023**. [CrossRef]
10. NurFarrahain, N.A.; Yew, H.W.; Nik, N.; Ghazali, N. A systematic review of fused deposition modeling process parameters. *Soft Sci.* **2022**, *2*. [CrossRef]
11. Gurrala, P.K.; Regalla, S.P. Part strength evolution with bonding between filaments in fused deposition modelling: This paper studies how coalescence of filaments contributes to the strength of final FDM part. *Virtual Phys. Prototyp.* **2014**, *9*, 141–149. [CrossRef]
12. Huang, B.; Singamneni, S. Raster angle mechanics in fused deposition modelling. *J. Compos. Mater.* **2015**, *49*, 363–383. [CrossRef]
13. Kumbhar, N.N.; Mulay, A.V. Post processing methods used to improve surface finish of products which are manufactured by additive manufacturing technologies: A review. *J. Inst. Eng. India Ser. C* **2018**, *99*, 481–487. [CrossRef]
14. Vanaei, H.R.; Khelladi, S.; Tcharkhtchi, A. Roadmap: Numerical-Experimental Investigation and Optimization of 3D-Printed Parts Using Response Surface Methodology. *Materials* **2022**, *15*, 7193. [CrossRef] [PubMed]
15. Chohan, J.S.; Singh, R. Pre and post processing techniques to improve surface characteristics of FDM parts: A state of art review and future applications. *Rapid Prototyp. J.* **2017**, *23*, 495–513. [CrossRef]
16. Hashmi, A.W.; Mali, H.S.; Meena, A. The surface quality improvement methods for FDM printed parts: A review. *Fused Depos. Model. Based 3D Print.* **2021**, 167–194.
17. Lalegani, D.M.; Mohd Ariffin, M.K.A.; Baharuddin, B.T.H.T. Experimental study of drilling 3D printed polylactic acid (PLA) in FDM Process. *Fused Depos. Model. Based 3D Print.* **2021**, 85–106.
18. Žigon, J.; Kariž, M.; Pavlič, M. Surface finishing of 3D-printed polymers with selected coatings. *Polymers* **2020**, *12*, 2797. [CrossRef]
19. Chamil, A.; Pimpisut, S.A.; Anura, F. Optimization of fused deposition modeling parameters for improved PLA and ABS 3D printed structures. *Int. J. Lightweight Mater. Manuf.* **2020**, *3*, 284–297.
20. Verstraete, G.; Samaro, A.; Grymonpré, W.; Vanhoorne, V.; Van Snick, B.; Boone, M.N.; Hellemans, T.; Van Hoorebeke, L.; Remon, J.P.; Vervaet, C. 3D printing of high drug loaded dosage forms using thermoplastic polyurethanes. *Int. J. Pharm.* **2018**, *536*, 318–325. [CrossRef]

21. Peng, W.; Bin, Z.; Hongchuan, X.; Shouling, D.; Chuanzhen, H. Effects of printing parameters of fused deposition modeling on mechanical properties, surface quality, and microstructure of PEEK. *J. Mater. Process. Technol.* **2019**, *271*, 62–74.
22. Shouling, D.; Bin, Z.; Peng, W.; Hongjian, D. Effects of nozzle temperature and building orientation on mechanical properties and microstructure of PEEK and PEI printed by 3D-FDM. *Polym. Test.* **2019**, *78*, 105948.
23. Heat, G.; Test, F. *Dtc-300 Thermal Conductivity Meter*; TA Instruments: New Castle, DE, USA, 2011.
24. *ASTM E1530-19*; Standard Test Method for Evaluating the Resistance to Thermal Transmission by the Guarded Heat Flow Meter Technique. ASTM International: West Conshohocken, PA, USA, 2019.
25. Filament2Print: Filament Aluminio 6061. Available online: https://filament2print.com/es/metalicos/1030-filamet-aluminio-6061.html (accessed on 20 October 2023).
26. 3DXTECH Advanced Materialsxtech.com: ThermaXTM PEEK 3D Printing Filament Physical. Available online: https://www.3dxtech.com/product/thermax-peek/?attribute_pa_diameter=1-75mm&attribute_pa_weight=250g&attribute_pa_color=natural (accessed on 20 October 2023).
27. Smart Materials. 3D Innovatefil: TPU Hardness. Available online: https://www.smartmaterials3d.com/innovatefil-tpu-hardness#/2-tamano-m_750g/11-color-true_black/26-diametro-175_mm (accessed on 20 October 2023).
28. Fillamentum Manufacturing Czech s.r.o.: ABS Extrafill. Available online: https://fillamentum.com/collections/abs-extrafill-filament/ (accessed on 20 October 2023).
29. Fillamentum Manufacturing Czech s.r.o.: PLA Extrafill. Available online: https://fillamentum.com/collections/pla-extrafill-filament/ (accessed on 20 October 2023).
30. Stratasys: ULTEM 1010 Resin. Available online: https://www.stratasys.com/en/stratasysdirect/materials/thermoplastics/ultem/ultem-1010-resin/ (accessed on 20 October 2023).
31. Stratasys: UltemTM 9085. Available online: https://www.stratasys.com/en/materials/materials-catalog/fdm-materials/ultem-9085/ (accessed on 20 October 2023).
32. Böğrekci, İ.; Demircioğlu, P.; Sucuoğlu, H.S.; Turhanlar, O. The Effect of the Infill Type and Density on Hardness of 3d Printed Parts. *Int. J. 3D Print. Technol. Digit. Ind.* **2019**, *3*, 212–219.
33. Chaidas, D.; Kitsakis, K.; Kechagias, J.; Maropoulos, S. The Impact of Temperature Changing on Surface Roughness of FFF Process. In *IOP Conference Series: Materials Science and Engineering, Proceedings of the 20th Innovative Manufacturing Engineering and Energy Conference (IManEE 2016), Kozani, Greece, 23–25 September 2016*; IOP Publishing: Bristol, UK, 2016; Volume 161.
34. Hanon, M.M.; Dobos, J.; Zsidai, L. The influence of 3D printing process parameters on the mechanical performance of PLA polymer and its correlation with hardness. *Procedia Manuf.* **2020**, *54*, 244–249. [CrossRef]
35. Si, H.; Yang, P. Uncertainty analysis of linear vertical bending moment in model tests and numerical prediction. *Mech. Syst. Sygnal Process.* **2022**, *178*, 109331. [CrossRef]

materials

MDPI

Article

From Transparent to Opaque: A Route towards Multifunctional Parts Injected with a Single Material

Luís D. Pedroso [1], António J. Pontes [1], António Alves [2], Fernando M. Duarte [1] and Olga S. Carneiro [1,*]

[1] Department of Polymer Engineering, Institute for Polymers and Composites (IPC), University of Minho, 4800-058 Guimarães, Portugal; luis_pedroso_09@hotmail.com (L.D.P.); pontes@dep.uminho.pt (A.J.P.); fduarte@dep.uminho.pt (F.M.D.)
[2] Cabopol—Polymer Compounds, 2480-049 Leiria, Portugal; antonio.alves@cabopol.com
* Correspondence: olgasc@dep.uminho.pt

Abstract: The technological, social and economic development observed in recent decades brought an exponential increase in consumption and inherent new challenges. Recycling is one of the best solutions to minimize the environmental impact of raw materials. However, multi-material components are difficult or even impossible to recycle. The present work focuses on the reduction in the number of different materials used in multifunctional components. In particular, it intends to assess the potential of injecting molding grades of polypropylene (PP) to produce parts with transparency (haze) gradients. Firstly, several polypropylene grades of different types were identified and injected under various thermal processing conditions, i.e., injection temperature and mold temperature, in order to vary the cooling rate, influencing the growth rate of the spherulites and eventually the presence/absence of α and β crystalline zones. The injected parts' optical properties were then characterized, and the most promising PP grades were identified and selected for subsequent work, namely grade DR 7037.01, showing the widest range of haze (from 29.2 to 68.7%). and PP070G2M, presenting the highest haze value (75.3%). Finally, in an attempt to understand the origin of the haze variations observed, the parts injected with the selected PP grades were further characterized through differential scanning calorimetry (DSC) and polarized light microscopy. It was concluded that the main factor causing the observed haze difference was, apart from the size of the spherulites, the presence of internal layers with different birefringence and, therefore, different refractive indices.

Keywords: graded property; transparency; haze; injection molding; polypropylene; circular economy; sustainability; recycling

Citation: Pedroso, L.D.; Pontes, A.J.; Alves, A.; Duarte, F.M.; Carneiro, O.S. From Transparent to Opaque: A Route towards Multifunctional Parts Injected with a Single Material. *Materials* **2023**, *16*, 6219. https://doi.org/10.3390/ma16186219

Academic Editors: Agnieszka Kijo-Kleczkowska and Adam Gnatowski

Received: 1 August 2023
Revised: 7 September 2023
Accepted: 8 September 2023
Published: 15 September 2023

1. Introduction

Plastics have changed our way of living and quality of life since the second half of the 20th century. Presently, the annual production of plastic products is of nearly 400 million tons per year, and Plastic Atlas predicts that by 2025 it will reach 600 million tons, which corresponds to an increase of more than 50% of the current production [1,2]. The unrestrained use of this type of materials, and its consequent increase as post-consumption material in the Earth's geological space, has led several scientists to classify our time as the plastic era [3]. Looking around, it is possible to find plastics in products/services of all application areas, from agriculture to aeronautics, packaging, sports, automotive industry, healthcare, civil construction, among many others [1,2,4]. The huge success of plastics can be attributed to two of their major characteristics: diversity and versatility [5]. In fact, there are thousands of polymeric materials, if the several existing grades and variants are considered, able to satisfy the most wide-ranging demands [6]. Their versatility comes not only from their easy processing, but also from their ability to be easily transformed into complex geometries [4,7–10]. Additionally, and as a general rule, when compared to other materials, plastics are cheaper, lighter, have a wide range of mechanical properties, and very good chemical resistance and thermal and electrical insulation properties [4,7,10].

The success of plastics has led to some severe problems. In fact, the colossal increase in their use in various areas of consumption and industry led to an unprecedented human development in the history of the planet [11]. However, this also led to an exponential increase in environmental pollution, with the main impact being on waterways and oceans, destroying habitats and endangering hundreds of marine species [1,11–13]. Additionally, the lack of economic resources and infrastructure in less developed countries led to the accumulation of these materials in inappropriate areas, which due to human or natural action are often involuntarily directed to waterways or are voluntarily directed to landfills and incineration [2]. None of these options is the most convenient, since materials should be reused or recycled at the end of the product's lifetime. The recycling figures in the European Union are quite encouraging [2,11], but are not similar across all countries. Recycling is not always possible when multi-material products such as multilayer co-extruded films or multi-material injection molded components are considered. In these cases, the materials' separation at the end of life and subsequent recycling, required for a circular economy, is much more difficult. Energy recovery is a popular method in these cases [14].

This paper investigates the feasibility of using a single material to produce injection-molded parts with a transparency gradient, thereby progressing towards mono-material products with property gradient. This aims to improve the end of life of these products, facilitating their recycling and, as a result, contributing to a circular economy.

There are several ways to modify the transparency of a plastic part, since light can be blocked from crossing the plastic and/or be scattered from its usual path via obstacles with dimensions greater than the wavelength of visible light [15–18]. These obstacles absorb the incident light and/or scatter it. Fillers, pigments, voids, or crystalline zones of semi-crystalline thermoplastics, among others, are examples of this kind of obstacles. Studies on the effect of thermoplastic crystallinity on optical properties revealed that the dimensions of the spherulites affect light transmission through the polymer. Transmission is higher in materials with very small spherulites, while haze is higher in parts with larger spherulites [15,17–21]. The crystalline structure can be found in two most common configurations: alpha and beta. The first one exists in large quantities in the polymer, whereas the second exists in very small quantities (sometimes not at all), unless the raw material is processed under the ideal conditions leading to its formation or specific nucleating agents are incorporated for this sake [15,22–24]. In order to vary the dimensions of the spherulites, nucleating and clarifying agents can be incorporated into the polymer matrix, which promote an increase in the number of crystallization active centers. Additives can also be added to speed up the crystallization process, favoring the appearance of alpha spherulites, which will be smaller and have less influence on the optical properties of the final part [19]. As a general rule, the lower the percentage of alpha crystallites, the better the optical properties of the final parts. For equal degrees of crystallinity, worse optical properties occur for the higher crystallite dimensions [15]. Another strategy consists of varying the injection temperature and mold temperature in order to affect the cooling rate [15,22,25]. Several recent studies on the crystallization kinetics of semi-crystalline thermoplastic materials concluded that very low cooling rates promote the growth of crystalline zones, allowing the spherulites to reach significant dimensions. On the other hand, high cooling rates inhibit their growth, leading to the development of many small spherulites [15,18,23,25–27]. In terms of crystalline structure, it has been demonstrated that processing at high shear rates promotes the appearance of the beta form; however, because in conventional processes it is not possible to process a material at high shear rates throughout the entire thickness of the part, this route is not an option [22]. Another way to promote the appearance of beta type spherulites is to slowly cool down the part near 120 °C; this is not feasible at the industrial level since high production rates are required [23]. The most feasible way to promote the beta form is through the incorporation of specific additives. In practical terms, materials with large amounts of beta phase have a lower stiffness, higher yield and rupture deformation, higher impact resistance and better optical properties (higher transparency) [15,24].

Polypropylene (PP) is one of the most widely used thermoplastic materials in the world, accounting for approximately 20% of total plastic material consumption [2]. This material is extremely versatile and inexpensive. It is a semi-crystalline polymer of the polyolefin family with low chemical reactivity, making it suitable for use in contact with strong acids and bases (even at high temperatures). This material is also an excellent candidate for food contact and medical devices. It is simple to color and to process with good thermal stability and low density [7,8,28]. As for the optical and mechanical properties of PP, they vary according to its nature (homopolymer, hPP, and copolymer, rPP) and according to the chemical structure of the polymer chains and polymerization conditions. Thus, PP can adopt three types of spatial configurations (tacticity) of the repetitive units: isotactic, syndiotactic and atactic [7,8,10,29]. This structure is related to the arrangement of the methyl group along the polymer chain, and the presence/absence of other monomers in the main chain, besides propene [28]. As a result, PP can be transparent, opaque, rigid or flexible, which makes it the ideal candidate for the present study [7,19,28,30].

In this work, several commercial PP grades were pre-selected in order to find the one that allows us to obtain greater ranges of haze, or turbidity, a property inversely related to transparency. The variation of turbidity in the injection-molded samples will be induced through the use of different cooling rates/thermomechanical environments since they are expected to result in different degrees of crystallinity. This will be later explored in a special mold enabling us to set/control different simultaneous local temperatures in the same cavity. The innovative character of this work is, therefore, the systematic research, performed with PP grades, with the objective of obtaining different degrees of transparency in the same injected component.

If the primary goal of this research is met, several advantages will result for the part design stage. In fact, the use of a single material with property gradient eliminates the existence of interfacing zones; also, the process and mold will be simpler since bi-injection will not be required. Additionally, it will facilitate the selection of materials, avoiding the need to select compatible or incompatible pairs of materials, according to the purpose of the part. Finally, it will also contribute to a circular economy and facilitate the recycling of components at the end of the product's useful life.

Potential applications of this research include opaque covers with a local transparent zone that allows for the viewing of an internal digital display, or opaque containers for liquids with a transparent strip that allows for the checking of the level of liquid, among others.

2. Materials and Methods

2.1. Materials

The polypropylene grades used in this study are listed in Table 1. All of the grades are commercially available and suitable for injection molding, the technology envisaged for future applications. These grades were chosen to have different material types (homopolymers and copolymers) and a wide range of melt flow indexes (MFI).

Table 1. Polypropylene grades used in the study.

Material Reference	Producer	Type	MFI (g/10 min) @230 °C, 2.16 kg
Isplen PP070G2M	Repsol	Homopolymer	12
Isplen PP099K2M	Repsol	Homopolymer	55
Isplen PR599C2M	Repsol	Random Copolymer	75
DR 7037.01	Braskem	Random Copolymer	23
DR 7051.01	Braskem	Random Copolymer	10
Inspire 364	Braskem	Random Copolymer	42
SB 520	Lotte Chemical	Random Copolymer	1.8

2.2. Part Geometry and Injection Molding

The injected part is a 60 mm long rectangle made up of two squares with 30 mm sides, designed for direct use in the equipment available for the evaluation of transmissivity. The mold consists of two interchangeable plates, with two cavity samples on each plate (see Figure 1a), and two different thicknesses on each injected sample, visible in Figure 1b. In one of the plates, one of the parts has 0.5 and 1.0 mm thicknesses, and the other part is 1.5 and 2.0 mm thickness. In the other plate the thicknesses of the two parts is 2.5 and 3.0 mm, and 3.5 and 4.0 mm. Therefore, this mold makes it possible to produce parts with thicknesses ranging from 0.5 mm to 4.0 mm, at 0.5 mm intervals, in order to investigate the effect of this parameter on transmissivity. The parts, illustrated in Figure 1b, were produced in an injection molding machine BOY, Berlin, Germany, model 22 A, with a clamping force of 22 tons.

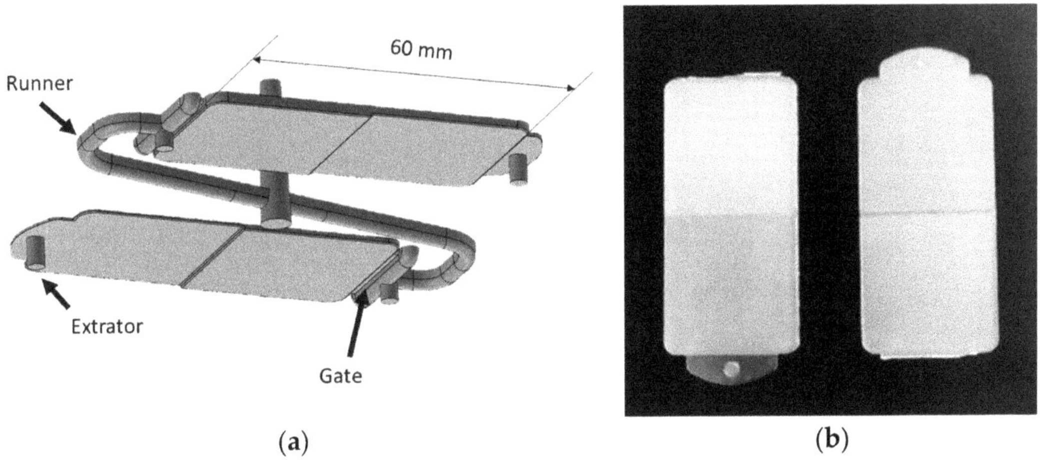

Figure 1. Injection molding of the samples: (**a**) representation of the complete molding; (**b**) examples of injected plastic samples.

Each material grade (Section 2.1) was injected with four different processing conditions by varying the injection temperature (Tinj) and the mold temperature (Tmold). The selected temperatures were 210 °C and 250 °C for Tinj, and 25 °C and 90 °C for Tmold, combined in four different processing conditions. For each material, the injection temperatures used are the materials' minimum and maximum limits, according to the datasheet information, in order to achieve the greatest possible variation of transparency. Table 2 shows the remaining reference processing conditions used. Considering the number of different processing conditions (four) and thicknesses (eight), 32 different types of samples were produced with each of the seven PP grades.

Table 2. Fixed processing conditions used in the injection molding of samples.

Sample Thickness (mm)	Injection Speed (mm/s)	Packing Pressure (MPa)	Packing Time (s)	Cooling Time (s)
0.5–2.0	80	25	6	22
2.5–4.0	65	20	6	22

The moldings were manually removed from the mold after the cooling stage, and their temperatures were monitored and recorded using a FLIR SC640 thermographic camera (FLIR, Hudson, NH, USA) throughout the subsequent cooling process. Frames were taken every 10s, allowing us to characterize the actual cooling process undergone after extraction. Measurements were taken on at least three samples of each condition, and the results were averaged.

2.3. *Samples Characterization*

2.3.1. Optical Properties

For the characterization of the injected samples transparency, haze has been considered. Haze is commonly referred to as the amount of visible light that is scattered at angles greater than 2.5° (2.5° < θ < 90°). A haze meter XL-211 Hazegard System, produced by BYK Gardner (BYK-Gardner, Geretsried, Germany), was used following the ASTM D1003 standard [31]. This device records values of the total transmission of light passing through the sample as well as the amount of light that is scattered. To obtain the final haze value, the errors associated with the measurement process were corrected and then the diffuse transmittance value was divided by the total transmittance value. For each condition, at least five samples were tested and the average value was calculated. Other optical properties, such as total, direct and diffuse transmittance, were compared. However, it was concluded that haze/turbidity is the one that best translates the set of values obtained into the real perception of the resulting transparency of the injected samples since it also takes into account the light scattered.

2.3.2. Morphology

Differential scanning calorimetry (DSC) was used to provide information about the degree of crystallinity of the injected samples. Only a first scan was performed, in order to preserve and assess the effect of the thermomechanical conditions in which processing occurred. For this purpose, a DSC 200 F3 Maia, from NETZSCH (Erich NETZSCH GmbH & Co, Selb, Germany) was used, heating the samples from 30 °C to 200 °C at a heating rate of 10 K/min, which is enough of a range to cover where the total melting of the crystalline zones occurs. Figure 2 shows the part regions where the samples were taken.

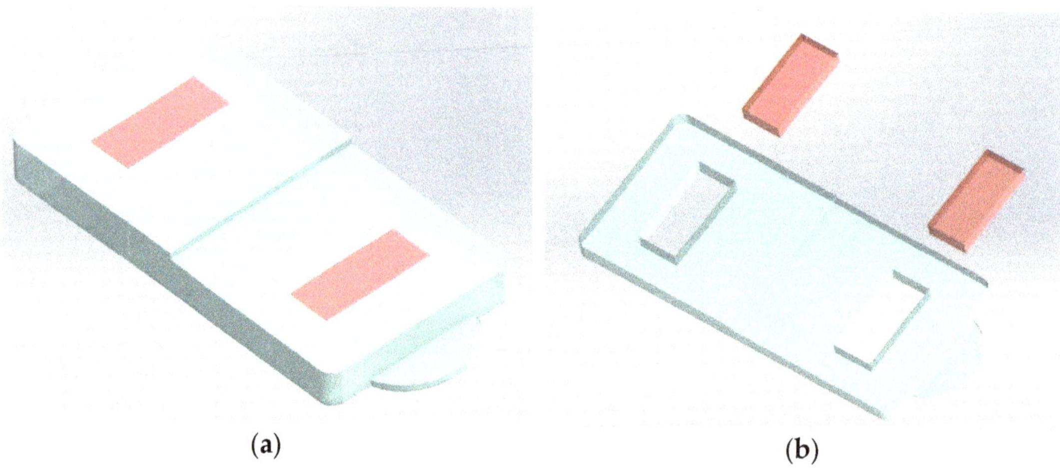

Figure 2. Injected parts and samples (orange blocks) collection for DSC: (**a**) injected part and location of the DSC samples; (**b**) part after the samples' removal.

The raw materials, as pellets, were also characterized under two different conditions: (i) heating at 10 K/min followed by cooling at 10 K/min; and (ii) heating at 10 K/min followed by cooling at 40 K/min using, at least, three samples for each condition. In this case, the first heating was used to eliminating the thermal history of the pellets.

Polarized light microscopy was used to characterize the internal structure of the injected samples in terms of the oriented shell/core ratio, as well as to measure the size of the spherulites. The samples were prepared using a microtome and fixed on a slide with the help of Canada balsam. Subsequently, these samples were observed in a transmission optical microscope LEICA DM 2500P (Leica-microsystems, Wetzlar, Hessen 35578, Germany) coupled to a polarizer. Images were taken using Leica Application Suite software (Leica-microsystems, Wetzlar, Hessen 35578, Germany) with magnification objectives of 2.5×, 4×, 10× and 20×, and an eyepiece of 1.63×.

These tests, DSC and polarized light microscopy, were only performed on the samples that showed the maximum absolute haze values injected with the PP070G2M grade, and also on the samples produced with the material that originated the wider range of haze, DR 7037.

3. Results and Discussion

3.1. Materials Screening

Figure 3 shows the haze values obtained for all tested materials, identified by different colors, and for all processing conditions (Tinj and Tmold combinations). Each point represents the experimental mean value. The standard deviation of the mean values is not shown because it is very low in the majority of cases (0.04–0.20%), reaching a maximum value of 2% for only two of the conditions.

As can be seen, for lower injection temperatures (210 °C, Figure 3a,b), a greater haze variation is observed among all of the materials. Additionally, a wide range of haze is obtained when only the material grade variable is considered, especially for lower thicknesses (from 0.5 mm to 2.5 mm). For the remaining thicknesses, the different material grades exhibited similar haze values, with the exception of grade SB 520. As for the parts injected at 250 °C, Figure 3c,d, there are two distinct behaviors regarding the haze values obtained, which vary according to the nature of the polypropylene, i.e., PP homopolymer and PP random copolymer. The light blue (PP070G2M) and orange (PP099K2M) curves represent the homopolymer grades, whose haze values were the highest for all the processing conditions. The grade that shows the lowest haze is the SB 520. However, the haze property of this grade is not very sensitive to the processing conditions, which does not make it a suitable candidate for the next phase of this work. The material grade with the highest absolute haze is the PP070G2M, which for typical injection molding thicknesses (from 0.5 mm to 1.5 mm), achieves absolute variations in the order of 10%, which, although small, are still interesting for this study. Table 3 shows the grade that presents the wider range of haze variation, as well as the corresponding processing conditions, for samples with a thickness of 3 mm. Note that this 39.5% absolute variation of the property corresponds to a relative variation of more than 100%.

According to the results obtained, the most promising polypropylene grades were selected for the next phase of the project. These were the DR 7037.01, the material with the widest range of haze variation, and the PP070G2M, which has a different nature from the first, has the highest absolute value of haze among all the materials tested (75.3% at Tinj 210 °C and Tmold 90 °C), and shows one of the widest ranges of haze for the lowest thickness (0.5 mm).

3.2. Optical Performance in Use

The injected samples were placed at various distances in front of a printed text to determine the effect of different haze values on their optical performance. Photographs of these experiments were taken and are shown in Figure 4. The left column, center column and right column correspond to sample–text distances of 0 mm, 4 mm and 8 mm, respectively. In the first line, the sample with thicknesses of 1.5 and 2 mm (as identified) was used, and in the remaining lines, specimens with thicknesses of 2.5 and 3 mm were used.

The effect of the thickness, and especially of the distance between the specimen and the text, is evident. These parameters have a significant impact on the clarity of the observation, which can play a key role in future applications. The differences in haze obtained with the different selected PP grades and with the different processing conditions are also remarkable.

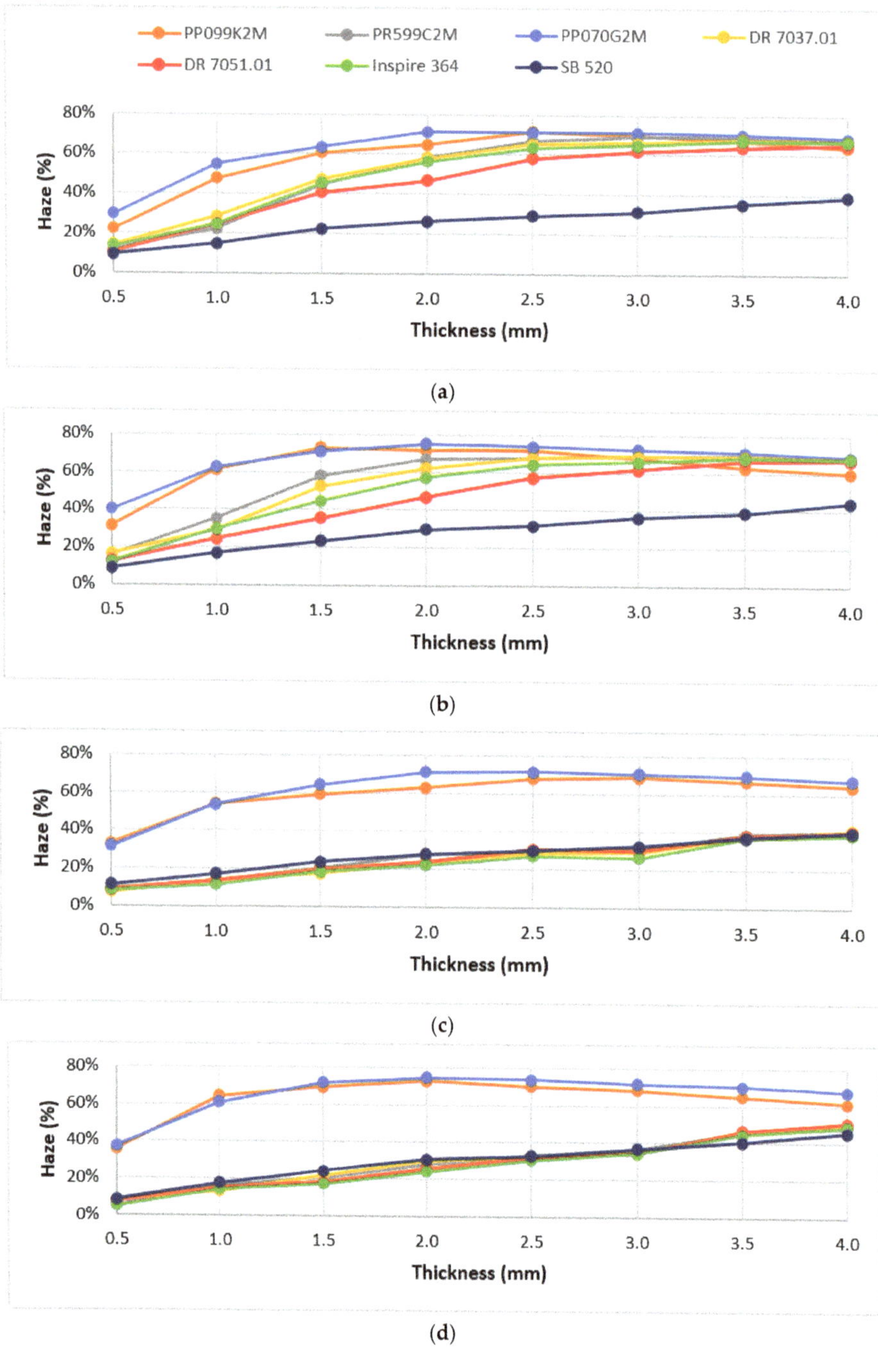

Figure 3. Haze as a function of samples thickness, for the seven PP grades and different processing conditions: (**a**) Tinj = 210 °C, Tmold = 25 °C; (**b**) Tinj = 210 °C, Tmold = 90 °C; (**c**) Tinj = 250 °C, Tmold = 25 °C; (**d**) Tinj = 250 °C, Tmold = 90 °C.

Table 3. Absolute haze ranges for the grade with wider range of haze variation (DR 7037.01) with a 3 mm thickness.

Processing Conditions	Minimum Haze (%)	Maximum Haze (%)	Haze Range (%)
Value	29.2	68.7	39.5
Tinj | Tmold (°C)	250 | 25	210 | 90	

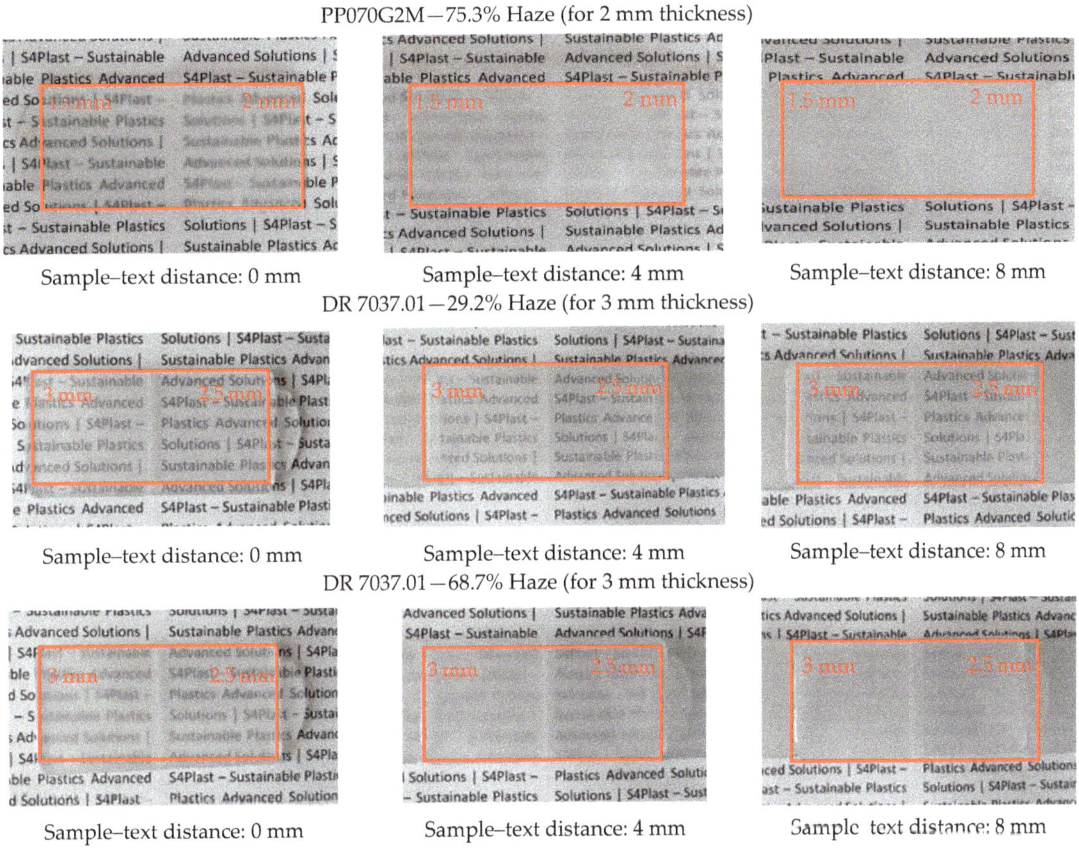

Figure 4. Selected PP grades: effect of the haze value on the optical performance of injected specimens, illustrated for various sample–text distances. Note: each specimen has two distinct thicknesses (see Section 2.2).

To investigate the causes of the haze variation, DSC tests and optical microscopy were used to characterize the morphology of the specimens for the selected grades. The results obtained using DSC are shown in Figure 5 and Table 4.

The curves presented in Figure 5 form two groups, each being representative of one of the two grades selected for further characterization. The first group, with the exothermic peak appearing at a lower temperature, corresponds to the copolymer polypropylene (rPP, DR 7037.01); the second one comprises the curves of the homopolymer polypropylene (hPP, PP070G2M). This difference, occurring in materials with the same chemical structure but with different nature, was expected and is due to the presence of a well-developed crystalline phase in hPP (material characterized by a more regular chain structure), which requires a greater amount of energy to melt.

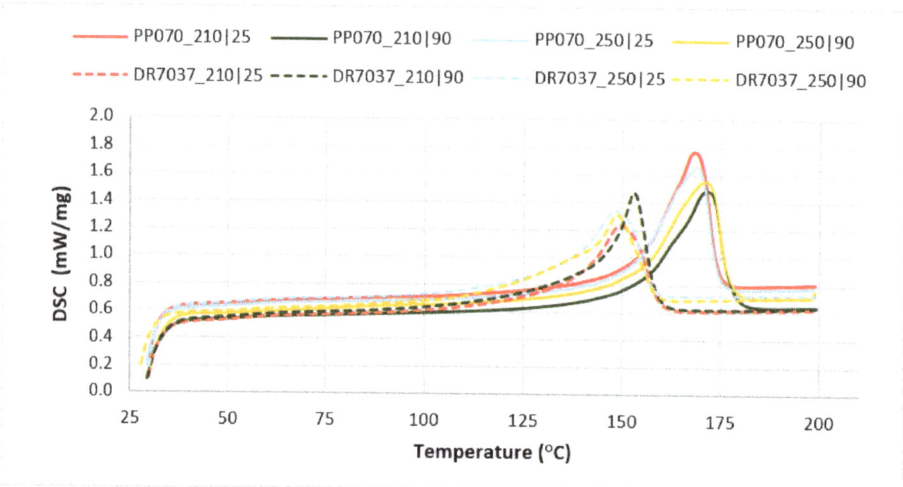

Figure 5. DSC curves obtained in this study (average values taken from the first scan of the injected samples).

Table 4. Thermal properties and degree of crystallization of the PP specimens. Note: this property was computed as the ratio of the specimen melting enthalpy and the melting enthalpy of 100% crystallized PP, i.e., 207 J/kg [32–34].

Material	Thickness (mm)	Injection Temperature (°C)	Mold Temperature (°C)	Melting Enthalpy (J/g)	Melting Temperature (°C)	Degree of Crystallization (%)
PP070G2M	2	210	25	90.01 ± 0.03	168.3 ± 0.09	43.48 ± 0.02
PP070G2M	2	210	90	94.82 ± 0.04	171.8 ± 0.15	45.80 ± 0.01
PP070G2M	2	250	25	87.77 ± 0.03	168.9 ± 0.02	42.40 ± 0.002
PP070G2M	2	250	90	102.25 ± 0.03	161.3 ± 0.34	49.40 ± 0.01
DR 7037.01	3	210	25	81.25 ± 0.06	150.34 ± 0.21	39.25 ± 0.04
DR 7037.01	3	210	90	90.78 ± 0.01	152.8 ± 0.08	43.86 ± 0.002
DR 7037.01	3	250	25	83.66 ± 0.11	148.3 ± 0.10	40.41 ± 0.06
DR 7037.02	3	250	90	85.51 ± 0.02	150.4 ± 1.73	41.31 ± 0.01

Analyzing the graphs shown in Figure 5, it can be highlighted that for a given grade, and regardless of the processing conditions, there are no major variation both in the melting temperature and the melting enthalpy (excluding the sample injected at 250 °C with a mold temperature of 90 °C). On the other hand, the homopolymer (PP070G2M) has a higher crystallinity than the copolymer grade (DR 7037.01), which was expected due to their different inherent propensity to crystallize. In any case, the differences in crystallinity, while significant, do not appear to be sufficient to justify the haze differences observed. DSC of the raw materials (granules) was also performed to better investigate this issue. Table 5 shows these results.

Table 5. Dependence of the thermal properties and degree of crystallization on the heating cycle of the raw materials (granules). Note: this property was computed as the ratio of the specimen melting enthalpy and the melting enthalpy of 100% crystallized PP, i.e., 207 J/kg [32–34].

Material	Heating Cycle	Melting Enthalpy (J/g)	Melting Temperature (°C)	Degree of Crystallization (%)
PP070G2M	1st	84.825 ± 0.01	168.9 ± 0.4	40.98 ± 0.02
PP070G2M	2nd	95.17 ± 0.02	168 ± 0.004	45.98 ± 0.01
PP070G2M	3rd	89.405 ± 0.02	168.4 ± 0.36	43.19 ± 0.01
DR 7037.01	1st	91.02 ± 0.01	152.8 ± 0.14	43.97 ± 0.04
DR 7037.01	2nd	87.015 ± 0.01	151.2 ± 0.02	42.04 ± 0.05
DR 7037.01	3rd	79.995 ± 0.01	149.8 ± 0.25	38.64 ± 0.05

The first heating is intended to clean the thermal history of the materials. The second and third heating cycles refer to heating at 10 K/min, after cooling at 10 K/min and 40 K/min, respectively. These results are comparable to those obtained from the injected samples. The observed differences can be attributed to the higher cooling rates occurring in injection molding. This supports the hypothesis that haze differences are determined by factors other than the crystallization degree.

In order to further investigate the causes of haze differences, the uncontrolled cooling that occurs after the parts' demolding was monitored with the help of a thermographic camera. Figure 6 shows a typical sequence of the images taken, and Figure 7 shows the evolution of temperature at location "1", corresponding to the thicker region of the sample, as a function of time.

Considering the left specimen, where the highest turbidity was registered, it can be observed that the values of temperature and its distribution varies along the cooling time. After demolding, the temperature of the part is almost homogeneous. After two minutes of demolding, the highest temperature zone is located in the central zone of the part, since natural convection cools down the external contour more efficiently. At the end of the monitored cooling time, and for the same reason, the maximum temperature continues to occur in the center of the parts. This behavior is common to all samples tested.

As shown in Figure 7, it takes more than five minutes for the temperature to drop to room temperature after demolding. Thus, cooling after demolding may be more important in terms of part crystallinity/turbidity than cooling that occurs in a controlled manner inside the mold.

In order to assess the morphological changes induced in the specimens (by processing and by the cooling in the environment), polarized light optical microscopy was used. This analysis intends to characterize the specimens' internal structure, in terms of the oriented shell/core ratio, as well as to check if there is any apparent variation in the size of the spherulites developed. The obtained micrographs are shown in Figures 8–10.

The presence of crystalline zones, evidenced by bright points, is not observed in the images shown in Figure 8. However, several bands along the cross section of the sample, identified with '3', can be seen in image (b). Image (a) clearly shows the presence of a very oriented and thin shell, '1', and an under-shell on both sides of the sample, '2'. The small marks visible along the width of the sample, visible in image (c), '4', were promoted by sectioning in the microtome during sample preparation. These images demonstrate what was expected for the polypropylene copolymer, that is, the lack of well-developed crystallites.

The micrographs shown in Figure 9, corresponding to the same material injected at a lower temperature (Tinj = 210 °C) in a mold at a higher temperature (Tmold = 90 °C), are those that better evidence distinct layers formed along the cross-section of the samples during cooling. One possible explanation for the formation of these layers is the successive reheating of the already solidified zones by the hotter inner zones during cooling outside the mold. This hypothesis was confirmed by measurements performed with the thermographic camera. For these processing conditions, there are also prominent white spots in the core of the samples, identified with '1', denoting the existence of well-developed spherulites. This is a consequence of the corresponding lower cooling rate that resulted from a high mold temperature.

The micrographs corresponding to the homopolymer PP070G2M are depicted in Figure 10. These reveal well-developed spherulites throughout all of the sample's cross-section. The presence of a very thin and oriented shell, '1', as well as a transition zone between this shell and the core of well-developed spherulites, '2', comprising a large number of smaller sized spherulites, are also noticeable.

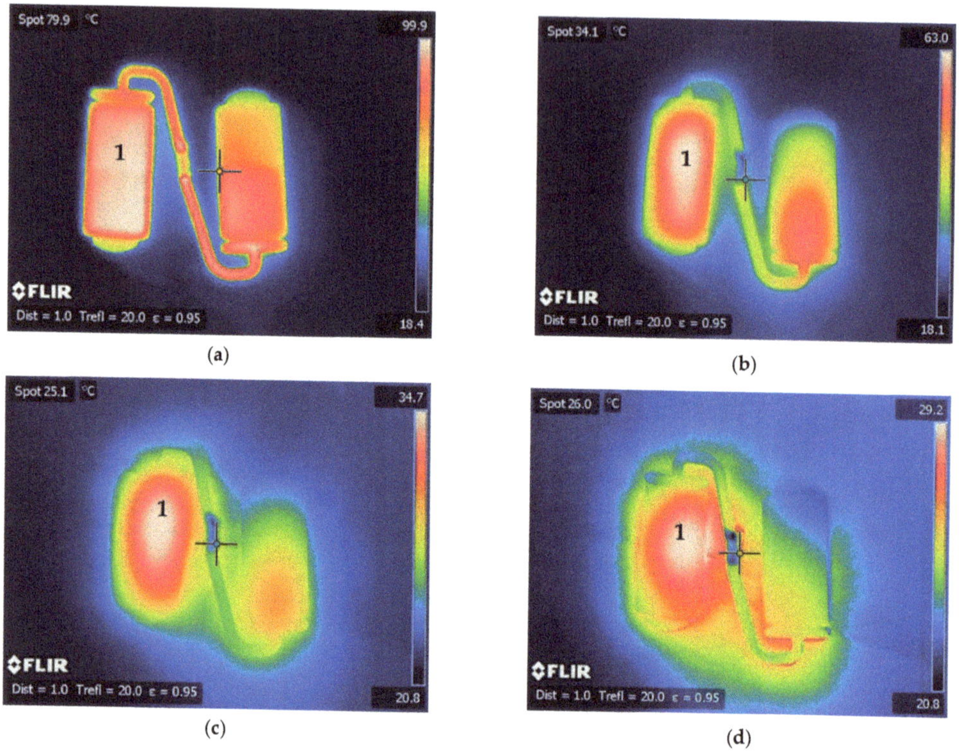

Figure 6. Thermograms corresponding to moldings injected with DR 7037.01 at 210 °C in a mold at 90 °C: (**a**) immediately after demolding; (**b**) 2 min after demolding; (**c**) 4 min after demolding; (**d**) after reaching a maximum temperature of 30 °C.

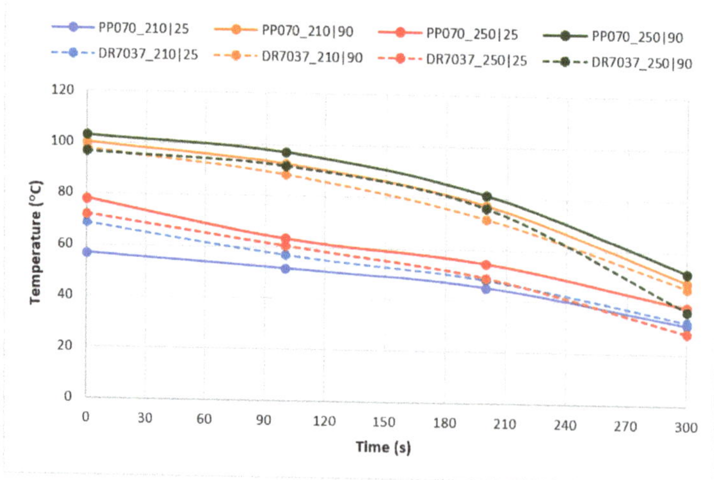

Figure 7. Part temperature variation with time, after demolding, at location "1" (identified in Figure 6).

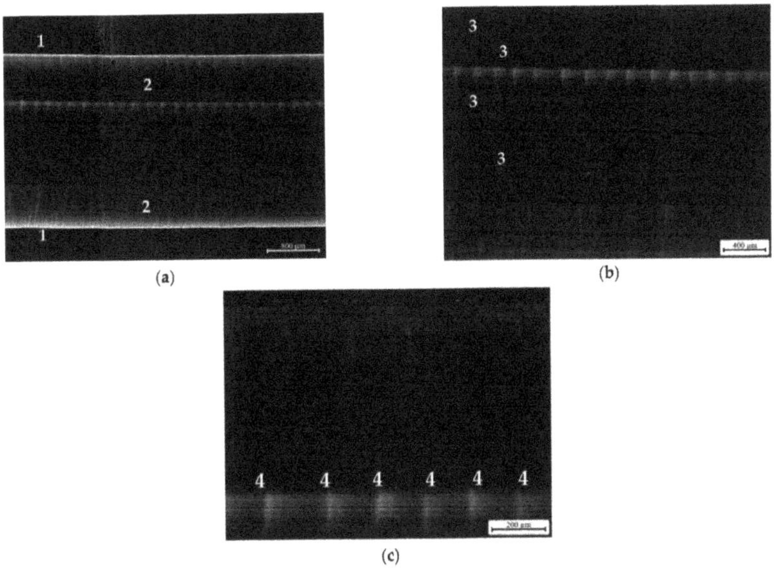

Figure 8. Polarized light microscopy images corresponding to the 3 mm thickness copolymer grade DR 7037.01 specimen, injected at Tinj = 250 °C and Tmold = 25 °C, observed at various magnifications: (**a**) complete sample; (**b**) magnification of the center region of the sample; (**c**) magnification of the bottom region of the sample. Notes: 1—oriented sample shell; 2—sample under-shell; 3—different bands observed; 4—marks promoted by the microtome.

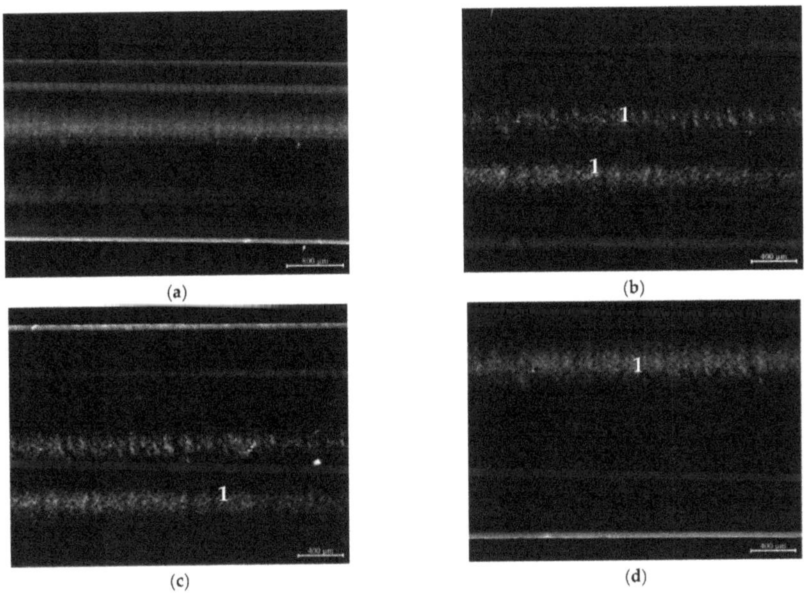

Figure 9. Polarized light microscopy images corresponding to the 3 mm thickness copolymer grade DR7037.01 specimen, injected at Tinj = 210 °C and Tmold = 90 °C: (**a**) complete sample; (**b**) magnification of the center region of the sample; (**c**) magnification of the upper region of the sample; (**d**) magnification of the bottom region of the sample. Note: number 1 identifies zones of well-developed spherulites.

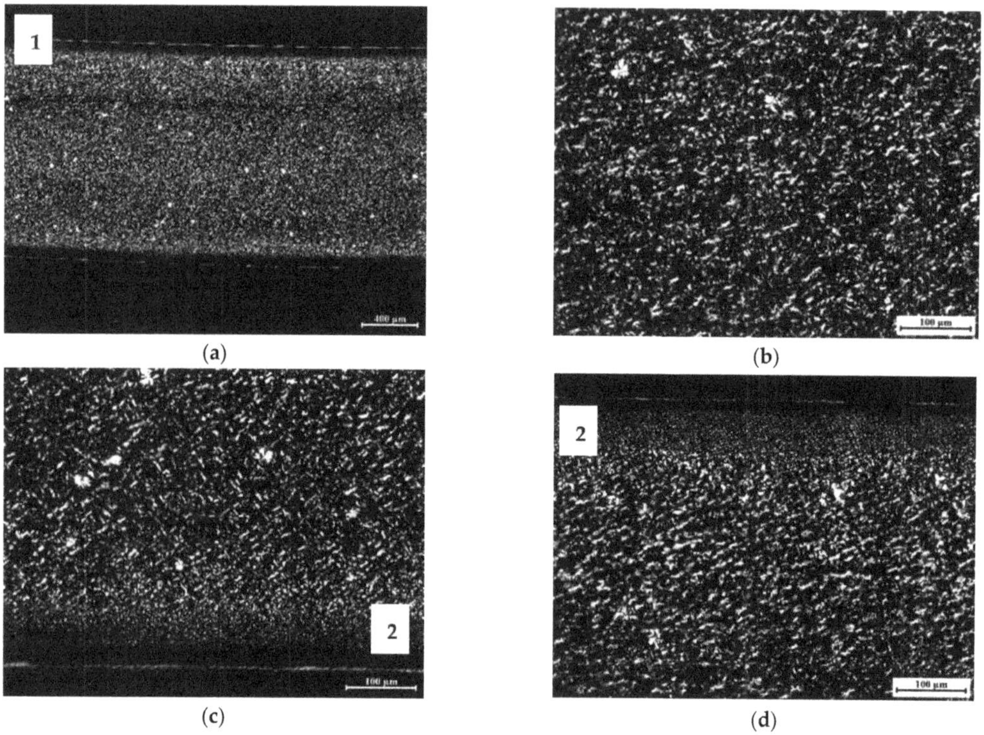

Figure 10. Polarized light microscopy images corresponding to the 2 mm thickness homopolymer grade PP070G2M specimen, injected at Tinj = 210 °C and Tmold = 90 °C: (**a**) complete sample; (**b**) magnification of the center region of the sample; (**c**) magnification of the bottom region of the sample; (**d**) magnification of the upper region of the sample. Notes: 1—oriented sample shell; 2—zone of well-developed spherulites.

The differences in the degree of crystallinity between the two PP grades observed in the micrographs are essentially apparent since they are not supported by the DSC results, presented in Table 4. Additionally, the considerable differences obtained for the haze (of the order of 39.5%) cannot be only attributed to differences in the degree of crystallinity. In fact, these differences may also be promoted by the layered zones, as revealed by the micrographs. These layers are expected to have different refractive indexes and, therefore, they deviate the light propagation during its passage through the injected specimens [15]. Furthermore, the higher the number of layers developed, the more times light is refracted, contributing to the increase in haze. Because it promotes light diffusion and reflection, the interface between layers can also contribute to an increase in haze.

4. Conclusions

This work assessed the potential of polypropylene injection molding grades to produce parts with varied transparency (haze) in order to obtain, in the future, mono-material products with a property gradient. Several polypropylene grades of different types were identified and the haze of parts injected under various thermal processing conditions was characterized. The results showed a significant dependence of haze on the tested materials, thermal processing conditions and thicknesses of the samples. The optical performance of the parts when used at various distances from a text (0, 4 and 8 mm) also revealed great differences. This feature has an interesting potential for future applications.

There were no significant differences in the degree of crystallinity measured via DSC capable to justify the differences observed in haze. However, microscopy results revealed the presence of several layers across the samples thickness, for copolymer grade DR 7037.01, or the presence of large spherulites, for homopolymer grade PP070G2M, that determined the samples' haze. These results should be related to the cooling rate of the parts, after demolding, which, under the studied conditions, takes about 5 min to reach a temperature of 30 °C.

The promising results obtained confirm the potential of using one single material to obtain parts with an optical property (haze) gradient. However, to accomplish this objective further developments and studies will be required. A mold with independent controlled temperatures in different cavity zones is already being designed for this purpose.

Author Contributions: Conceptualization and methodology, L.D.P., F.M.D. and O.S.C.; experimental work, L.D.P.; formal analysis, L.D.P., A.A., F.M.D. and O.S.C.; resources, A.J.P. and A.A.; writing—original draft preparation, L.D.P.; writing—review and editing, A.J.P., A.A., F.M.D. and O.S.C.; supervision, A.J.P., F.M.D. and O.S.C.; funding acquisition, A.J.P. All authors have read and agreed to the published version of the manuscript.

Funding: Conducted under the project "S4Plast—Sustainable Plastics Advanced Solutions" (POCI-01-0247-FEDER-046089) financed by European Regional Development Fund (ERDF), through the Incentive System to Research and Technological development, within the Portugal2020 Competitiveness and Internationalization Operational Program. IPC researchers acknowledge also funding by National Funds through FCT-Portuguese Foundation for Science and Technology, References UIDB/05256/2020 and UIDP/05256/2020.

Conflicts of Interest: The authors declare no conflict of interest.

References

1. Böll, H. *Plastic Atlas 2019: Facts and Figures about the World of Synthetic Polymers*; Heinrich Böll Foundation: Cologne, Germany, 2019.
2. Plasctics Europe. *Plastics—The Facts 2021: An Analysis of European Plastics Production, Demand and Waste Data*; Plasctics Europe: Brussels, Belgium, 2021.
3. Carrington, D. After Bronze and Iron, Welcome to the Plastic Age, Say Scientists. The Guardian. 4 September 2019. Available online: https://www.theguardian.com/environment/2019/sep/04/plastic-pollution-fossil-record (accessed on 4 July 2022).
4. Culter, J.D.; Selke, S.E.M. *Plastics Packaging Book*, 3rd ed.; Hanser Publishers: Munich, Germany, 2016.
5. Thompson, R.C.; Swan, S.H.; Moore, C.J.; Vom Saal, F.S. Our plastic age. *Philos. Trans. R. Soc. B Biol. Sci.* **2009**, *364*, 1973–1976. [CrossRef] [PubMed]
6. Wypych, G. *Handbook of Polymers*, 2nd ed.; ChemTec Publishing: Toronto, ON, Canada, 2016.
7. Maier, C.; Calafut, T. *Polypropylene The Definitive User's Guide and Databook*, 1st ed.; Plastics Design Library: Cambridge, UK, 1998.
8. Osswald, T.A.; Menges, G. *Material Science of Polymers for Engineers*, 3rd ed.; Hanser Publishers: Munich, Germany, 2012.
9. Encyclopaedia Britannica. *The Composition, Structure, and Properties of Plastics*, Encyclopaedia Britannica, Inc.: Edinburgh, UK, 2020; pp. 1–20. Available online: https://www.britannica.com/print/article/4636841 (accessed on 2 May 2023).
10. Harper, C.A. *Handbook of Plastics, Elastomers, and Composites*, 4th ed.; McGraw-Hill Education: Forest Hill, MD, USA, 2002.
11. Bakas, I.; Ferna, R.; Reichel, A.; Trier, X.; Zeiger, B. *Plastics, the Circular Economy and Europe's Environment—A Priority for Action*; European Environment Agency: Copenhagen, Denmark, 2021. [CrossRef]
12. Pattanaik, S.S.; Sangavi, S. Marine habitat destruction: An anthropogenic way towards the end of life in the ocean. *Vigyan Varta* **2020**, *1*, 19–22.
13. Sheavly, S.B.; Register, K.M. Marine debris & plastics: Environmental concerns, sources, impacts and solutions. *J. Polym. Environ.* **2007**, *15*, 301–305. [CrossRef]
14. Korniejenko, K.; Kozub, B.; Bak, A.; Balamurugan, P.; Uthayakumar, M.; Furtos, G. Tackling the Circular Economy Challenges-Composites Recycling: Used Tires, Wind Turbine Blades, and Solar Panels. *J. Compos. Sci.* **2011**, *5*, 243. [CrossRef]
15. De Santis, F.; Pantani, R. Optical properties of polypropylene upon recycling. *Sci. World J.* **2013**, *2013*, 354093. [CrossRef] [PubMed]
16. Molnár, J.; Sepsi, Ö.; Erdei, G.; Lenk, S.; Ujhelyi, F.; Menyhárd, A. Modeling of light scattering and haze in semicrystalline polymers. *J. Polym. Sci.* **2020**, *58*, 1787–1795. [CrossRef]
17. Lin, Y.; Bilotti, E.; Bastiaansen, C.W.M.; Peijs, T. Transparent semi-crystalline polymeric materials and their nanocomposites: A review. *Polym. Eng. Sci.* **2020**, *60*, 2351–2376. [CrossRef]
18. Gahleitner, M.; Grein, C.; Kheirandish, S.; Wolfschwenger, J. Nucleation of Polypropylene Homo-and Copolymers. *Int. Polym. Process.* **2011**, *26*, 2–20. [CrossRef]

19. Luo, S.; Wei, L.; Sun, J.; Huang, A.; Qin, S.; Luo, H.; Gao, C.; Zheng, Y.; Shen, J. Crystallization behavior and optical properties of isotactic polypropylene filled with α-nucleating agents of multilayered distribution. *RSC Adv.* **2019**, *10*, 387–393. [CrossRef] [PubMed]
20. Zia, Q.; Androsch, R.; Radusch, H.J. Effect of the structure at the micrometer and nanometer scales on the light transmission of isotactic polypropylene. *J. Appl. Polym. Sci.* **2010**, *117*, 1013–1020. [CrossRef]
21. Menyhárd, Z.H.A.; Doshev, P.; Gahleitner, M.; Vörös, G.; Varga, J.; Pukánszky, B. Effect of the molecular structure of the polymer and nucleation on the optical properties of polypropylene homo-and copolymers. *ACS Appl. Mater. Interfaces* **2014**, *6*, 7456–7463. [CrossRef]
22. Tordjeman, P.; Robert, C.; Marin, G.; Gerard, P. The effect of alpha, beta crystalline structure on the mechanical properties of polypropylene. *Eur. Phys. J. E* **2001**, *4*, 459–465. [CrossRef]
23. Gradys, A.; Sajkiewicz, P.; Minakov, A.; Adamovsky, S.; Schick, C.; Hashimoto, T.; Saijo, K. Crystallization of polypropylene at various cooling rates. *Mater. Sci. Eng. A* **2005**, *413*, 442–446. [CrossRef]
24. Papagiorgiou, D.G.; Chrissafis, K.; Bikiaris, D. β-Nucleated Polypropylene: Processing, Properties and Nanocomposites. *Polym. Rev.* **2011**, *55*, 596–629. [CrossRef]
25. Lamberti, G. Isotactic polypropylene crystallization: Analysis and modeling. *Eur. Polym. J.* **2011**, *47*, 1097–1112. [CrossRef]
26. Nielsen, A.S.; Pyrz, R. The effect of cooling rate on thermal residual strains in carbon/polypropylene microcomposites. *Sci. Eng. Compos. Mater.* **1998**, *7*, 1–22. [CrossRef]
27. Brucato, V.; Piccarolo, S.; La Carrubba, V. An experimental methodology to study polymer crystallization under processing conditions. The influence of high cooling rates. *Chem. Eng. Sci.* **2002**, *57*, 4129–4143. [CrossRef]
28. Tripathi, D. *Practical Guide to Polypropylene*, 1st ed.; Rapra Technology Limited: Shawbury, UK, 2002.
29. Maddah, H.A. Polypropylene as a Promising Plastic: A Review. *Am. J. Polym. Sci.* **2016**, *6*, 1–11. [CrossRef]
30. Ahmed, A.K.; Atiqullah, M.; Pradhan, D.R.; Al-Harthi, M.A. Crystallization and melting behavior of i-PP: A perspective from Flory's thermodynamic equilibrium theory and DSC experiment. *RSC Adv.* **2017**, *7*, 42491–42504. [CrossRef]
31. *ASTM D1003-21*; Standard Test Method for Haze and Luminous Transmittance of Transparent Plastics. ASTM: West Conshohocken, PA, USA, 2021.
32. van der Meer, D.W. Structure-Property Relationships in Isotactic Polypropylene. Ph.D. Thesis, University of Twente, Enschede, The Netherlands, 2003.
33. Ehrenstein, G.W.; Theriault, R.P. *Polymeric Materials: Structure, Properties, Applications*; Hanser: Munich, Germany, 2001.
34. Ruiz-Orta, C.; Fernandez-Blazquez, J.P.; Anderson-Wile, A.M.; Coates, G.W.; Alamo, R.G. Isotactic polypropylene with (3,1) chain-walking defects: Characterization, crystallization, and melting behaviors. *Macromolecules* **2011**, *44*, 3436–3451. [CrossRef]

materials

MDPI

Article

Research on Waste Combustion in the Aspect of Mercury Emissions

Agnieszka Kijo-Kleczkowska [1], Adam Gnatowski [2,*], Barbara Tora [3], Krzysztof Kogut [4], Krzysztof Bytnar [4], Jaroslaw Krzywanski [5] and Dorota Makowska [4]

1 Department of Thermal Machinery, Faculty of Mechanical Engineering and Computer Science, Czestochowa University of Technology, 42-201 Czestochowa, Poland; a.kijo-kleczkowska@pcz.pl
2 Department of Technology and Automation, Faculty of Mechanical Engineering and Computer Science, Czestochowa University of Technology, 42-201 Czestochowa, Poland
3 Faculty of Civil Engineering and Resource Management, AGH University of Science and Technology, 30-059 Cracow, Poland; tora@agh.edu.pl
4 Faculty of Energy and Fuels, AGH University of Science and Technology, 30-059 Cracow, Poland; kogut@agh.edu.pl (K.K.); bytnar@agh.edu.pl (K.B.); makowska@agh.edu.pl (D.M.)
5 Faculty of Science & Technology, Jan Dlugosz University in Czestochowa, 42-201 Czestochowa, Poland; j.krzywanski@ujd.edu.pl
* Correspondence: adam.gnatowski@pcz.pl

Abstract: The topic of waste combustion/co-combustion is critical, given the increasingly restrictive legal regulations regarding its environmental aspects. In this paper, the authors present the test results of selected fuels of different compositions: hard coal, coal sludge, coke waste, sewage sludge, paper waste, biomass waste and polymer waste. The authors conducted a proximate and ultimate analysis of the materials and mercury content in them and their ashes. An interesting element of the paper was the chemical analysis of the XRF of the fuels. The authors conducted the preliminary combustion research using a new research bench. The authors provide a comparative analysis of pollutant emissions—especially mercury emission—during the combustion of the material; this is an innovative element of this paper. The authors state that coke waste and sewage sludge are distinguished by their high mercury content. The value of Hg emission during the combustion depends on the initial mercury content in the waste. The results of the combustion tests showed the adequacy of mercury release compared to the emissions of other compounds considered. Small amounts of mercury were found in waste ashes. The addition of a polymer to 10% of coal fuels leads to a reduction in mercury emissions in exhaust gases.

Keywords: waste; polymer; combustion; emission; mercury

Citation: Kijo-Kleczkowska, A.; Gnatowski, A.; Tora, B.; Kogut, K.; Bytnar, K.; Krzywanski, J.; Makowska, D. Research on Waste Combustion in the Aspect of Mercury Emissions. *Materials* **2023**, *16*, 3213. https://doi.org/10.3390/ma16083213

Academic Editor: Andrea Petrella

Received: 23 March 2023
Revised: 13 April 2023
Accepted: 17 April 2023
Published: 19 April 2023

1. Introduction

The issues of waste and its cataloguing can be found in legislation [1,2]. Depending on the source of its generation, waste is divided into groups, including waste generated during the extraction of minerals, waste from agriculture, waste from paper and cardboard processing, waste from thermal processes, waste from the mechanical surface treatment of metals and plastics and municipal waste.

According to the Announcement of the Marshal of the Polish Parliament [3], waste is understood as "any substance or object which the holder discards, intends to dispose of or is obliged to dispose of". Conversely, a waste incineration plant is intended for thermal processing, with or without the recovery of the generated thermal energy, including installations and devices used to control and monitor the aforementioned process, together with the treatment of exhaust gases released to the atmosphere.

The information on the exact amount and types of waste from plants that produce more than 1000 tonnes of waste per year and the ones that produce 1 million tonnes or more (excluding municipal waste) is presented in Ref. [4].

According to [5], regarding sludge from industrial and municipal wastewater treatment plants, sludge production per year changed as follows (in thousands of tonnes of dry matter): 1063.1 in 2000, 1124.4 in 2005, 895.1 in 2010, 951.5 in 2015, 989.5 in 2020 and 1025.8 in 2021. Meanwhile, their storage per year changed as follows (in thousands of tonnes of dry matter): 474.5 in 2000, 399.1 in 2005, 16.9 in 2010, 131.5 in 2015, 63.9 in 2020 and 84.0 in 2021. Furthermore, their thermal processing per year changed as follows (in thousands of tonnes of dry matter): 34.1 in 2000, 37.4 in 20015, 66.4 in 2010, 165.4 in 2015, 219.4 in 2020 and 221.8 in 2021. A clear reduction in sewage sludge storage with increasing thermal utilization in recent years can be observed.

The issue of waste storage requirements was examined in Ref. [6]. According to Ref. [7], one of the reference methods for testing soils on which municipal sewage sludge has been applied is the determination of the content of heavy metals—lead, cadmium, mercury, nickel, zinc, copper and chromium—by atomic absorption spectrometry (AAS) or by optical emission spectrometry with inductively coupled plasma (ICP-OES) with their prior digestion using strong acids. According to the criteria for admitting waste, coded 19 08 05 (stabilized municipal sewage sludge) [8], the acceptable limits are: loss on ignition (LOI): 8% dry weight; total organic carbon (TOC): 5% dry weight; calorific value: 6 MJ/kg dry weight.

There is an insufficient load of incineration plants in Poland, which could handle sewage sludge if only it were properly prepared for incineration, mainly in terms of a level of dewatering of >90% DS [9]. Among the available sludge drying technologies, including thermal dryers, low-temperature dryers using heat pumps and solar and hybrid dryers, solar energy has the greatest use.

One of the strategic goals presented in Ref. [10] is to eliminate the production of municipal sewage sludge as waste, which, due to its quality, poses problems with its management (in accordance with applicable regulations). This is possible due to thermal processing.

With reference to the above, the thermal utilization of sewage sludge, the need for continuous research aimed at identifying the properties and course of the waste incineration and co-incineration process have become extremely important. Given that there are 11 incineration plants operating in Poland using sewage sludge as energy fuel, further development of these technologies in Poland should be expected while taking into account the benefits of their co-combustion with other fuels. Sewage sludge, due to its properties, can be a valuable energy fuel, which has been emphasized in a number of research works, e.g., Refs. [11–13]. The similar calorific value of dried sewage sludge to lignite justifies the possibility of using this waste as fuel.

It should be emphasized that 86% of electricity generated in Poland is based on hard coal and lignite. The main components of coal are organic matter, mineral matter and water. The mineral substance originates from plant material from which coal seams were formed, as well as from mineral deposits mixed with plant material. Organic matter, on the other hand, mainly consists of carbon, hydrogen, oxygen, sulphur and nitrogen, as well as trace amounts of phosphorus and other compounds. Ash is a non-combustible part of the fuel, consisting mainly of a mineral substance. Together with moisture, it forms fuel ballast, reducing its quality. Thus, the calorific value of the fuel decreases with the increase in ash content. In the heating process, the organic and mineral substance of the fuel is decomposed. Moisture, numerous gases and vapours of organic substances are then released, leading to the formation of char, composed of the remains of organic matter and mineral substances of the fuel changed as a result of heating [14,15].

As the quality requirements of the coals burned in power plants increase, the issue of the combustion of highly watered fuels becomes more and more important. In order to meet the expectations of power engineers, hard coal mines were forced to expand and modernize coal enrichment plants. This causes a continuous increase in waste in the form of flotation sludge. The best way to dispose of these sludges is to incinerate them as slurries and to co-incinerate them with other materials and fuels. The issue of the thermal treatment of coal sludge has been addressed, e.g., in Refs. [16,17].

In high-temperature conditions, materials undergo thermal decomposition, which is also accompanied by a change in their physical and chemical properties. Coal fuels, in the presence of oxygen, undergo combustion and a series of stages of this process: heating, the evaporation of moisture, the degassing and combustion of volatiles and char burning [18]. These steps may be consecutive or overlapping. In the case of coal, char burnout is the stage that determines the course of the fuel combustion process. Char combustion is, in its simplest form, a high-temperature oxidation of elemental carbon to carbon dioxide [14,15].

Comparing the energy properties of biomass to coals, one can notice the same qualitative elemental composition of these fuels. However, there is a difference in quantity. The advantage of biomass is definitely the lower content of sulphur and ash compared to fossil fuels. The basic features of biomass distinguishing it from other types of fuels and having a direct impact on the course of the combustion process are the high content of volatiles, resulting in the high reactivity of the fuel, low ignition temperature, high and variable moisture content, differentiated grain size distribution of the fuel and alkali content, including potassium, sodium, chlorine and sulphur. Due to the high reactivity, the burning rate of the biomass char is definitely higher compared to that of coal. For example, the thermal decomposition of wood begins at a temperature of about 220 °C, and its individual components decompose at different temperatures, e.g., hemicellulose: 220–320 °C; cellulose: 320–370 °C; lignin: 320–500 °C. The most important gaseous components emitted from wood include CO, CO_2, CH_4, C_2H_4, C_2H_6 and H_2, and the main liquid elements of the thermal decomposition of this type of fuel are water, methanol, acetic acid, acetaldehyde and tar. In general, the process of burning wood is similar to burning young coals, especially brown coals. A very large share of volatiles in this type of fuel causes over 60–70% of the calorific value of wood to be released in the degassing process. Due to the high reactivity, the burning rate of wood char is much higher compared to that of coal. The mechanism of oxidation of the wood coke residue mainly depends on the combustion temperature. At high temperatures, oxidation to CO dominates, which is then burnt in the exhaust gas. At low temperatures, however, the oxidation process to CO_2 takes place, with the participation of oxygen sorption on the char surface [19].

The behaviour of fuels during the high-temperature process depends on their composition and the thermal and flow conditions, which also affect the emission of gases into the atmosphere [13,20–26]. For example, in Ref. [13], the authors emphasized that an important issue in sewage sludge incineration is the emission of pollutants such as heavy metals, mercury, dioxins, furans, NO_x, N_2O, SO_2, HCl, HF and C_xH_y. The high nitrogen content in the sludge affects the emission of N_2O and NO_x. Heavy metals can be present in sewage sludge in the form of hydroxides, carbonates, phosphates, silicates and sulphates. Dioxins and furans mainly refer to compounds containing chlorinated dibenzo-p-dioxins and chlorinated dibenzofurans.

Heavy metals are monitored and controlled in combustion plants. In sewage sludge, zinc, copper, lead, chromium, nickel, manganese, cobalt, cadmium and arsenic may be present in significant amounts. Dried sewage sludge contains about 50% of ash (the main elements are Si, Fe, Al, Ca and P) and a higher proportion of volatile elements (including Na, K and P) in comparison to coal. According to the authors of Ref. [25], the behaviour of heavy metals during fuel combustion is crucial. The classification of selected metals under combustion conditions was made, with cadmium, cadmium chloride, lead chloride, lead oxide, zinc, zinc chloride, arsenic and arsenic chloride being classed as volatile; mercury and mercury chloride being classed as very volatile; cadmium oxide, chrome copper and nickel being classed as non-volatile; and lead being classed as indirect.

Table 1 shows an example share of heavy metals in fuels [26]. In relation to other materials, sewage sludge and municipal waste are characterized by a high proportion of mercury.

Table 1. The share of heavy metals in fuels, in mg/kg [26].

Parameter	Sewage Sludge (Dry)	Municipal Waste	Beech Wood	Willow	Lignite	Hard Coal
cadmium	1.4–20	5.95	1.2	719	0.4–1.0	0.4–2.0
copper	80–800	45	1530	3.8	8–44	12–60
nickel	16–50	3	605	2.01	3–38	6–48
lead	20–50	279	185	2.91	3–24	6–50
zinc	2400–6100	663	550	135.96	3–73	20–420
mercury	2–2.5	1.22	<5	0.03	—	0.004–0.13

In Poland, there are no legal regulations regarding the permissible content of heavy metals in fuels/waste. The emission standards for waste combustion and co-combustion equipment and installations are presented in Ref. [27]. The amounts of heavy metals in exhaust gases include 0.05 mg of (cadmium + thallium)/m^3, 0.05 mg of mercury/m^3 and mg (antimony + arsenic + lead + chromium + cobalt + copper + manganese + nickel + vanadium)/m^3.

According to Ref. [28], from the point of view of the management or storage of ashes, it is extremely important to analyse them in terms of the content of heavy metals. These compounds are characterized by a very long shelf life in the environment, which may result in their bioaccumulation in plants and inclusion in the biological cycle. Among all metals, the following elements should be particularly distinguished:

- cadmium destroys the kidneys, causes hypertension and influences reproductive functions, and it should also be emphasized that it poses a huge toxicological risk because it easily penetrates into ground and underground waters;
- lead and antimony destroy bones, soft tissues, liver, brain and bone marrow; and
- mercury attacks brain cells and the entire nervous system and causes paralysis [29,30].

According to Ref. [31], heavy metals can be divided with environmental protection being taken into consideration as: As, B, Cd, Hg, Mo, Pb and Se being elements of the greatest concern; Cr, Cu, Ni, V and Zn being elements of moderate concern; Ba, Co, Ge, Li, Mn, Sb and Sr being elements of less concern; Rn, Th and U radioactive elements; and Be, Sn, Te and TI being elements present in only very small concentrations.

In Ref. [32], the authors analysed the average concentrations of the heavy metals in coal, slag and fly ash samples. They found that the average concentrations of the heavy metals in the materials were:

- arsenic (As): 39.2 (17.7–56.8) mg/kg, 44.5 (37.4–50.5) mg/kg, 119.2 (73.5–217.9) mg/kg;
- lead (Pb): 10.9 (7.8–16.9) mg/kg, 25.9 (23.0–29.1) mg/kg, 40.3 (23.9–77.4) mg/kg;
- mercury (Hg): 1.3 (<0.8–1.4) mg/kg, 1.8 (1.2–2.2) mg/kg, 3.0 (1.7–3.8) mg/kg;
- chromium (Cr): 68.9 (52.3–82.2) mg/kg, 181.6 (160.0–199.1) mg/kg, 189.2 (148.4–214.3) mg/kg;
- iron (Fe): 13,445 (11,530–15,830) mg/kg, 34,839 (31,530–37,400) mg/kg, 35,211 (27,370–40,600) mg/kg;
- zirconium (Zr): 32.5 (24.4–41.8) mg/kg, 118.2 (105.3–129.7) mg/kg, 103.8 (69.8–125.2) mg/kg;
- cobalt (Co): 13.9 (<3.0–20.5) mg/kg, 25.8 (22.7–31.5) mg/kg, 22.5 (12.9–32.9) mg/kg;
- zinc (Zn): 120.2 (55.2–156.3) mg/kg, 328.2 (262.8–397.2) mg/kg, 367.6 (239.8–602.1) mg/kg;
- copper (Cu): 18.8 (13.6–23.1) mg/kg, 42.3 (39.5–45.6) mg/kg, 45.3 (32.9–52.5) mg/kg;
- nickel (Ni): 63.5 (47.1–79.0) mg/kg, 156.7 (143.2–170.0) mg/kg, 155.4 (117.3–173.3) mg/kg;
- manganese (Mn): 111.8 (73.3–141.8) mg/kg, 226.0 (207.0–254.9) mg/kg, 272.8 (214.5–312.8) mg/kg;
- vanadium (V): 145.3 (114.4–182.3) mg/kg, 278.5 (222.7–320.7) mg/kg, 374.7 (299.0–420.4) mg/kg;
- titanium (Ti): 133.1 (1132.0–1522.0) mg/kg, 3519.7 (3164.0–3982.0) mg/kg, 3568.0 (2739.0–3959.0) mg/kg.

The authors of Ref. [33] stated that the content of heavy metals in wood pellet ash produced by biomass combustion depends on the type and quality of wood biomass, the production process, the use of additives, the characteristics of the furnace, the temperature of the process, etc., and that the mean concentrations of the heavy metals they analysed are ordered as follows: Fe > Mn > Zn > Cu > Pb > Ni > Cr > Cd > Co.

In Ref. [34], the authors showed that the trend of metal concentrations was as follows: Zn > Cr > Cu > Pb > Ni > Cd for wastewater treatment plant 1; Zn > Pb > Cr > Ni > Cu > Cd for wastewater treatment plant 2; Zn > Cu > Ni > Cr > Pb > Cd for wastewater treatment plant 3; Zn > Pb > Cu > Cr > Ni > Cd for wastewater treatment plant 4; and Zn > Cu > Pb > Cr > Ni > Cd for wastewater treatment plant 5. The results were compared to the literature: according to Ref. [35], the concentrations of particular heavy metals in sewage sludge are ordered as follows: Zn > Cu > Cr > Ni > Pb > Cd.

The authors of Ref. [36] stated that the concentration of heavy metals in leaves was found to decrease in the order of Zn > Ni > Cr > Pb > Mn > Cu > Co > Cd, while in the roots, this was in the order of Zn > Ni > Cr > Pb > Mn > Cu > Co > Cd.

The concentration of selected heavy metals—chromium (Cr), manganese (Mn), iron (Fe), nickel (Ni), copper (Cu) and zinc (Zn)—in 5-year-old *Populus trichocarpa* wood was studied by Krutul et al. [37]. The average concentration of all the tested elements that can inhibit the bioethanol production process (chromium, nickel, copper and iron) was quite low in native wood and ranged between 10.0 ppm and 54.0 ppm. The average concentration of manganese and zinc in the native wood was higher, at the level of 118.0 ppm and 155.0 ppm, respectively.

The study 'Distribution and Accumulation of Heavy Metals in Red Cedar (*Cedrela odorata*) Wood Seedling Grown in Dumpsite Soil', by Akintola, O.O. and Bodede [38], assessed the ability of *Cedrela odorata* to accumulate and distribute heavy metals such as Cu, Pb, Zn, Cd and Co in their roots and shoots planted in dumpsite soil by determining the distribution factors and enrichment coefficients between soils and plant parts. The heavy metal concentrations before planting were 48.01–356.71 mg/kg of Cu, 28.42–26.48 mg/kg of Pb, 39.99–437.88 mg/kg of Zn, 0.69–9.59 mg/kg of Cd and 16.88–29.22 mg/kg of Co, while their concentrations after planting were 8.12–226.56 mg/kg of Cu, 11.22–227.41 mg/kg of Pb, 7.66–321.51 mg/kg of Zn, 0.31–4.78 mg/kg of Cd and 3.21–14.11 mg/kg of Co. Heavy metal concentrations (mg/kg) in roots were 9.93–20.11 mg/kg of Cu, 7.26–15.21 mg/kg of Pb, 9.05–22.35 mg/kg of Zn, 0.11–0.99 mg/kg of Cd and 4.56–6.11 mg/kg of Co and their concentration shoots of the plant were 18.01–35.22 mg/kg of Cu, 9.01–17.51 mg/kg of Pb, 18.66–37.86 mg/kg of Zn, 0.15–1.32 mg/kg of Cd and 6.45–8.01 mg/kg of Co. The study of the distribution characteristics of hazardous heavy metals in ginseng and wood-cultivated ginseng was carried out by Yang et al. [39]. Samples of ginseng and wood-cultivated ginseng were collected from 14 and 5 regions of Korea, respectively. The cultivated ginseng peels contained 40.3% of Pb, 25.9% of Cd, 47.6% of As, and 89.9% of Al. Meanwhile, heavy metals consisting of 27.2% of Pb, 28.2% of Cd, 48.3% of As and 56.8% of Al were distributed in the peels of the wood-cultivated ginseng.

The issue of mercury release from fuels and waste was addressed, for example, in Refs. [40–50]. In the Ref. [40], the authors analysed various types of waste: paper waste, cardboard, textiles, plastics, plastic film, sewage sludge, RDF and car tires. The mercury content in the materials was from 5 to 764 μg Hg/kg (paper waste: 10 μg Hg/kg; cardboard: 18 μg Hg/kg; textiles: 16μg Hg/kg; plastics and plastic film: 5 μg Hg/kg; car tires: 68 μg Hg/kg; RDF: 764 μg Hg/kg; sewage sludge: 564 μg Hg/kg). They revealed that mercury content in the samples results from the production process and the fractional composition. The authors characterized the behaviour of mercury in the thermal process of waste. For example, for paper waste, cardboard, RDF and sewage sludge, mercury release starts at 150 °C, and it is removed almost completely at 350 °C. The mercury in coal and biomass can be released in low-temperature processes [41,42].

According to Ref. [43], the average content of mercury in coals from Polish power plants is 112.9 μg Hg/kg, with a variety ranging from 30 to 321 μg Hg/kg. The authors stated that, along with increasing ash content in fuels, mercury content is increasing. In a mineral matter of coals, mercury is associated with pyrite and sulphates [44].

During the combustion of fuels in power plants, more than 99% of mercury is present at flame temperatures as Hg^0; then, as the flue gas cools, some of the Hg^0 mercury can be oxidized to Hg^{2+}, combining with other flue gas components, such as SO_2 and Cl_2, forming the $HgCl_2$ compounds Hg_2Cl_2 and HgS. Some of the resulting compounds can be absorbed by the fly ash that forms. Hg^0 mercury is insoluble in water, which makes it persist in the atmosphere for up to two years. Hg^{2+} mercury is easily soluble in water, which means that it can be relatively easily removed from flue gas by means of a wet flue gas desulfurization installation [45,46].

The issue of mercury content in various fuels is also discussed in Refs. [47–49]. In Ref. [49], it was emphasized that during the combustion of fuels, mercury might be emitted in three main forms: as elemental mercury Hg^0, divalent Hg^{2+} and mercury bound to ash particles.

The authors of Ref. [50] emphasized that combustion processes, in particular, those of hard and brown coal, constitute one of the main sources of ecotoxic elements emissions to the atmosphere.

The emission of ecotoxic elements into the atmosphere from combustion depends on the type of materials, their chemical composition and the process technology used [51,52].

The composition of gases generated during the combustion process is very important. Ref. [53] presents the results of experimental studies of the concentrations of the main gases after the pyrolysis process and combustion of various materials. The concentrations of gas mixture components were determined for each material at three temperatures, corresponding to the ignition, average combustion temperature and typical process temperature. In order to predict the formation of nitrogen oxides during the incineration of plastic wastes such as polyamide, it is important to quantify the composition of the gases that are generated during the primary thermal degradation [54]. In Refs. [55,56], modelling of the thermal degradation of polyamide 6 was performed. The thermal and thermooxidative degradation of polyamide 6 was investigated. The influence of oxygen in the degradation mode was demonstrated. Moreover, it explains the role of modifiers in the improvement of the fire behaviour of the blends, which reduces the flow rate of the fuel feeding the flame. Polyethylene combustion studies have been carried out in many works. In Ref. [57], the thermal degradation of pure, high-density polyethylene was studied in a cone calorimeter under ventilated conditions with a piloted ignition. The concentration and type of flue gases emitted during the combustion process were tested. Flammability and fire-resistance tests of HDPE were carried out in Refs. [58,59], and flammability and fire-resistance tests of the compounds formed during combustion were carried out in Refs. [60–63]. In Ref. [60], the thermal degradation of polyethylene was studied under a wide range of oxidative degradation conditions. The test techniques described are designed to isolate mixtures of products to minimize secondary reactions. In general, both pyrolytic and oxidative decomposition products were obtained. The pyrolysis products included a range of saturated and unsaturated hydrocarbons, which did not vary greatly in product ratio with conditions. In Ref. [61], pollutants emitted from steady-state, steady-flow gasification and combustion of polyethylene in a two-stage furnace were examined. The powdered polymer was first subjected to pyrolysis at 1000 °C and then to the combustion of its gaseous pyrolysates after mixing with air at high temperatures (900–1100 °C). The motivation for this indirect type of PE combustion was to achieve nominally pre-mixed combustion of pyrolysis gases with air, thus achieving lower pollutant emissions than those from the direct combustion of solid PE polymer. In Ref. [62], chromatographic analysis shows that, at high decomposition temperatures, pyrogas mainly consists of hydrogen, methane, ethylene and ethane, whereas at low decomposition temperatures, it mainly consists of ethylene, ethane, methane, hydrogen, propane and higher hydrocarbons. The thermal and catalytic pyrolysis of virgin high-density polyethylene (HDPE) was carried out in Ref. [63]. The result of the thermal pyrolysis showed that the product contained significant amounts of hydrocarbons. The result suggests that the oil produced by the catalytic pyrolysis of plastic waste has the potential to burn and produce combustion compounds as an alternative fuel.

Due to the harmfulness of the mercury emitted from combustion processes and the limits on such emissions in waste incineration and co-incineration facilities, the analysis of Hg emissions from the combustion of different types of waste is an important aspect of operating such installations.

2. Materials and Methods

The behaviour of waste during combustion depends on its composition and the conditions of the process. Coal, coal sludge, biomass waste, paper waste, coke waste, sewage sludge and polymer waste were examined during the research.

The proximate and ultimate analysis of materials, shown in Table 2, was carried out for samples with a grain size of below 200 μm, according to the Polish Standard PN-G-04502:2014-11 [64]. The tests of waste heat combustion were carried out with the use of LECO AC 500 (St. Joseph, MI, USA) apparatus, in accordance with the PN-ISO 1928:2002 [65] standards. The tests of the ash, volatile matter and moisture content in waste were carried out with the use of TGA Thermostep Eltra (Haan, Germany) apparatus, in accordance with Polish Standard PN-G-04511:1980 [66] and Polish Standard PN-G-04512:1980 [67]. The tests of the coal, hydrogen, and sulphur content in waste were carried out with the use of CHS-580 (Eltra) apparatus, using high-temperature combustion and detection in the infrared method of total sulphur content determination, in accordance with Polish Standards: PN-G-04584:2001 [68] and PN-G-04571:1998 [69].

Table 2. Proximate and ultimate analyses of waste; content of mercury in materials.

Parameter/ Material	W^{ad}	V^{ad} [%]	A^{ad} [%]	C^{ad} [%]	H^{ad} [%]	S^{ad} [%]	HHV [kJ/kg]	Hg [μg/kg]
Hard coal	12.07	29.49	6.26	51.63	10.48	1.17	21,506	216.93
Coal sludge	10.66	16.32	55.92	19.26	3.83	0.47	16,039	133.51
Coke waste	14.85	11.62	36.25	24.10	6.24	2.22	12,609	523.16
Sewage sludge	13.81	21.81	61.02	15.00	5.59	0.88	5062	527.81
Paper waste	4.33	54.80	36.69	30.19	5.20	0.64	10,740	179.78
Biomass waste	13.59	64.67	2.64	44.19	11.93	0.13	17,026	7.43
Polymer waste *	1.31	98.60	0.01	62.40	9.07	0.01	30,496	2.74

* proximate and ultimate analyses made by an external company, ^{ad}—air-dried basis.

The tests of mercury content in waste were carried out with absorptive atomic spectrometry with cold vapour (CV-AAS) generation in an automated mercury analyser: DMA-80 Milestone (Sorisole, Italy).

An interesting cognitive element was the chemical analysis of XRF, i.e., X-ray fluorescence analysis. This is a technique that allows the quantitative analysis of a material's elemental composition. The method consists of irradiating the test sample with X-rays. The tests were carried out using Panalytical Epsilon 3XLE (Malvern Panalytical, Malvern, UK) apparatus.

For measuring the continuous combustion gas emissions during the combustion of waste, an automatic exhaust gas analyser, ECO 3000 (MRU Messgeräte für Rauchgase und Umweltschutz GmbH. Neckarsulm, Germany), was used, enabling the assessment of the emissions of carbon dioxide, carbon monoxide, sulphur dioxide, hydrogen sulphide and nitrogen oxides.

The mercury concentration in the flue gases was measured using an original bespoke measurement system. To determine the total mercury content, the following solutions were used: 10% $SnCl_2$ to reduce Hg^{2+} to Hg^0 and 10% KOH to remove acidic components. At the measurement path's end was an EMP-3 (Nippon Instruments Corporation, Osaka, Japan) mercury detector.

The tests of the materials' combustion were carried out using a test bench (Figure 1). An interesting cognitive element was the study of the combustion of a mixture of coal fuels (90%) with polymer waste (10%).

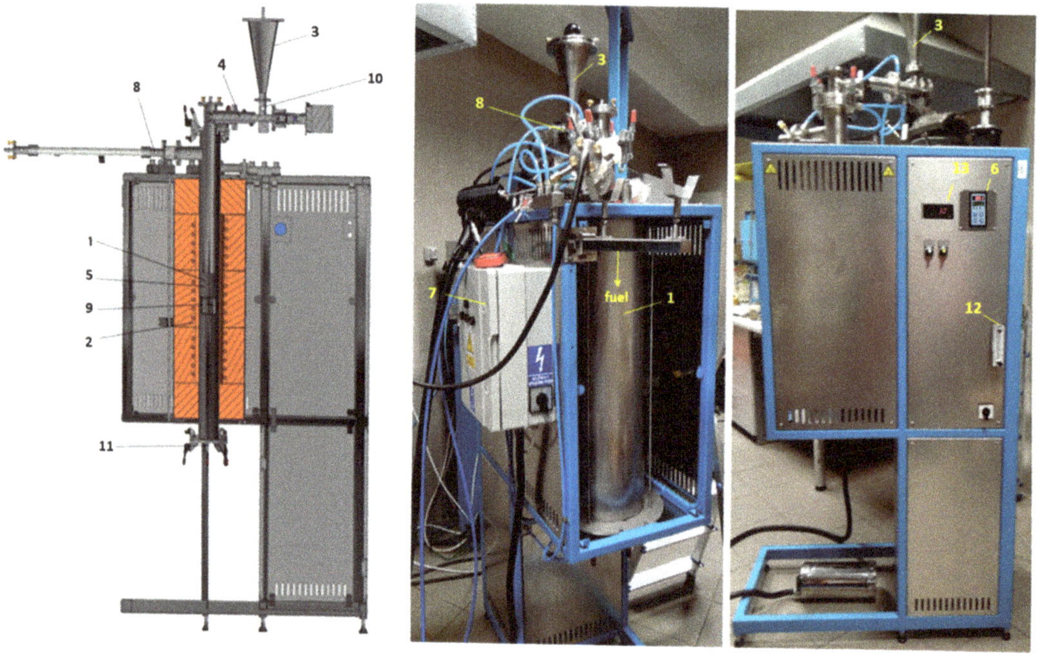

Figure 1. The scheme and photo of the research bench. The main element of the research bench is the combustion chamber (1), which is a pipe with a diameter of 76.1 mm, on which a three-sequence heating spiral (2) with a total power of 3.4 kW was wrapped. The pipe was thermally insulated to ensure maintenance of the desired temperature in the combustion chamber: 850 °C. Fuel/waste with a granulation of below 0.2 mm and a weight of about 30 g is served once to the charging tank (3) and then, using a screw feeder (4), to the combustion chamber. To adjust the temperature in the place of combustion (at the height of 737 mm from the feeder axis), the Ni-NiCr thermoelement was installed (5). This thermoelement is connected to the temperature regulator (6) and power controller (7). In addition, the temperature in the combustion chamber at the measurement site is controlled using the second Ni-NiCr thermoelement. The burning of waste along the height of the combustion chamber is continuous and concurrent with primary air. A connector (8) measures the gas emissions from the tested waste during combustion at the combustion chamber outlet. A container from the bottom of the combustion chamber enables the ash collection (9) in order to analyse its composition further. The sealing of detached elements is cooled using water. The total air for combustion (10 L/min) is supplied to the combustion chamber as primary air (10) and secondary air (11). The secondary air enables the burning of waste in a container of ash. The total air is measured using the rotameter (12). The spigot located behind the rotameter evenly separates the primary and secondary air. The fuel/waste is delivered to the combustion chamber using the screw feeder (4) with a rotational speed of approx. 0.63 turnover/min (13).

3. Results and Discussion

Table 2 presents the proximate and ultimate analyses of the materials used in the research and the content of Hg in waste before combustion. It shows the diversity in their composition and heat of combustion. Polymer and biomass waste have a high content of volatile matter, while hard coal, coke waste, sewage sludge and paper waste have a high sulphur content. This is reflected in the results of the XRF analysis and the SO_3 content in the waste samples (Table 3).

Table 3. The chemical analysis of XRF of materials' samples.

Material/Chemical Compounds [%]	Hard Coal	Coal Sludge	Coke Waste	Sewage Sludge	Paper Waste	Biomass Waste	Polymer Waste
SiO_2	25.61	44.61	17.19	28.59	20.47	18.26	17.50
Al_2O_3	14.31	22.77	10.46	—	11.34	10.51	10.59
Fe_2O_3	5.13	3.82	28.75	7.02	5.07	0.27	0.28
CaO	1.90	1.36	1.28	5.36	33.34	2.13	0.60
MgO	0.87	0.87	0.83	—	0.86	0.86	0.86
K_2O	0.79	2.34	0.12	1.04	0.53	1.42	0.06
P_2O_5	0.20	0.10	4.80	15.93	2.16	0.50	0.04
SO_3	6.88	0.87	2.96	5.88	2.53	0.10	—
MnO	0.06	0.05	0.10	0.08	0.14	0.12	0.01
SrO	0.04	0.04	0.04	0.06	0.12	0.02	0.01
TiO_2	0.49	1.00	0.09	0.40	0.01	—	2.50

For the fuel/waste combustion tests carried out using a research bench (Figure 1), materials were used which were in the air-dry state, had a grain size of below 200 μm and had the properties presented in Table 1. Before starting the combustion measurements, the combustion chamber was heated to 850 °C, and an air stream of 10 L/min was set on the rotameter. The fuel was fed by the setting on the screw feeder controller. The average mass flow rate of each fuel/waste was about 0.067 g/s. During the combustion process, flue gas composition was measured using the ECO 3000 analyser and mercury DMA-80. The resulting ash was collected to analyse mercury content.

Based on the results of the pollutant emissions (Figures 2–5), it can be indicated that the maximum values of mercury emission during the combustion of the considered materials are adequate in terms of their release of other considered gaseous compounds (CO, CO_2, SO_2, H_2S and NO_x). The authors also noted the adequacy of the composition of the materials in terms of the emission of gaseous compounds during combustion.

Higher NO_x emission values during the combustion of coal fuels–polymer mixtures result from the high nitrogen content in the considered polymer, compared to coal or coal sludge. In this paper, the authors did not determine the nitrogen content in fuels/wastes. However, the authors found nitrogen content in the considered polymer waste (N^{ad} = 15.17). It was observed that a high nitrogen content distinguishes this polymer waste from the example coal fuels and biomass (N^{ad} = 1.04 for hard coal; N^{ad} = 0.41 for coal sludge; N^{ad} = 0.53 for wheat straw [16]). This may be the reason for higher NO_x emissions in the combustion of a coal fuels–polymer mixture compared to the combustion of coal or coal sludge. According to Refs. [11,16], the polymer is also distinguished from coal fuels and biomass by its significant amount of hydrogen.

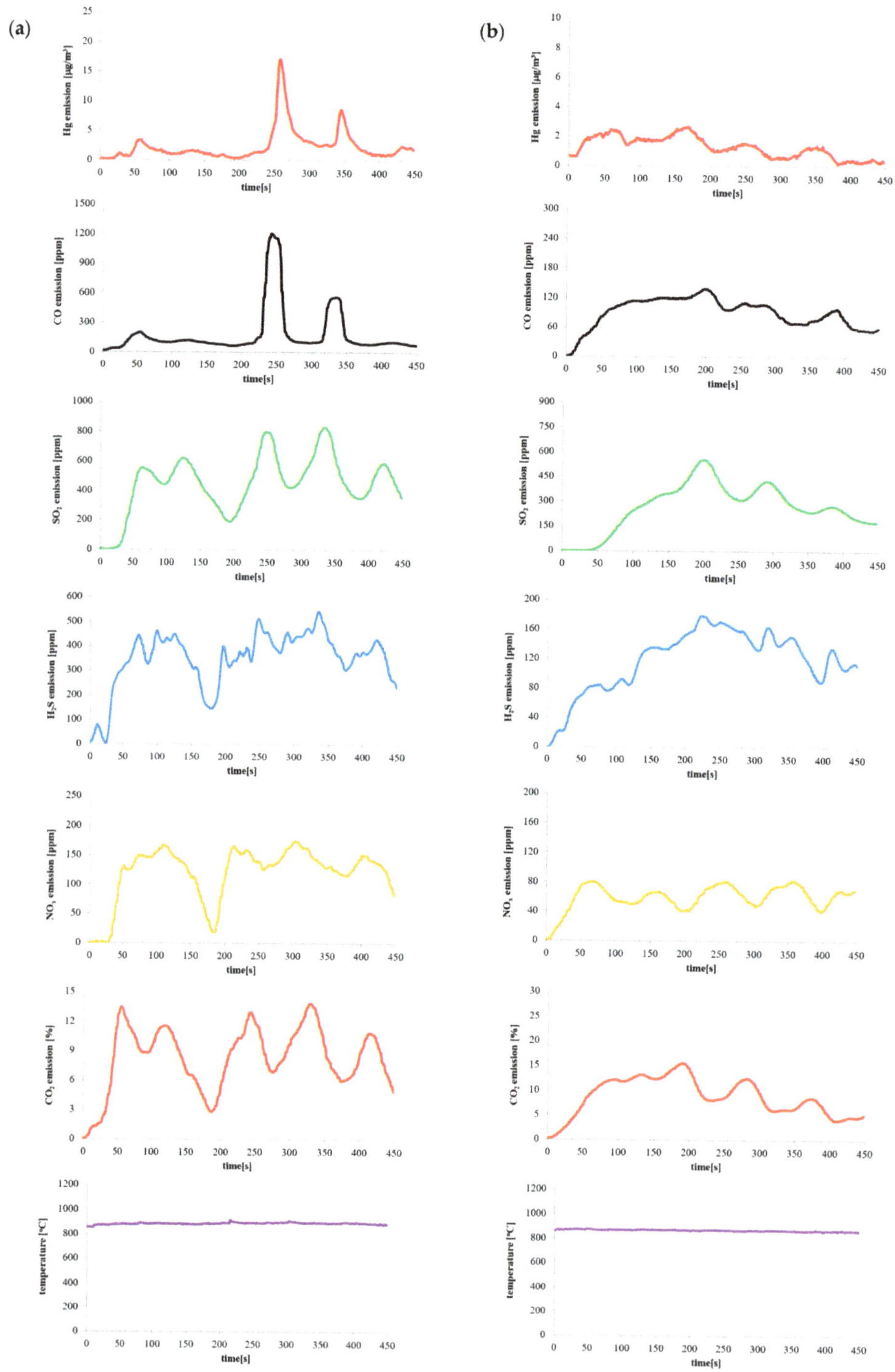

Figure 2. Emission of pollutants during (**a**) hard coal and (**b**) coal sludge combustion.

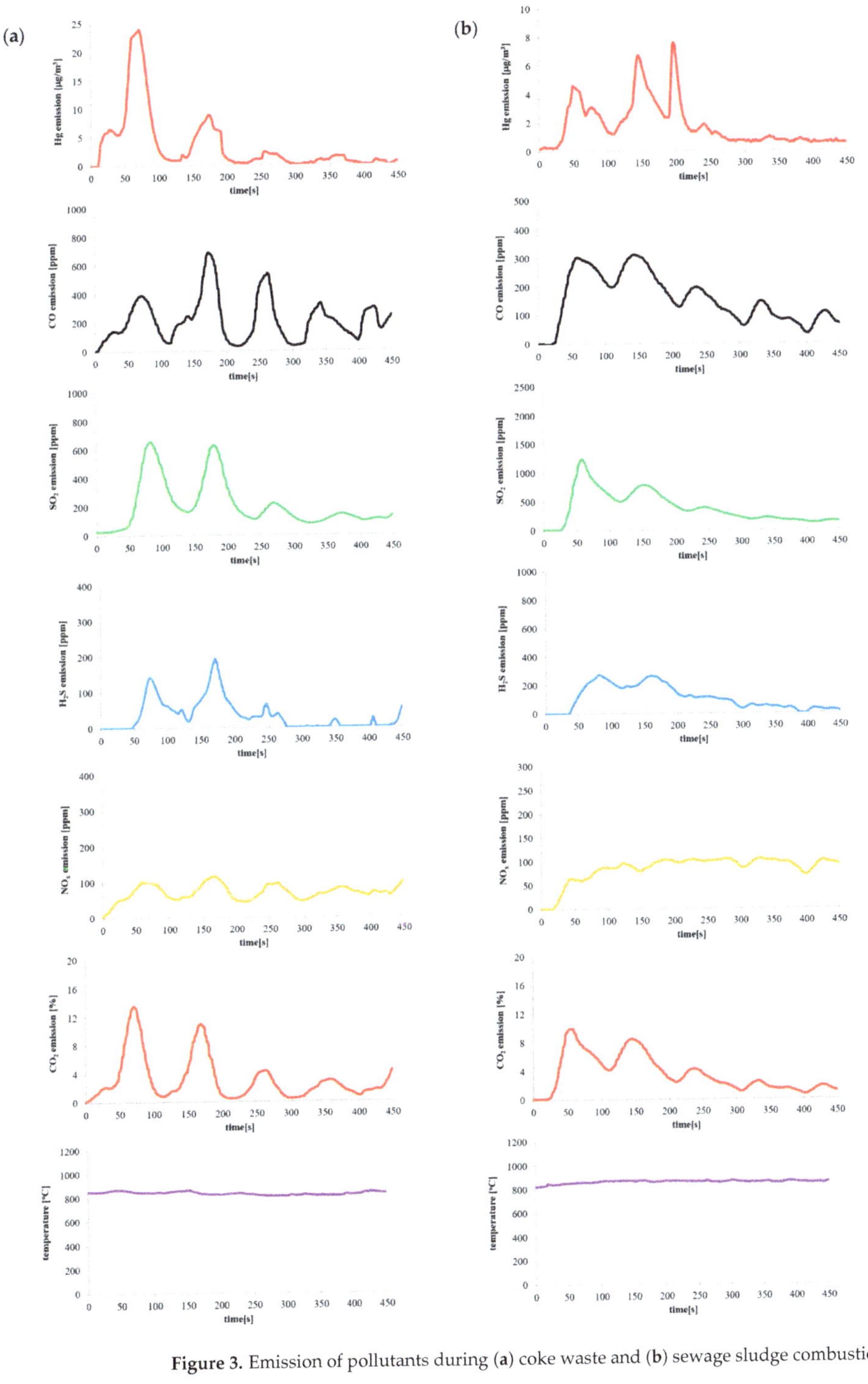

Figure 3. Emission of pollutants during (**a**) coke waste and (**b**) sewage sludge combustion.

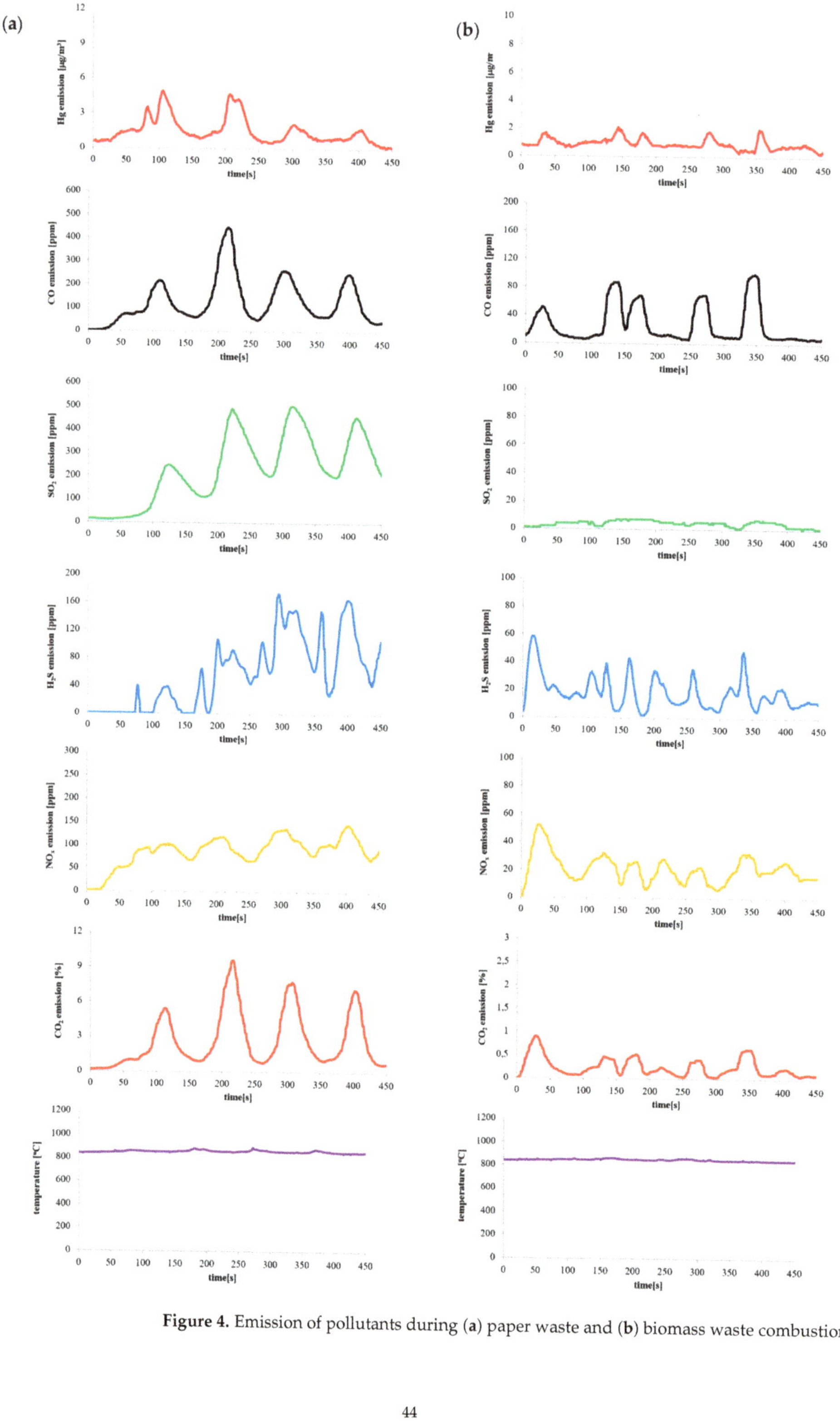

Figure 4. Emission of pollutants during (**a**) paper waste and (**b**) biomass waste combustion.

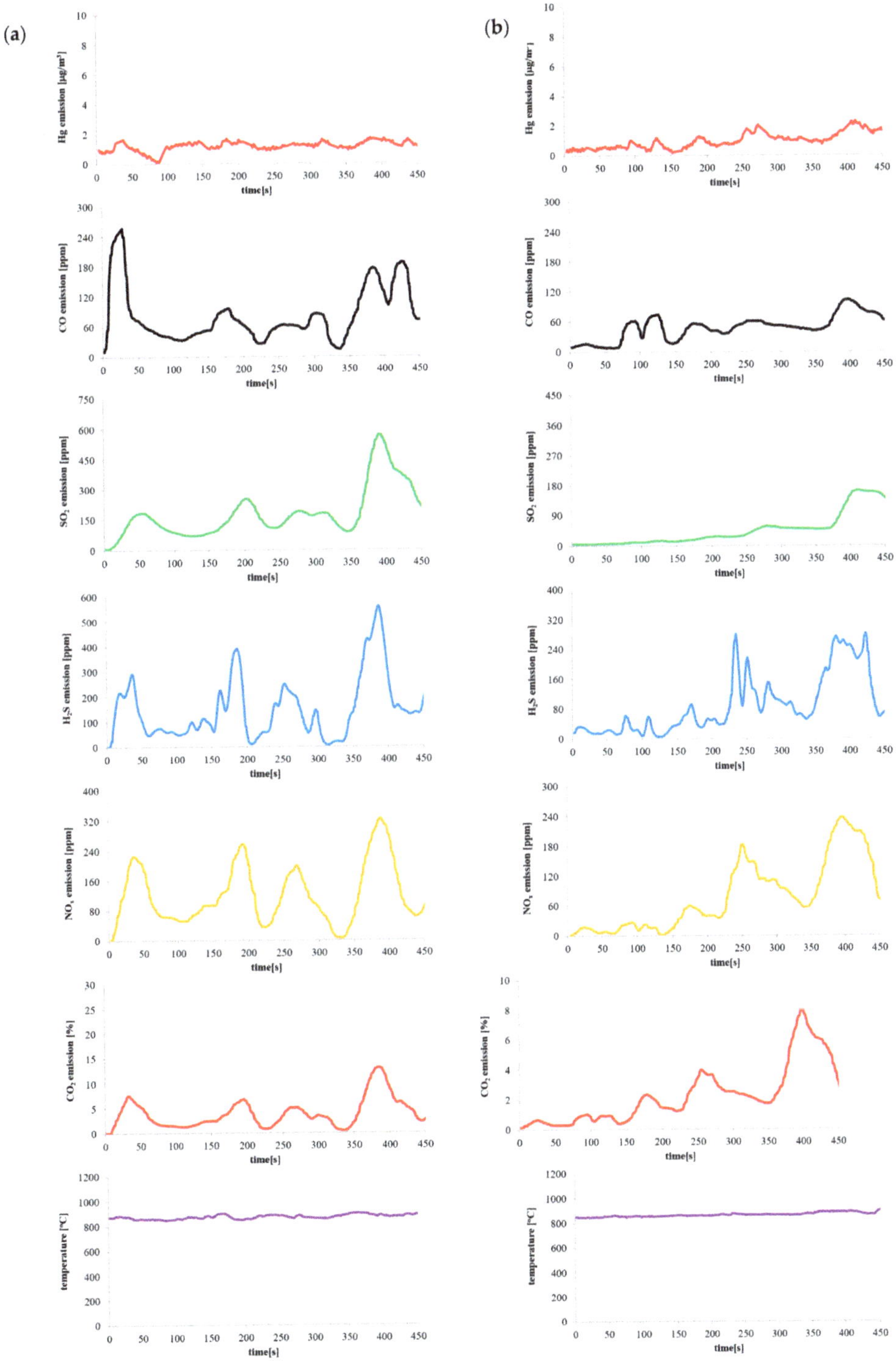

Figure 5. Emission of pollutants during (**a**) hard coal (90%) + polymer waste (10%) and (**b**) coal sludge (90%) + polymer waste (10%) combustion.

In this paper, during coal and coal sludge combustion, the maximum NO_x values were 169 ppm and 81 ppm, respectively. In the case of the combustion of a mixture of hard coal and polymer waste, this value was 319 ppm, and in the case of the combustion of a mixture of coal sludge and polymer waste this value was 235 ppm. Ref. [70] presents NO_x emission amounts from the combustion of various plastics in fluidized bed conditions; they found different values, even up to 233.8 ppm, which confirms the high amount of NO_x emissions during the incineration of polymer waste.

The temperature shown in Figures 2–5 is the temperature in the combustion chamber during the combustion process of the presented materials. The task of the temperature controller was to constantly strive to obtain the desired temperature in the combustion chamber.

As mentioned earlier, the rotational speed of the screw feeder was kept constant throughout the measurement. However, this rotational speed was so low that it caused small masses of the tested materials to fall from the edge of the screw feeder. A similar phenomenon was observed by the authors of the paper during the measurements of a large industrial installation at very high fuel flow rates.

Table 4 presents the content of Hg in waste ash.

Table 4. Analysis of material samples during the combustion process.

Parameter/ Material	Content of Hg in Materials' Ash [μg/kg]
Hard coal	6.02
Coal sludge	2.24
Coke waste	38.90
Sewage sludge	35.08
Paper waste	1.45
Biomass waste	6.47
Hard coal (90%) + polymer waste (10%)	4.10
Coal sludge (90%) + polymer waste (10%)	1.52

Depending on the tested material, we discovered that the mercury content in ashes is from 1.45 to 38.90 μg/kg Hg. We stated that coke sludge and municipal sewage sludge have the highest mercury content, both in the samples before incineration and in the ashes. In accordance with Table 5, the average emission of mercury during the combustion process relates to the content of Hg in a material's sample before the combustion. The addition of polymer waste to coal or coal sludge decreases the average mercury emissions.

Table 5. The average value of emission of pollutants for the analysed materials.

Gas Emissions/Material	Hard Coal	Coal Sludge	Coke Waste	Sewage Sludge	Paper Waste	Biomass Waste	Hard Coal (90%) + Polymer Waste (10%)	Coal Sludge (90%) + Polymer Waste (10%)
Hg [μg/m^3]	2.42	1.29	2.73	1.79	1.46	0.96	1.12	0.95
CO [ppm]	167.20	90.38	213.10	153.54	126.61	27.43	81.14	47.68
SO_2 [ppm]	448.68	267.87	206.37	385.77	229.35	4.12	178.26	44.56
H_2S [ppm]	348.43	118.07	33.87	108.46	57.49	18.34	148.12	90.25
NO_x [ppm]	119.73	59.58	69.52	84.60	87.01	20.84	120.54	78.73
CO_2 [%]	8.31	8.77	3.05	3.51	2.71	0.25	3.83	2.28

4. Conclusions

1. Due to the increasingly restrictive legal regulations regarding environmental aspects, it is extremely important to constantly recognize the course of coal and waste combustion, also from the point of view of the emissions of gases released into the atmosphere.

2. The mercury content of materials varies and this depends on their origin.
3. Among the materials considered, coke waste and sewage sludge are distinguished by their high mercury content: 523.16 μg/kg and 527.81 μg/kg, respectively.
4. The value of Hg emission during the combustion process depends on the initial mercury content in the waste.
5. The addition of a polymer to 10% of coal fuels leads to a reduction in mercury emissions in exhaust gases. This reduction is 53.72% in the case of hard coal (90%) + polymer waste (10%) and 26.36% in the case of coal sludge (90%) + polymer waste (10%).
6. The results of the conducted tests show the adequacy of mercury release during the considered materials' combustion in terms of the emissions of the other compounds considered (CO, CO_2, SO_2, H_2S and NO_x).
7. The higher NO_x emission values during the combustion of coal fuels–polymer mixtures result from the high nitrogen content compared to coal or coal sludge in the considered polymer.
8. Under the considered process conditions, small amounts of mercury in waste ashes were found, including 6.02 μg/kg for hard coal ash, 1.45 μg/kg for paper waste ash and 6.47 μg/kg for biomass waste ash.

The tests of the considered materials' combustion presented in the paper are preliminary studies conducted with the use of the introduced research bench. We will continue and expand our research on this aspect, including the analysis of ash.

Model research of mercury emissions can also be carried out in the future, e.g., via a fuzzy logic-based approach [71–74], which constitutes one of the leading artificial intelligence methods.

Author Contributions: Conceptualization, A.K.-K.; fuels' samples preparing, A.K.-K., A.G., B.T., K.K. and K.B.; methodology, A.K.-K., B.T., K.K., K.B. and D.M.; analysis, A.K.-K.; investigation, A.K.-K., B.T., K.K. and K.B.; writing—original draft preparation, A.K.-K. and A.G.; writing—review and editing, A.K.-K., A.G., B.T., K.K., K.B., J.K. and D.M. All authors have read and agreed to the published version of the manuscript.

Funding: The paper was carried out under BS/PB-1-100-3011/2023/P (Czestochowa University of Technology, Faculty of Mechanical Engineering and Computer Science). The tests were conducted using the research apparatus of AGH as part of the co-financing of the statutory activity of AGH University of Science and Technology, Faculty of Energy and Fuels (16.16.210.476).

Institutional Review Board Statement: Not applicable.

Informed Consent Statement: Not applicable.

Data Availability Statement: Data are contained within the article.

Conflicts of Interest: The authors declare no conflict of interest.

References

1. Ustawa z Dnia 14 Grudnia 2012 r. o Odpadach. Available online: https://eli.gov.pl/eli/DU/2013/21/ogl/pol (accessed on 8 January 2013). (In Polish)
2. Rozporządzenie Ministra Klimatu z Dnia 2 Stycznia 2020 r. w Sprawie Katalogu Odpadów. Available online: https://eli.gov.pl/eli/DU/2020/10/ogl/pol (accessed on 3 January 2020). (In Polish)
3. Obwieszczenie Marszałka Sejmu Rzeczypospolitej Polskiej z Dnia 3 Marca 2022r. w Sprawie Ogłoszenia Jednolitego Tekstu Ustawy o Odpadach. Available online: https://isap.sejm.gov.pl/isap.nsf/download.xsp/WDU20220000699/U/D20220699Lj.pdf (accessed on 12 December 2013). (In Polish)
4. Rocznik Statystyczny Przemysłu. GUS, 2021. Available online: https://stat.gov.pl/obszary-tematyczne/roczniki-statystyczne/roczniki-statystyczne/rocznik-statystyczny-przemyslu-2021,5,15.html (accessed on 28 February 2022). (In Polish)
5. *Ochrona Środowiska 2022*; GUS: Warszawa, Poland, 2022. (In Polish)
6. Rozporządzenie Ministra Klimatu z Dnia 11 Września 2020 r. w Sprawie Szczegółowych Wymagań Dla Magazynowania Odpadów. Available online: https://eli.gov.pl/api/acts/DU/2020/1742/text/O/D20201742.pdf (accessed on 11 September 2020). (In Polish)

7. Obwieszczenie Ministra Klimatu i Środowiska z Dnia 18 Listopada 2022 r. w Sprawie Ogłoszenia Jednolitego Tekstu Rozporządzenia Ministra Środowiska w Sprawie Stosowania Komunalnych Osadów Ściekowych. Available online: https://eli.gov.pl/api/acts/DU/2023/23/text.pdf (accessed on 4 January 2023). (In Polish)
8. Rozporządzenie Ministra Gospodarki z Dnia 16 Lipca 2015 r. w Sprawie Dopuszczania Odpadów do Składowania na Składowiskach. Available online: https://eli.gov.pl/eli/DU/2015/1277/ogl (accessed on 1 September 2015). (In Polish)
9. Suszarnie Niskotemperaturowe Czy Technologia Granulacji Osadów? Available online: https://seidel-przywecki.eu/2020/09/03/suszarnie-niskotemperaturowe-czy-technologia-granulacji-osadow/ (accessed on 3 September 2020). (In Polish).
10. *Strategia Postępowania z Komunalnymi Osadami Ściekowymi na lata 2019–2022*; Ministerstwo Środowiska: Warsaw, Poland, 2018. (In Polish)
11. Kijo-Kleczkowska, A.; Środa, K.; Kosowska-Golachowska, M.; Musiał, T.; Wolski, K. Experimental research of sewage sludge with coal and biomass co-combustion, in pellet form. *Waste Manag.* **2016**, *53*, 165–181. [CrossRef] [PubMed]
12. Kijo-Kleczkowska, A.; Środa, K.; Kosowska-Golachowska, M.; Musiał, T.; Wolski, K. Mechanisms and kinetics of granulated sewage sludge combustion. *Waste Manag.* **2015**, *46*, 459–471. [CrossRef] [PubMed]
13. Werther, J.; Ogada, T. Sewage sludge combustion. *Prog. Energy Combust. Sci.* **1999**, *25*, 55–116. [CrossRef]
14. Tomeczek, J. *Spalanie Węgla*; Wydawnictwo Politechniki Śląskiej: Gliwice, Poland, 1992.
15. Chomiak, J. *Podstawowe Problemy Spalania*; PWN: Warszawa, Poland, 1977. (In Polish)
16. Kijo-Kleczkowska, A.; Szumera, M.; Gnatowski, A.; Sadkowski, D. Comparative Thermal Analysis of Coal Fuels, Biomass, Fly Ash and Polyamide. *Energy* **2022**, *258*, 124840. [CrossRef]
17. Nadziakiewicz, J.; Kozioł, M. Co-combustion of sludge with coal. *Appl. Energy* **2003**, *75*, 239–248. [CrossRef]
18. Kijo-Kleczkowska, A.; Środa, K.; Kosowska-Golachowska, M.; Musiał, T.; Wolski, K. Combustion of pelleted sewage sludge with reference to coal and biomass. *Fuel* **2016**, *170*, 141–160. [CrossRef]
19. Kordylewski, W. *Spalanie i Paliwa*; Oficyna Wydawnicza Politechniki Wrocławskiej: Wrocław, Poland, 2005. (In Polish)
20. Kosowska-Golachowska, M.; Luckos, A.; Kijo-Kleczkowska, A.; Musiał, T.; Wolski, K.; Środa, K. Analysis of pollutant emissions during circulating fluidized bed combustion of sewage sludge. *J. Phys. Conf. Ser.* **2019**, *1398*, 012010. [CrossRef]
21. Li-Ming, S.; Shi-Suo, F.; Hua, Z.; Qi-Sheng, Y.; Pin-Jing, H. SO_2 and NO_x emissions from sludge combustion in a CO_2/O_2 atmosphere. *Fuel* **2013**, *109*, 178–183.
22. Saènger, M.; Werther, J.; Ogada, T. NO_x and N_2O emission characteristics from fluidised bed combustion of semi-dried municipal sewage sludge. *Fuel* **2001**, *80*, 167–177. [CrossRef]
23. Shimizu, T.; Toyono, M.; Ohsawa, H. Emissions of NO_x and N_2O during co-combustion of dried sewage sludge with coal in a bubbling fluidized bed combustor. *Fuel* **2007**, *86*, 957–964. [CrossRef]
24. Leckner, B.; Åmand, L.-E.; Lücke, K.; Werther, J. Gaseous emissions from co-combustion of sewage sludge and coal/wood in a fluidized bed. *Fuel* **2004**, *83*, 477–486. [CrossRef]
25. Hartman, M.; Pohořelý, M.; Trnka, O. Behaviour of inorganic constituents of municipal sewage sludge during fluidized-bed combustion. *Chem. Pap.* **2007**, *61*, 181–185. [CrossRef]
26. Nadziakiewicz, J. *Spalanie Stałych Substancji Odpadowych*; Wyd. Gnome: Katowice, Poland, 2001. (In Polish)
27. Rozporządzenie Ministra Klimatu z Dnia 24 Września 2020 r. w Sprawie Standardów Emisyjnych Dla Niektórych Rodzajów Instalacji, Źródeł Spalania Paliw Oraz Urządzeń Spalania Lub Współspalania Odpadów. Available online: https://isap.sejm.gov.pl/isap.nsf/DocDetails.xsp?id=WDU20200001860 (accessed on 22 October 2020). (In Polish)
28. Sushil, S.; Batra, V.S. Analysis of fly ash heavy metal content and disposal in three thermal power plants in India. *Fuel* **2006**, *85*, 2676–2679. [CrossRef]
29. Piecuch, T. Termiczna utylizacja odpadów. *Rocz. Ochr. Sr.* **2000**, *2*, 11–37. (In Polish)
30. Wilk, M.; Gworek, B. Metale ciężkie w osadach ściekowych. *Ochr. Sr. I Zasobów Nat.* **2009**, *3*, 40–59. (In Polish)
31. Vejahati, F.; Xu, Z.; Gupta, R. Trace elements in coal: Associations with coal and minerals and their behavior during coal utilization—A review. *Fuel* **2010**, *89*, 904–911. [CrossRef]
32. Altıkulaç, A.; Turhan, S.; Kurnaz, A.; Gören, E.; Duran, C.; Hançerlioğulları, A.; Uğur, F.A. Assessment of the Enrichment of Heavy Metals in Coal and Its Combustion Residues. *ACS Omega* **2022**, *7*, 21239–21245. [CrossRef]
33. Pazalja, M.; Salihović, M.; Sulejmanović, J.; Smajović, A.; Begić, S.; Špirtović-Halilović, S.; Sher, F. Heavy metals content in ashes of wood pellets and the health risk assessment related to their presence in the environment. *Sci. Rep.* **2021**, *11*, 17952. [CrossRef]
34. Kowalik, R.; Latosínska, J.; Gawdzik, J. Risk Analysis of Heavy Metal Accumulation from Sewage Sludge of Selected Wastewater Treatment Plants in Poland. *Water* **2021**, *13*, 2070. [CrossRef]
35. Shrivastava, S.K.; Banerjee, D.K. Speciation of metals in sewage sludge and sludge—Amended soils. *Water. Air. Soil. Pollut.* **2004**, *152*, 219–232. [CrossRef]
36. Hussain, A.; Priyadarshi, M.; Dubey, S. Experimental study on accumulation of heavy metals in vegetables irrigated with treated wastewater. *Appl. Water Sci.* **2019**, *9*, 122. [CrossRef]
37. Krutul, D.; Szadkowski, J.; Antczak, A.; Drożdżek, M.; Radomski, A.; Karpiński, S.; Zawadzki, J. The concentration of selected heavy metals in poplar wood biomass and liquid fraction obtained after high temperature pretreatment. *Wood Res.* **2021**, *66*, 39–48. [CrossRef]
38. Akintola, O.O.; Bodede, I.A. Distribution and accumulation of heavy metals in Red Cedar (Cedrela odorata) wood seedling grown in dumpsite soil. *J. Appl. Sci. Environ. Manag.* **2019**, *23*, 811–817. [CrossRef]

39. Yang, S.H.; Lee, T.W.; Lee, J.I.; Choi, H. Distribution Characteristics of Hazardous Heavy Metals in Ginseng and Wood-cultivated Ginseng. *J. Food Hyg. Saf.* **2019**, *34*, 325–333. [CrossRef]
40. Dziok, T.; Bury, M.; Burmistrz, P. Mercury release from municipal solid waste in the thermal treatment process. *Fuel* **2022**, *329*, 125528. [CrossRef]
41. Dziok, T.; Kołodziejska, E.K.; Kołodziejska, E.K. Mercury content in woody biomass and its removal in the torrefaction process. *Biomass Bioenergy* **2020**, *143*, e105832. [CrossRef]
42. Guo, S.; Zhang, L.; Niu, X.; Gao, L.; Cao, Y.; Wei, X.-X.; Li, X. Mercury release characteristics during pyrolysis of eight bituminous coals. *Fuel* **2018**, *222*, 250–257. [CrossRef]
43. Burmistrz, P.; Kogut, K. Mercury in bituminous coal used in polish power plants. *Arch. Min. Sci.* **2016**, *61*, 473–488. [CrossRef]
44. Zhang, C.; Chen, G.; Yang, T.; Guoging, L.; Mak, C.; Kelly, D.; Xu, Z. An Investigation on Mercury Association in an Alberta Sub-bituminous Coal. *Energy Fuels* **2007**, *21*, 485–490. [CrossRef]
45. Senior, C.; Helble, J.; Sarofim, A. Emissions of mercury, trace elements, and fine particles from stationary combustion sources. *Fuel Process. Technol.* **2000**, *65*, 263–288. [CrossRef]
46. Wichliński, M. Emisja rtęci z polskich elektrowni w świetle konkluzji BAT. *Polityka Energetyczna* **2017**, *20*, 79–88. (In Polish)
47. Kogut, K.; Gorecki, J.; Burmistrz, P. Opportunities for reducing mercury emissions in the cement industry. *J. Clean. Prod.* **2021**, *293*, 126053. [CrossRef]
48. Dziok, T.; Bury, M.; Bytnar, K.; Burmistrz, P. Possibility of using alternative fuels in Polish power plants in the context of mercury emissions. *Waste Manag.* **2021**, *126*, 578–584. [CrossRef]
49. Wichliński, M.; Kobyłecki, R.; Bis, Z. Emisja rtęci podczas termicznej obróbki paliw. *Polityka Energetyczna* **2011**, *14*, 191–202. (In Polish)
50. Makowska, D.; Wierońska, F.; Dziok, T.; Strugała, A. Emisja pierwiastków ekotoksycznych z procesów spalania paliw stałych w świetle regulacji prawnych. *Polityka Energetyczna Energy Policy J.* **2017**, *20*, 89–102. (In Polish)
51. Lorenz, U. *Skutki Spalania Węgla Kamiennego Dla Środowiska Przyrodniczego i Możliwości Ich Ograniczania;* Szkoła Eksploatacji Podziemnej—Sympozja i Konferencje nr 64: Kraków, Poland, 2005; pp. 97–112. (In Polish)
52. Burmistrz, P.; Kogut, K.; Marczak, M.; Zwaździak, J. Lignites and subbituminous coals combustion in Polish power plants as a source of anthropogenic mercury emission. *Fuel Process. Technol.* **2016**, *152*, 250–258. [CrossRef]
53. Kropotova, S.S.; Kuznetsov, G.V.; Strizhak, P.A. Identifying products of pyrolysis and combustion of materials at incipient stages of fires. *Fire Saf. J.* **2022**, *132*, 103643. [CrossRef]
54. Leichtnam, J.N.; Schwartz, D.; Gadiou, R. The behaviour of fuel-nitrogen during fast pyrolysis of polyamide at high temperature. *J. Anal. Appl. Pyrolysis* **2000**, *55*, 255–268. [CrossRef]
55. Siat, C.; Bourbigot, S.; Bras, M. Thermal behaviour of polyamide-6-based intumescent formulations—A kinetic study. *Polym. Degrad. Stab.* **1997**, *58*, 303–313. [CrossRef]
56. Dabrowski, F.; Bourbigot, S.; Delobel, R.; Bras, M. Kinetic modelling of the thermal degradation: Of polyamide-6 nanocomposite. *Eur. Polym. J.* **2000**, *36*, 273–284. [CrossRef]
57. Luche, J.; Mathis, E.; Rogaume, T.; Richard, F.; Guillaume, E. High-density polyethylene thermal degradation and gaseous compound evolution in a cone calorimeter. *Fire Saf. J.* **2012**, *54*, 24–35. [CrossRef]
58. Zhang, S.; Chen, H.; Zhang, Y.; Zhang, Y.-M.; Kan, W.; Pan, M. Flame Retardancy of High-Density Polyethylene Composites with P,N-Doped Cellulose Fibrils. *Polymers* **2020**, *12*, 336. [CrossRef] [PubMed]
59. Shi, L.; Gong, J.; Huang, D.; Liu, X.; Zhang, G. Numerical Study on the Spontaneous Combustion of High-density Polyethylene. *Procedia Eng.* **2018**, *211*, 621–628. [CrossRef]
60. Hodgkin, J.H.; Galbraith, M.N.; Chong, Y.K. Combustion Products from Burning Polyethylene. *J. Macromol. Sci. Part A Chem.* **2006**, *17*, 35–44. [CrossRef]
61. Gonçalves, C.K.; Tenório, J.; Levendis, Y.A.; Carlson, J.B. Emissions from the Premixed Combustion of Gasified Polyethylene. *Energy Fuels* **2008**, *22*, 372–381. [CrossRef]
62. Frolov, S.M.; Shamshin, I.O.; Kazachenko, M.V.; Aksenov, V.S.; Bilera, I.V.; Ivanov, V.S.; Zvegintsev, V.I. Polyethylene Pyrolysis Products: Their Detonability in Air and Applicability to Solid-Fuel Detonation Ramjets. *Energies* **2021**, *14*, 820. [CrossRef]
63. Anene, A.F.; Fredriksen, S.B.; Sætre, K.A.; Tokheim, L.-A. Experimental Study of Thermal and Catalytic Pyrolysis of Plastic Waste Components. *Sustainability* **2018**, *10*, 3979. [CrossRef]
64. *PN-G-04502:2014-11;* Węgiel Kamienny i Brunatny—Pobieranie i Przygotowanie Próbek do Badań Laboratoryjnych—Metody Podstawowe. Polish Committee for Standardization: Warsaw, Poland, 2014. (In Polish)
65. *PN-ISO 1928:2002;* Paliwa Stałe—Oznaczanie Ciepła Spalania Metodą Spalania w Bombie Kalorymetrycznej i Obliczanie Wartości Opałowej. Polish Committee for Standardization: Warsaw, Poland, 2002. (In Polish)
66. *PN-G-04511:1980;* Paliwa Stałe—Oznaczanie Zawartości Wilgoci. Polish Committee for Standardization: Warsaw, Poland, 1980. (In Polish)
67. *PN-G-04512:1980;* Paliwa Stałe—Oznaczanie Zawartości Popiołu Metodą Wagową. Polish Committee for Standardization: Warsaw, Poland, 1980. (In Polish)
68. *PN-G-04584:2001;* Paliwa Stałe—Oznaczanie Zawartości Siarki Całkowitej i Popiołowej Automatycznymi Analizatorami. Polish Committee for Standardization: Warsaw, Poland, 2001. (In Polish)

69. *PN-G-04571:1998*; Paliwa Stałe—Oznaczanie Zawartości Węgla, Wodoru i Azotu Automatycznymi Analizatorami—Metoda Makro. Polish Committee for Standardization: Warsaw, Poland, 1998. (In Polish)
70. Jankowski, D.; Żukowski, W. Procesy termiczne jako metody redukcji odpadów z tworzyw sztucznych—Badania wstępne w reaktorze ze złożem fluidalnym. *CHEMIK* **2011**, *65*, 1041–1048.
71. Krzywanski, J.; Czakiert, T.; Zylka, A.; Nowak, W.; Sosnowski, M.; Grabowska, K.; Skrobek, D.; Sztekler, K.; Kulakowska, A.; Ashraf, W.M.; et al. Modelling of SO_2 and NO_x Emissions from Coal and Biomass Combustion in Air-Firing, Oxyfuel, iG-CLC, and CLOU Conditions by Fuzzy Logic Approach. *Energies* **2022**, *15*, 8095. [CrossRef]
72. Krzywanski, J. Heat Transfer Performance in a Superheater of an Industrial CFBC Using Fuzzy Logic-Based Methods. *Entropy* **2019**, *21*, 919. [CrossRef]
73. Krzywanski, J.; Urbaniak, D.; Otwinowski, H.; Wylecial, T.; Sosnowski, M. Fluidized Bed Jet Milling Process Optimized for Mass and Particle Size with a Fuzzy Logic Approach. *Materials* **2020**, *13*, 3303. [CrossRef] [PubMed]
74. Krzywanski, J.; Czakiert, T.; Nowak, W.; Shimizu, T.; Zylka, A.; Idziak, K.; Sosnowski, M.; Grabowska, K. Gaseous emissions from advanced CLC and oxyfuel fluidized bed combustion of coal and biomass in a complex geometry facility: A comprehensive model. *Energy* **2022**, *251*, 123896. [CrossRef]

materials

MDPI

Article

Research into Specific Mechanical Properties of Composites Produced by 3D-Printing Additive Continuous-Fiber Fabrication Technology

Peter Pokorný [1], Daynier Rolando Delgado Sobrino [1,*], Štefan Václav [1], Jana Petru [2] and Rafał Gołębski [3]

1 Institute of Production Technologies, Faculty of Materials Science and Technology in Trnava, Slovak University of Technology in Bratislava, Ulica Jána Bottu 25, 917 24 Trnava, Slovakia
2 Faculty of Mechanical Engineering, VŠB—Technical University of Ostrava, 70800 Ostrava, Czech Republic
3 Faculty of Mechanical Engineering and Computer Science, Department of Technology and Automation, Częstochowa University of Technology, 42-201 Czestochowa, Poland
* Correspondence: daynier_sobrino@stuba.sk; Tel.: +421-908-674-132

Abstract: This paper introduces novel research into specific mechanical properties of composites produced by 3D printing using Continuous-Fiber Fabrication (CFF). Nylon (Onyx) was used as the composite base material, while carbon constituted the reinforcement element. The carbon fiber embedment was varied in selected components taking values of 0°, 45°, 90°, and 135° for parts undergoing tensile testing, while one specific part type was produced combining all angles. Carbon-fiber-free components with 100% and 37% fillings were also produced for comparison purposes. Parts undergoing the Charpy impact test had the fibers deposited at angles of 0° and 90°, while one part type was also produced combining the four angles mentioned before. Carbon-fiber-free parts with 100% and 37% fillings were also produced for comparison purposes as with the first part. The Markforged MARK TWO 3D printer was used for printing the parts. These were subsequently scanned in the METROTOM 1500 computed tomography and submitted to the tensile and impact tests. The results showed that adding carbon fiber to the base material increased the volume of defects in the samples as a result of the porosity increase. Although the tensile testing manifested an overall increase in tensile strength Rm of up to 12 times compared to the sample without reinforcement, it was proven that an improper fiber orientation significantly diminished the strength and that combining the four selected angles did not lead to the highest strength values. Finally, the impact tests also showed that fiber-reinforced parts implied up to 2.7 times more work to fracture, and that an improved fiber orientation also led to strength reduction.

Keywords: composites; CFF technology; mechanical properties; computer tomography; 3D printing

Citation: Pokorný, P.; Delgado Sobrino, D.R.; Václav, Š.; Petru, J.; Gołębski, R. Research into Specific Mechanical Properties of Composites Produced by 3D-Printing Additive Continuous-Fiber Fabrication Technology. *Materials* **2023**, *16*, 1459. https://doi.org/10.3390/ma16041459

Academic Editor: Silvia Farè

Received: 17 January 2023
Revised: 30 January 2023
Accepted: 6 February 2023
Published: 9 February 2023

1. Introduction

Usually, a material that consists of two or more components with different properties and a clear interface or limit between the individual components is referred to as a composite material. It is well known that these composites have different properties arising from the same components residing in them. Composite materials are used in various industries such as construction, shipbuilding, automotive, and aerospace. In the engineering industry, they are commonly used in the form of measuring and welding elements, in different grippers' jaws, as components of diverse fixtures and clamping devices for CNC machining, as functional end parts or tools with high mechanical strength, and in the prototyping of parts as replacements for those obtained in more conventional machining processes [1].

In this regard, it has been documented in the states of the art and practice that the 3D printing production process has a great influence on the properties of composites. Similarly, it has also been discussed that the choice of materials that make it possible to meet given performance requirements in a final part or component also implies a series of trade-off decisions and/or compromises between several factors that may end in having a pronounced impact on the obtained composite. In this sense, there has been much relevant research; see, for example, [2], where the mechanical properties of a given composite in a combination of polyamide and continuous carbon fibers have been properly boarded. In their paper, these authors created a hierarchical structure of criteria enabling and facilitating the selection of the proper material while also taking into account physical, economic, environmental, and social impacts. They applied the results of the analyses to the construction of a high-speed train, and in consequence, their research confirmed that such an approach to composite material selection could be an effective tool for designers in general. Similarly, the study presented in [3] examined in detail 3D-printed composite parts by analyzing and comparing the mechanical properties of different composites as well as various production techniques used for their production. This work also provided a detailed review of the scientific literature including current trends and areas of application of composite materials. On the other hand, the authors in [4] studied a composite made of nylon, which was reinforced with fiberglass fiber. They concluded that such composites could be used for complex parts in industry, while also reaching with their analysis a dimensional accuracy of ± 0.15 mm and a surface roughness of Ra = 5.3 μm. This research also recommended that in order to shorten the production process, the density of the filling was to be reduced. In a study presented in [5], the authors also focused on the development of fiber-reinforced materials, while mentioning some shortcomings of composite materials in relation to their mechanical properties, and also pointing out clear gaps in the current states of the art and practice when it came to predicting the performance of composites.

On the other hand, it is generally well known that composite fibers can be oriented in different directions in a 3D printer. There are usually three different approaches when laying down the fiber into the plastic matrix (see [6–8]), i.e., (1) depositing the fiber before the printing process, (2) incorporating the feeder into the same printing head, and (3) laying it down directly on the plastic matrix. The authors in [9] built their own 3D printing device using an experimental print head. This printer was capable of extruding carbon-fiber-reinforced composites, and during this study, they used an ABS matrix reinforced with 10% of continuous carbon fiber. The results of this research also showed that the flexural and tensile strengths of the composite presented higher values when compared to those obtained in a classical injection molding process, where the fiber content is the same. In particular, the flexural strength values increased to a total 127 MPa, while the tensile strength values increased to 147 MPa, leading these authors to logically conclude that the 3D printing of fiber-reinforced components had a great potential in the production of composite parts in practice. In the case of the research presented in [10], several samples were printed using the free-hanging printing method. The printed samples had diverse shapes, i.e., quadrilateral, pyramidal, circular grid, and others. The study compared printing accuracy; examined print morphology, stresses and deformations, relative density, and fiber volume fraction; compared these values with existing well-documented lattice structures. The authors concluded that by placing a continuous fiber in a polymer matrix, it was possible to significantly improve the functional properties of the parts, while also stating that other advantages lied in lower costs for the production of parts from composite materials and in the possibility of optimizing the topology of these.

Similarly, the study in [11] compared two types of fibers used for the additive production of parts; it presented price differences and uses of both carbon and glass fibers, while it also listed areas of industry where both of these fibers were commonly used because of low-weight and high-strength requirements (aerospace, aviation, and sport industries). The parts presented in the paper were produced using a unidirectional placement of the fibers at an angle of 0°, while the authors also analyzed the microstructure, density, and porosity of such parts, and performed tensile and bending tests as well as a quasi-static indentation force test. At the end of this study, the authors identified the advantages, possibilities, but also limitations of reinforcing plastic components with carbon and glass fibers, and more specifically, they also came to the conclusion that it was possible to produce high-performance parts with these composites due to, mainly, their high tensile strength if compared to unreinforced parts. In addition, the research presented by [12] investigated parts in which 70% of their volume was a polymer and the remaining 30% contained a continuous natural fiber. In this case, parts investigated in the paper were printed with XZ and YZ orientations with respect to the 3D printer platform. At the end of the research, the authors concluded that the porosity of the parts was around 4%, which, if compared to glass- and carbon-fiber-reinforced parts (approximately 15%), was considered as a relatively small value.

The study presented in [13] dealt with the research of carbon- and glass-fiber-reinforced ABS parts. In the production of experimental parts, the authors used an approach that lied in depositing the reinforcing fiber after the nozzle. Tensile strength tests and Young's modulus were performed on the parts, leading to an increase of 17% and 21%, respectively, with a simple fiber volume in the parts of 0.6%. On the other hand, the authors in [14] used in their experiment a nylon matrix (Onyx) reinforced with Kevlar fiber. In this study, they proposed and investigated several Kevlar fiber deposition strategies, i.e., triangular, rectangular, and hexagonal types of fills. The samples were made on a Markforged Mark Two 3D printer, while the Markforged Eiger software (https://markforged.com/software, accessed on 7 June 2022)was used to prepare the assembly and simulate fiber placement. After extrusion, the samples underwent the Charpy impact test and a ballistic test using a gas weapon. In the Charpy test, the sample was oriented in two directions, i.e., Z and XY, which were determined by the position V of the groove on the sample. The authors found that the sample oriented in the Z direction had better properties than the sample in XY. They also concluded in the study that the reinforcement had no effect if the fibers were placed parallel to the V-groove, while on the other hand, the best results were obtained with samples reinforced with Kevlar fibers using a rectangular type of filling. Closely related to the previous research, the study in [15] also presented research into composite materials with a carbon-fiber-reinforced Onyx plastic matrix. The produced samples were tested by tensile and bending tests, and the results showed delamination failure as the main cause of component failure in the tensile tests. These findings led the authors to identify a problem in the bonds between subsequent layers and adjacent fiber layers, while they also calculated tensile and flexural strengths of about 560 MPa and 271 MPa respectively.

In the works presented by [16,17], the authors identified and presented a key step forward in the creation of composite materials that are reinforced with carbon fibers, by introducing patented technologies for the production of such composites. They divided these patents in two directions, i.e., patents for new production methods and patents for new structures of composite materials. During their work, these authors also reported an improvement in the mechanical properties when using continuous fibers over short fiber in the production of the composites. At the same time, they also concluded that the strength and stiffness of continuous-fiber-reinforced composites (produced by 3D printing) were still lower than the ones found in conventionally produced components, which is known to be a disadvantage of these materials, limiting their use in industry where tough requirements are placed on their mechanical properties. This study also identified the reason for the low strength, which lied in the same creation of cavities during 3D printing; in other words, they concluded that if fibers are incorporated into the matrix, the porosity increases, making the interfacial bond between the polymer and the fiber weak as well. Similarly, the authors in [18] also investigated the mechanical properties of fiber composite materials produced by additive technology. They compared them with composites made by conventional pressing. In this study, an ABS carbon-fiber-reinforced matrix was used to make the samples. The production of samples was realized on three different 3D printers, and the results showed different mechanical properties in the extruded parts as a result and/or consequence of the specific printer used. In addition, the manufactured parts had differently oriented fibers in the ABS matrix, and as it is well known that the mechanical properties of the parts also depend on the same direction of fiber placement, it was also possible for the authors to conclude that when comparing their 3D-printed extruded parts to parts made by conventional pressing, the extruded ones showed worse properties in each case.

In the same area of research, the paper in [19] had as a goal to investigate the mechanical properties of carbon-fiber-reinforced polymer matrix parts. Tensile and bending tests were performed, while the fiber volume in the matrix was varied. One type of part was reinforced with fibers of different orientations. The fiber volume variation in the matrix was performed by changing the number of layers of fibers. Next, the authors extruded a part with a carbon fiber Onyx matrix. Subsequently, a pressure of 50 bar was applied to the part at a constant temperature (130 °C) for 1 h, leaving the orientation of the fibers along the sample to bear the greatest load. It was found that by applying a pressure and temperature for 1 h, the structure between the layers improved as a result of lower porosity. The authors also found a significant increase in tensile strength of up to 768 MPa and modulus of elasticity of up to 80 GPa compared to non-pressed samples. Other research (see [20]) also made parts by 3D printing, investigating their resulting mechanical properties. In this case, the base material was also Onyx, which was reinforced with carbon fibers. The properties investigated were flexural strength and flexural modulus. Both of these works discussed above allowed the authors of this paper to reinforce their opinion once more on the fact that the distribution of fibers in the matrix is very important because it may lead to an uneven distribution of stresses in the part. For example, if the direction of fiber deposition is conceived in the direction of the force, the structure is significantly strengthened. Knowledge like this allows designers to position and orient the fibers into precisely defined areas of the parts in order to achieve the best possible properties in the end product.

An extensive study of composite materials was also carried out in paper presented in [21]. The authors shared key knowledge about production techniques used for the production of composite materials. The last chapter of this publication is dedicated to the application areas in which composite materials are used. The researchers managed to improve the functional properties of the materials, and as mentioned before, these composite materials were identified to be of great value and use in industries such as aerospace, chemical, sports, and automotive.

As for the classification of composite materials based on the individual components of the composite, the works in [22–24] offer a good approach to this. The authors based their classification with regard to (1) the size (nanocomposites), (2) the reinforcement (fiber, particles, shaping), (3) the matrix material (polymer, ceramics, metal), and (4) biocomposites. Composite materials with different matrix materials (epoxy resin, polyester, Cu, PLA, etc.) are clearly presented and discussed in [25–31]. In these publications, the matrices were reinforced with various materials (sheep wool fibers, TiO2 particles, jute fibers, etc.). In addition, such composites proved to improve properties such as impact resistance, heat resistance, tensile and flexural strength, and wear resistance. In this last classification of composites, it is also important to highlight the very matured and complex research presented by [32] in the field of high performance and/or more novel thermosetting resins such as the cyanate ester/benzoxazine one. These authors clearly proved the effect and relevance of silane in terms of treating the carbon fibers for the creation of composites using the mentioned resins. Likewise, it is also important to mention the research carried out by [33,34]; in the first one, the authors clearly investigated structural, morphological, mechanical, and thermal properties of newly designed polymeric materials using, again, high-performance hybrid fibers to reinforce the polybenzoxazine resins. In this paper, the authors clearly demonstrated and drew key conclusions on the progressive enhancement in the thermal and mechanical properties in the hybrid composites as a result of their treated fibers. As for the case of [34], the authors presented interesting research that boarded hyperbranched-polymer-coated carbon-fiber-reinforced DCBA/BA-a composites, while conducting detailed microscopic analysis (SEM) of their structural, morphological, and mechanical properties, allowing them to verify positive improvements in terms of modulus, bending, and impact strength in the analyzed specimens as a result of a better dispersion and adhesion of the coated carbon fibers within the resin matrix.

In general terms, it is possible to state that the improved properties of composites usually make them useful materials in various application areas such as mortars and plasters in tribological applications, in heat exchangers, or in the construction industry, where several improvements have been made in recent years; see, for example, [35], where the authors implemented short alkali-resistant glass fibers into fine-grained concrete, leading to an improvement in several mechanical properties, or the work presented in [36] where the authors focused on using agricultural waste in the creation of eco-friendly cementitious composite materials. Of course, the success of composites greatly depends on the individual production techniques used in their production, i.e., cold pressing, injection molding, and pressing. However, nowadays, there is no doubt that a key current trend in the production of parts relies more on additive technologies, which are mainly used for rapid prototyping of complex shapes, and where polymers or fiber-reinforced polymers are used as the base material. At present, metals are also used for the additive production of parts; for example, the work in [37] properly investigated some of their properties. However, it is still safe to say that additive technologies remain a tool primarily used with polymers, although there are even authors who have stated (see [38,39]) that fiber-reinforced polymeric composite materials (of various types and compositions) are in competition with some metals, e.g., aluminum itself.

As covered up until this point in this paper, it is clear that some significant research has been already performed and implemented in the area of fiber-reinforced polymeric composite materials. However, the authors of this paper have found that there are still gaps and poor coverage in the states of the art and practice in terms of analyzing how the selection of the proper fiber deposition orientation angles influences the resulting composite, especially if compared to other unreinforced materials. In addition, the consulted literature has also shown that there are only few studies that have used non-destructive tests in the investigation of composites and, hence, the importance and scientific and practical soundness in the use of computer tomography in combination with classic tests as they are performed are presented in this research.

In this sense, the present paper focuses on the analysis of selected mechanical properties of parts produced by 3D continuous-fiber fabrication technology, which, in the case of polymers, is usually referred to in the scientific literature as CFRPs (Carbon-Fiber-Reinforced Polymers). The basic material for the research of the mechanical properties is nylon filled with micro-carbon fibers (Onyx); this material is the base for the Markforged composite parts and is reinforced with continuous carbon fibers. In the individual parts undergoing tensile tests, the positioning or carbon fiber orientation is changed at different parts of the process. In the specific case of one single sample, the fiber placement is performed in a combination of angles with values of 0°, 45°, 90°, and 135°. In the case of other samples, fibers are placed at angles of 0° and 90°. For comparison purposes, carbon-fiber-free parts with 100% and 37% fillings are also produced. For the Charpy impact test, parts with a fiber orientation of 0° and 90° are produced as well. As with the tensile test part, another sample is made in a combination of all the same four angles mentioned before. In order to compare the reinforced and unreinforced parts, carbon-fiber-free parts with 100% filling and with 37% filling are also produced for this case. The production of parts takes place on a Markforged (Watertown, MA, USA) MARK TWO printer. The components are subsequently scanned using a highly precise Zeiss (Oberkochen, Germany) METROTOM 1500 computed tomography system. The non-destructive tomography system allows for observing in detail the distribution of fibers in the matrix of the base material, previous to the realization of tensile and impact tests. This will allow the authors to better understand the whole experimentation process afterward and draw relevant conclusions about it.

Taking into consideration all of what has been mentioned in the above elements, the main aim of the research is to analyze how a controlled orientation of the reinforcing element during the design and production process may influence and possibly lead to the improvement of the mechanical properties of the resulting composites. This is to be performed and/or complemented by properly comparing these properties with those of unreinforced materials while making use what will allow the author to draw further conclusions from the study. This is a research space that has not been sufficiently investigated in the states of the art and practice, especially if taken into account that this paper aims to use either classical destructive or non-destructive tests for these purposes at the same time. All these elements mentioned above enhance and constitute the practical and scientific novelty of this research.

2. Materials and Methods

As mentioned in the previous section, the material to be used as matrix was Onyx, which is a nylon that contains micro-carbon fibers in its structure, and that is characterized by high strength, toughness, and chemical resistance [40]. The flexural strength is stated to be 71 MPa, which happens to be 1.4 times higher than that of ABS. In addition, Onyx is a material that can be even further fiber-reinforced. In this study, a continuous carbon fiber was used as the reinforcing material; when Onyx is reinforced with this fiber, it has been proved to achieve up to six times higher strength and eight times higher stiffness, and, thus, this composite has found application in the production of parts that are to replace aluminum in many cases, which enhances its relevance and its use potential for many industrial sectors [41].

For the purposes of programming the test samples, the Eiger software was used within this research. This software allows for setting composite printing parameters and then generating print paths. On the other hand, for the purposes of producing the parts, a Markforged Mark Two 3D printer pertaining to the company ADMASYS SK s.r.o. (Šaľa, Slovakia) was used.

As also mentioned before, in order to scan the parts and observe the deposition of fibers in the matrix, the Zeiss METROTOM 1500 computed tomography system was used. As it is well known and has been stated in [42], computed tomography is one of the best non-destructive methods for evaluating the internal structure of materials. In addition, at present, this measuring method is increasingly used and is compared with coordinate measuring machines and their techniques; for example, the authors in [43–45] clearly pointed out the benefits of applying computed tomography measurements over other methods.

On the other hand, it is also well known that mechanical tests are used to determine the mechanical properties of materials, in which a test specimen made is exposed to the action of mechanical forces. The force applied to the test specimen may be static, and in consequence, it is possible to break any material by tensile stress under extreme load conditions.

In this context, the Charpy impact test was used to evaluate the amount of energy absorbed by a material during fracture. The test was performed on a pendulum hammer and the energy absorbed by the parts was, in consequence, determined (see [46]).

3. Results and Discussion

The modeling of the experimental samples was the first step in experimental research; these samples had the same shape as the tensile test ones. Similarly, samples for the Charpy impact test were also modeled. Subsequently, these models were allocated on the 3D printer's workspace in the environment of the Eiger software (see Figure 1).

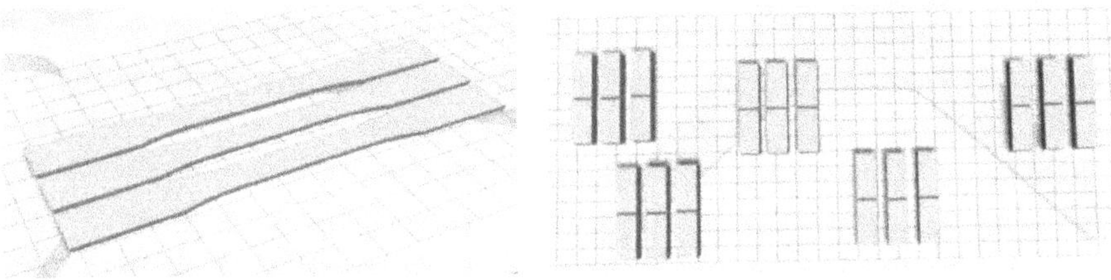

(a) (b)

Figure 1. Allocation of test sample models on the 3D printer's workspace in the environment of the Eiger software: (**a**) tensile test samples; (**b**) Charpy impact test samples.

The main monitored parameter was the distribution and/or orientation of the carbon fibers in the base material Onyx; this is something that can be set in the same Eiger software. The experimentation continued with the creation of four different types of composite samples. The first type created was a sample where the orientation of the carbon fibers changed after each layer, which resulted in a sample combining the fiber orientation at angles of 0°, 45°, 90°, and 135° in a single specimen (Type A). In particular, the first 4 layers were only of Onyx, the next 12 layers were Onyx + carbon fiber with the 5th layer having a 0° orientation, the 6th layer 45°, the 7th layer 90°, the 8th layer 135°, the 9th layer 0°, the 10th layer 45°, the 11th layer 90°, the 12th layer 135°, the 13th layer 0°, the 14th layer 45°, the 15th layer 90°, and the 16th layer having an orientation of 135°. The last 4 layers had no carbon fiber. Thus, in total, the whole set of angles (0°, 45°, 90°, 135°) was used 3 times. The last 4 layers were also made of only Onyx as well. This is an approach that has not been properly explored in the consulted literature and, given the many positives that seem to be involved in it, the authors wanted to fully investigate this course of action.

Next, a sample with a fiber orientation at an angle of 0° (Type B) was made (see Figure 2). On the other hand, the next sample (Type C) had the fibers oriented at an angle of 90° (see Figure 3). In order to compare the samples, carbon-fiber-free samples were also made out of Onyx material only. These samples were made as: Full Core Solid Fill 100% (Type D) and Infill Triangle Network 37% (Type E).

Figure 2. Carbon fiber oriented at an angle of 0°—sample for tensile test.

Figure 3. Carbon fiber oriented at an angle of 90°—sample for tensile test.

In Figure 4, it is possible to see a summarized take on the Charpy impact test. In this regard, it is convenient to specify that the parts for the Charpy test were made of 80 layers. The carbon fiber content was 20%, so 16 layers contained carbon fiber. Carbon fibers began to be deposited only in the first layer under the notch. The notches were printed.

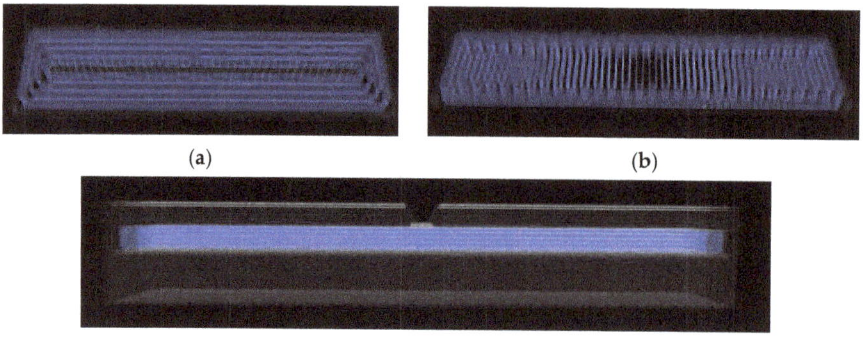

(a) (b) (c)

Figure 4. Carbon fiber orientation: Samples for the Charpy impact test: (**a**) fiber orientation at an angle of 0°; (**b**) fiber orientation at an angle of 90°; (**c**) side view of the sample and the beginning of the carbon fibers under the notch.

3.1. Scanning of the Printed Samples

After printing the samples that were to undergo the tensile and the impact tests, these proceeded to be scanned first. In this case, the VGStudio MAX 3.0 software was used to evaluate the results of the scanning process. A total of three pieces were made from each type of sample. Figure 5 showcases a scan example of a type A sample.

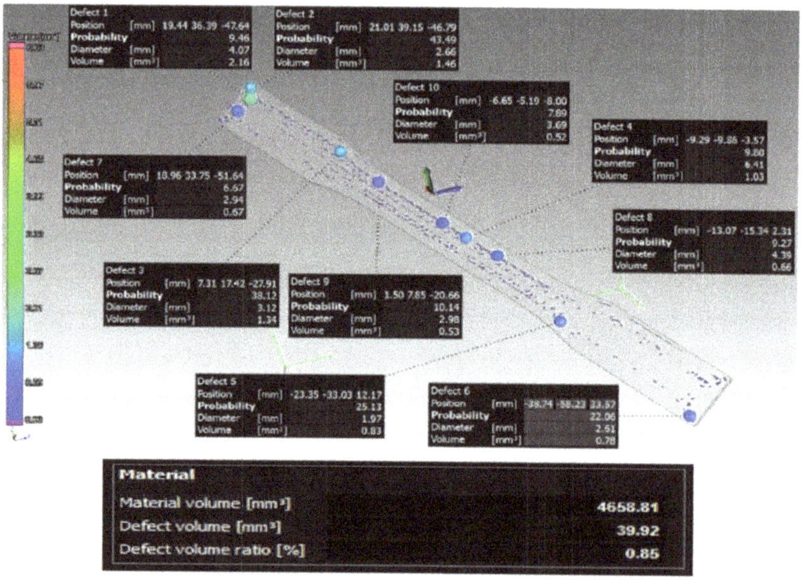

Figure 5. Distribution of the defects and their volume in the case of the type A sample.

The percentage of defects in the individual samples of type A was A1—0.85%, A2—0.47%, and A3—0.71%; these percentages relate to the whole sample volume.

Figure 6 shows a scan of the type B sample.

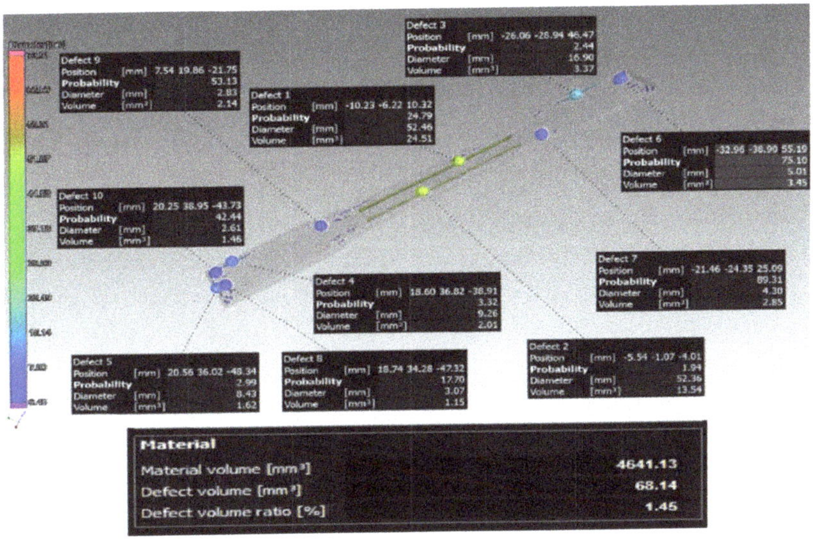

Figure 6. Distribution of the defects and their volume in the case of the type B sample.

Similarly, the percentage of defects of the sample volume in individual samples of type B was: B1—1.45%, B2—1.22%, and B3—1.05%. Here, a clearly visible defect is shown in yellow on both sides of the sample. This phenomenon was caused by a large gap between the carbon fibers in the program generated by the Eiger software. This gap was present because the diameter of the inserted carbon fiber was 0.125 mm and because it was not possible to insert the fibers evenly in the sample, due to its shape and size. In this regard, the Eiger software calculated and determined the gap with respect to the carbon fiber diameter (the fiber could not be inserted in this gap), and in consequence, this appeared to be a defect in the scans, when in fact, it was not.

Figure 7 presents a tomographic scan of a type C sample.

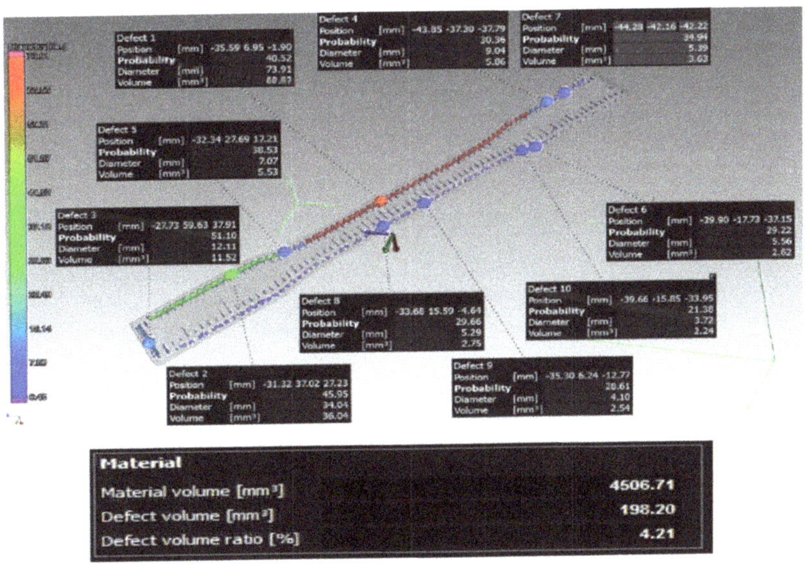

Figure 7. Distribution of the defects and their volume in the case of the type C sample.

This type of carbon fiber storage insertion caused the largest number of defects compared to the previous types of carbon fiber deposition. The percentage of defects of the sample volume in individual samples was: C1—4.21%, C2—2.69%, and C3—2.67%. The defects in this case were distributed around the perimeter of the sample and, at the point of occurrence of defects, the carbon fiber generally bended by a total of 90°. It was precisely in such places where it was not possible to fill the space with the Onyx base material, which led to the appearance of such defects.

Figure 8 shows scans of the type D and type E samples where it is possible to see the distribution of defects. These were carbon-fiber-free comparative samples.

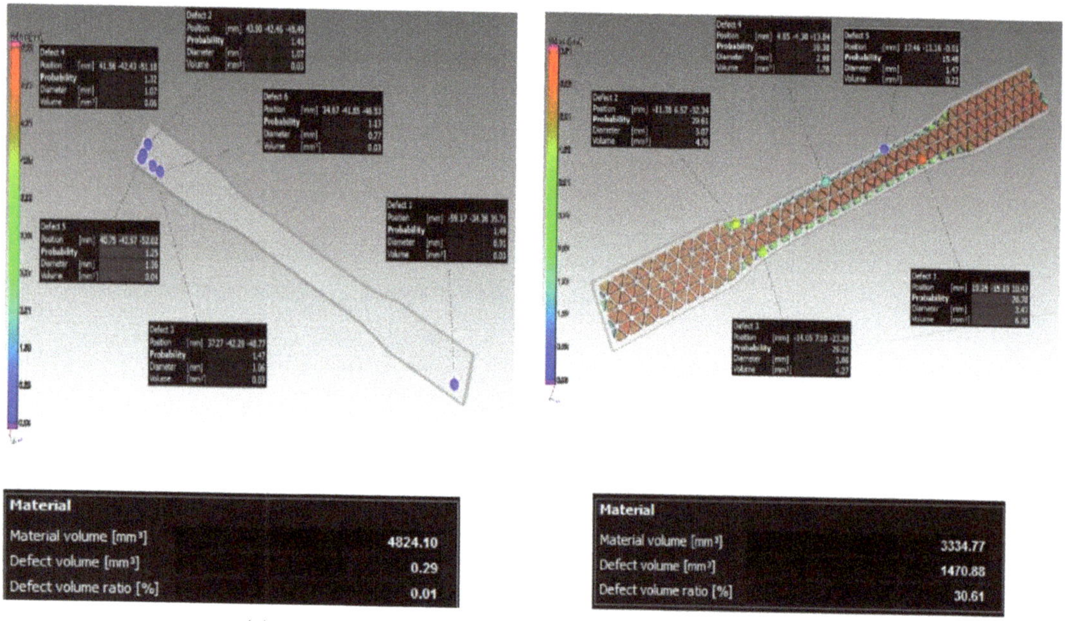

(**a**) (**b**)

Figure 8. A general view on the distribution of the defects and their volume in the case of the samples without carbon fibers: (**a**) type D sample; (**b**) type E sample.

Similarly, as for the case of the percentage of defects of the sample volume in individual samples of type D, the results were: D1—0.01%, D2—0.05%, and D3—0.06%, while in the case of sample type E, the percentage of defects in individual samples was: E1—30.61%, E2—30.56%, and E3—30.53%.

In a similar way, the samples for undergoing the Charpy impact test were also scanned. All of the samples were made by the triangular infill strategy with the exception of the type D sample. For this reason, the detected defect values in volume were also high. Due to the use of this type of strategy, it was not possible to evaluate the effect of fiber deposition on the volume of defects in the impact test specimens. Figure 9 shows a type B sample scan before the sample was submitted to the impact test.

An example of the porosity in the type D samples can be seen in Figure 10. As mentioned before, in this case, carbon fibers were not used as reinforcement for the Onyx base matrix, and the whole sample was extruded with a solid core. For each of the individual samples, the values of porosity with respect to the sample volume reached were as follows: D1—0.74%, D2—1.09%, and D3—1.17%.

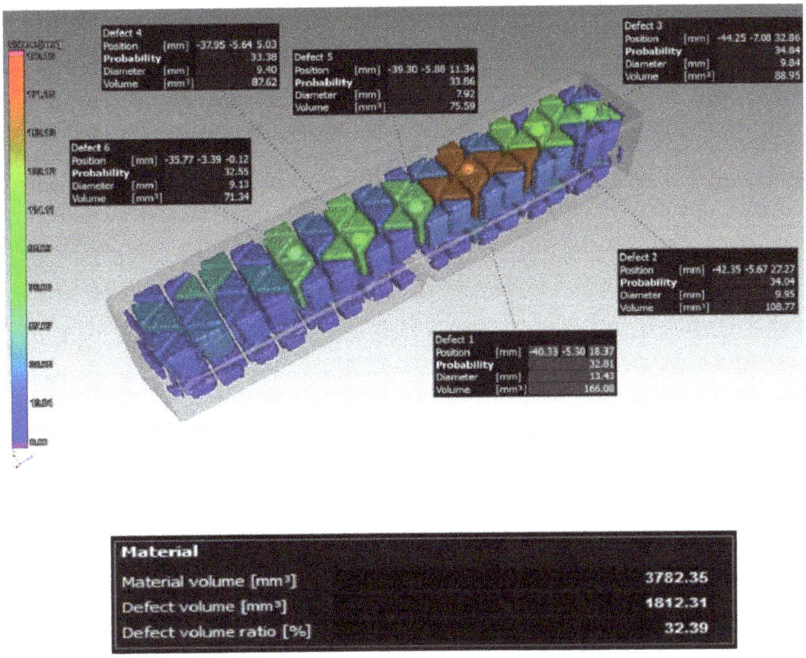

Figure 9. Distribution of the defects and their volume in the case of the type B sample for the purposes of the bending impact test.

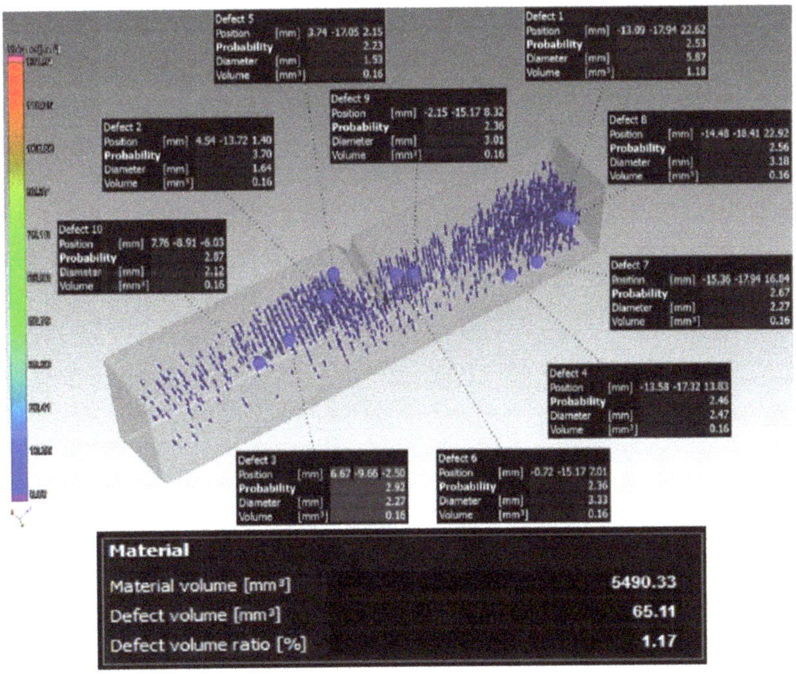

Figure 10. Distribution of the defects and their volume in the case of the type D sample for the purposes of the bending impact test.

From the analysis of the porosity of the samples for the tensile test, the authors found that when combining the layers' deposition angles at values of 0°, 45°, 90°, and 135° in one sample (Type A), the defects were distributed over the entire sample volume. In addition, the authors could also observe a higher number of defects in the area where sample narrowed. On the other hand, in the case of the type B sample, defects were caused by an uneven fiber distribution, and in the type C sample, defects were observed at 90° bending points of the fiber. In addition, if compared to the type D sample without carbon fiber reinforcement, the authors could confirm and, thus, state that by adding reinforcing fibers to the base material, the porosity of the samples increased, and this is something that has also been proven in studies such as [47,48], where the authors also investigated fiber-reinforced composite materials. Thus, after this conclusion, it is certain and verified for the authors of this paper that the reinforcement in the base material causes the formation of pores and has an effect on the interlaminar strength.

On the other side of things, according to Handwerker [49], carbon-fiber-reinforced parts have or may have better properties than parts made of aluminum. However, the main disadvantage in the production of parts from reinforced composites produced by additive technology is their usually weak interlaminar strength and high porosity. In addition, in the case of the type B sample, the authors also observed a gap in the volume of the part created during the deposition of the fiber; however, it is necessary to state that due to the fiber diameter, the software did not allow the fiber to be evenly distributed. One of the reasons for this effect was the shape of the manufactured part, given that the geometry and corner radii on the parts are often limiting conditions for the deposition of the continuous filament in the matrix. This, in consequence, often leads to the formation of defects in components of more complex shapes, as also stated in [50,51].

3.2. Tensile Test

The tensile test was performed according to the ASTM D638 standard. The device on which the test was performed was a universal testing machine, i.e., the LaborTech LabTest 5.250 SP1. The speed of movement of the crossbar during the test was set at 5 mm.min^{-1}.

Figure 11 shows the individual sample types A and B after the tests.

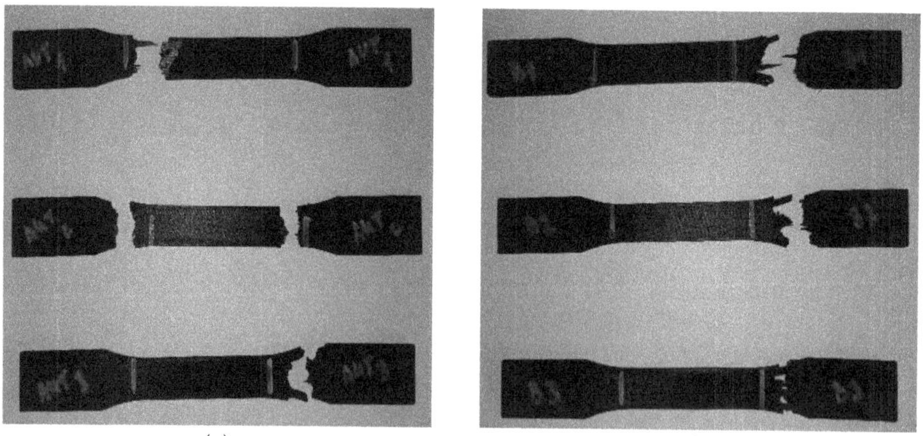

(a) (b)

Figure 11. A take on the ruptured samples: (**a**) type A sample; (**b**) type B sample.

Figure 12 shows the individual sample types C, D, and E after the tests.

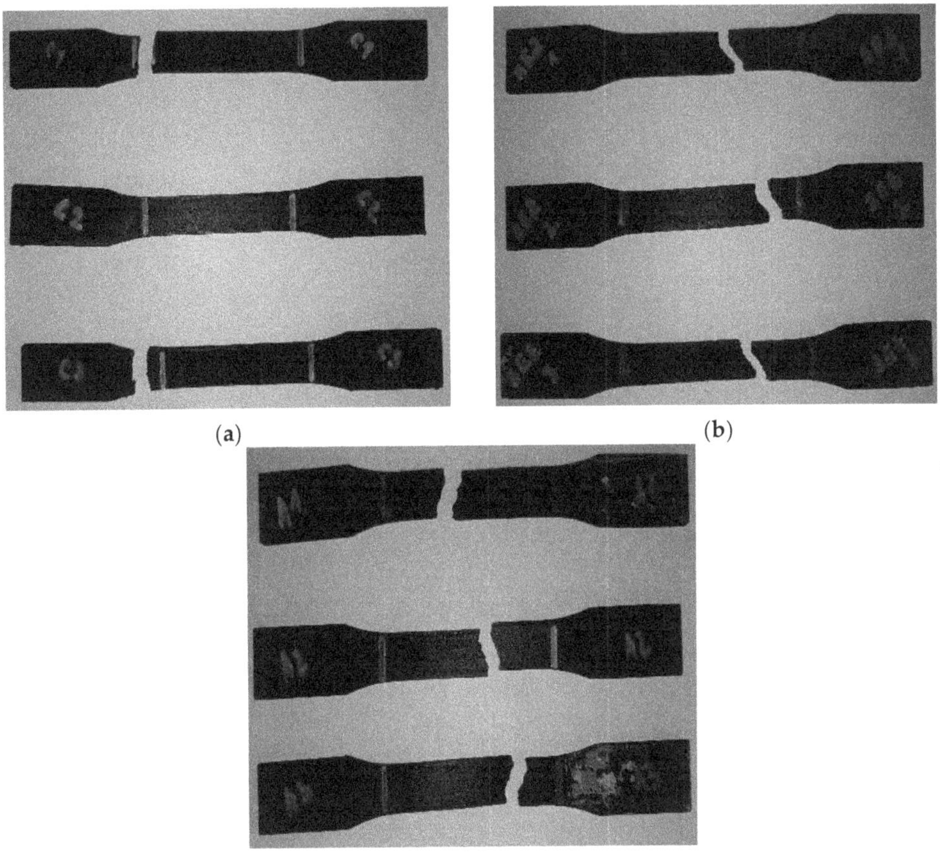

Figure 12. A take on the ruptured samples: (a) type C sample; (b) type D sample; (c) type E sample.

As it can be appreciated in the whole of Figure 11 and in Figure 12a, the samples did not usually break in the narrowest part of the sample between the two yellow lines. This phenomenon was caused by the placement of the filament in the matrix during or when executing the calculations in the Eiger software. More specifically, this was caused when setting the printing parameters. When calculating the paths, the software is not always able to place fiber in all spots, due to the same shape of some parts, and then an uneven distribution of the fibers can, in consequence, take place. This issue may lead to having places without reinforcement in the part, which, of course, results in the weakening of these areas or places and, thus, in the subsequent failure of the sample.

As for Figure 12b,c, the type D and E samples broke where they were expected to break. The slight offset from the center of the part could have been caused by various factors, e.g., layer thickness, melting temperature, printing speed, and infill geometry. Another factor is the same fact that the Onyx material is a micro-carbon-fiber-filled nylon.

On the other hand, Figure 13 shows examples of tensile diagrams for sample types B and D.

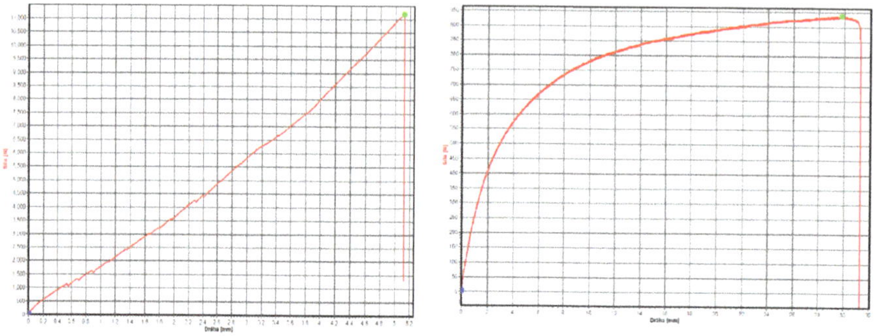

(**a**) (**b**)

Figure 13. A general view of the tensile diagrams for samples: (**a**) type B sample; (**b**) type D.

Table 1 shows the measured values of the tensile strength Rm (MPa) obtained in the static tensile test for individual sample types.

Table 1. Strength values Rm for the individual sample types.

Sample	**Number**			**Average R_m Value (MPa)**
	1	**2**	**3**	
Type A	287.91	311.12	304.89	301.31 ± 12.01 MPa
Type B	400.85	389.12	403.15	397.67 ± 7.57 MPa
Type C	42.86	35.95	37.19	38.67 ± 3.68 MPa
Type D	33.89	33.76	31.49	33.05 ± 1.73 MPa
Type E	21.88	24.32	22.88	23.03 ± 1.23 MPa

A summary of the results of the measured values is shown in Figure 14.

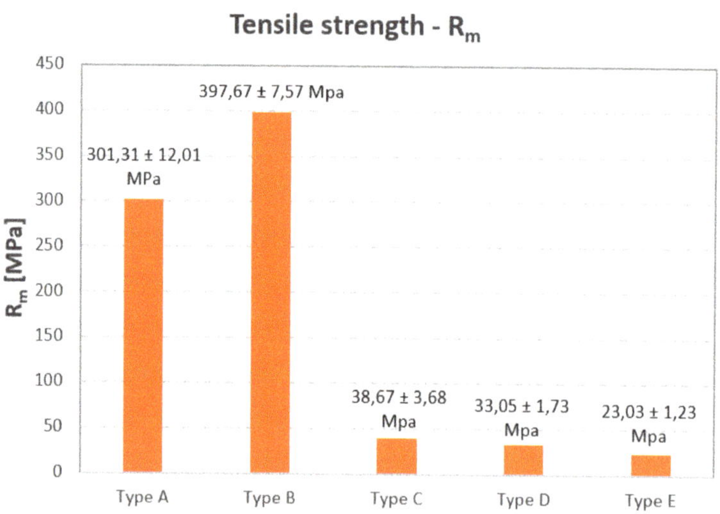

Figure 14. Resulting values of the tensile strength Rm (MPa) obtained in the static tensile test.

In Figure 14, it is possible to see that there is an increase in the tensile strength R_m of the type B sample; this value is approximately 12 times higher if compared to the Type D sample without reinforcement. Similar results have also been confirmed by other similar studies (see [52–57]). In the production of the type A sample, the authors used fiber deposition angles with values of 0°, 45°, 90°, and 135°, and, thus, each subsequent layer had a different fiber orientation. In this sense, the authors assumed that this sample would have the best tensile strength results; however, this did not prove to be true. This lower strength of the samples may be explained by the limitations that often occur during fiber deposition; for example, the fiber is either unevenly distributed in the volume of the matrix, or defects occur at the 90° bending point of the fiber, which, as mentioned before, leads to a higher number of pores in the composite. Other factors that affect the strength are also possible; for example: the layer thickness, the amount of fiber in the matrix, and also the fiber deposition itself [58,59]. In the case of the type C sample, the strength was also low. This was caused by an inappropriate orientation of the fiber with respect to the direction of the load, i.e., the fiber was oriented perpendicular to its direction, which usually leads to inferior strength values. Therefore, when designing composites, it is necessary to determine the direction of the load on the part as a whole; thus, based on this, it is then possible to define the correct deposition direction of the reinforcing fibers.

As it is already known, by changing the orientation of the fibers, it is possible to change the mechanical properties of the components. This is even possible without changing the volume fraction of the fiber in the composite [60]. Composites where the fiber is oriented perpendicular to the direction of the load have the lowest strength, and this is related to the surface roughness of the part [61]. In the case of the type E sample (Infill 37%), this showed a decrease in strength compared to the type D sample (Solid Fill 100%). This was due to the lower volume of material, i.e., the filling of the sample, which was produced by the triangular mesh strategy with a filling of 37%.

3.3. Charpy Impact Test

The test was performed according to the ASTM E23 standard. The device on which the test was performed is an instrumented pendulum impact hammer of the type LaborTech CHK-300. Table 2 shows the measured values of the KV [J] work required for the individual sample types to break.

Table 2. Values of work in KV [J] required to break the test specimen.

Sample	Number			Average Work Values KV [J]
	1	2	3	
Type A	5.13	5.34	5.23	5.23 ± 0.11 J
Type B	4.52	4.52	4.52	4.52 ± 0 J
Type C	1.82	1.67	1.67	1.72 ± 0.09 J
Type D	1.82	2.05	1.97	1.95 ± 0.12 J
Type E	1.44	1.71	1.80	1.65 ± 0.19 J

The above-mentioned ASTM E23 standards define standard test methods for testing metal materials by impact using the Charpy impact test. During these tests, among other parameters, the amount of energy absorbed by a material during fracture is determined. Despite the fact that this standard is used for metallic materials, the authors decided to implement it for the case of the carbon-fiber-reinforced composite presented in this paper. It was the intention of the authors to monitor whether the fibers and their orientation affected the amount of work required to break the sample. From the results in Figure 15, it is clear that the fiber reinforcement did affect the work required to break the sample. In the case of the type A sample, a work of 5.23 J was required to break the specimen; this represented an increase of approximately 2.7 times in terms of the work required for the same purposes if compared to the Type D non-fiber sample.

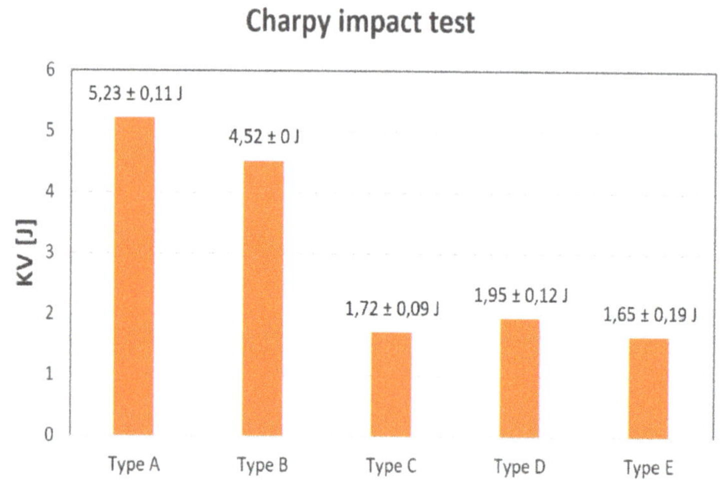

Figure 15. Resulting work values from the Charpy impact test.

The Type B sample also saw approximately a 2.3 times increase in the work required to break. As in the tensile test, the authors can state and conclude that the fiber-reinforcement and their orientation did indeed affect the mechanical properties of the composite. On the other hand, in the case of the type C sample, a decrease of 0.23 J in terms of work was observed. This, like in the case of the tensile test specimens, was due to the unsuitable orientation of the fibers with respect to the direction of the applied force. This, in addition, may also be related to the adhesive characteristics of the fiber and the base material in the production of the samples themselves, which may have led and may indeed lead to delamination as also covered in [62]. Similarly, the decrease in work in the case of the type E sample (Infill 37%) was due to the different fill density compared to the type D sample (Solid Fill 100%). Figure 16 offers a general take on the tested samples of individual types A–E.

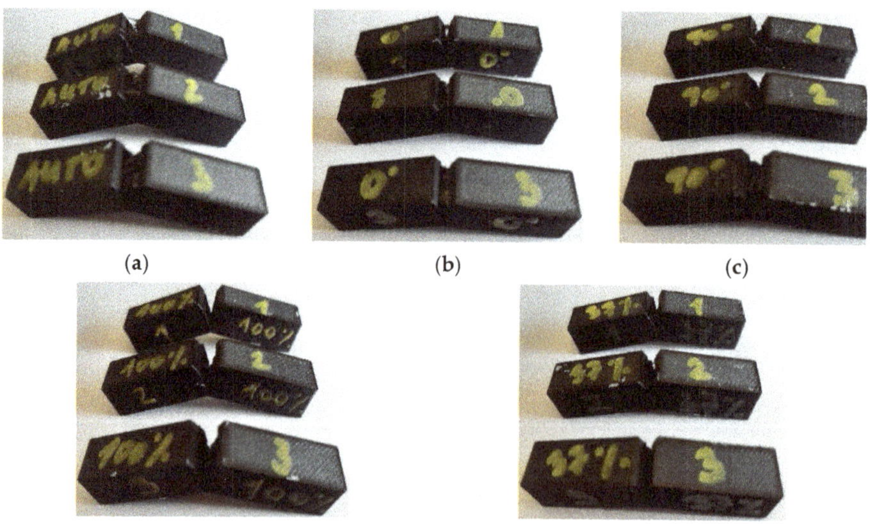

(a) (b) (c) (d) (e)

Figure 16. A general take on the tested samples: (**a**) Type A sample; (**b**) Type B sample; (**c**) Type C sample; (**d**) Type D sample; (**e**) Type E sample.

The higher value of impact energy for the case of sample A can be explained by the fact that layers with different fiber orientations were placed on top of each other. In consequence, each of these layers became essentially stressed in a different direction during impact, which provided a certain advantage in this sense. Overall, this setup or configuration of all layers required higher energy to break the sample.

4. Conclusions

The aim of the research presented in this paper was to investigate the improvement of mechanical properties of composite materials with a controlled placement and/or orientation of the reinforcing element (carbon fiber). Reaching and verifying this goal also implied comparing these properties with those of unreinforced materials. In this sense, the achieved results and conclusions of this research can be summarized in the following three areas:

(1) Porosity of reinforced composite material. In this case, the addition of carbon fiber to the base material was proven to increase the volume of defects in the sample. This came as a result of a porosity increase, which affected the same adhesive capacity of the fiber and the base material and had the potential to lead and indeed led to delamination of the individual layers. All of these facts are more evident and prone to occur in the case of more complex shape components.

(2) As for the mechanical properties—tensile tests, the authors recorded an increase in the tensile strength Rm of up to 12 times higher compared to the sample without reinforcement. In this case, although the combination of several fiber orientations in one sample showed an increase in strength, these results did not achieve the highest strength among the tested samples, which was due to an improper placement of the fiber or their orientation, as it was discussed in detail in the respective section. This fact was exhibited by a sample with a fiber orientation at an angle of 0° (parallel to the direction of the load), i.e., a lower volume of material in the sample causes a decrease in strength, and this was also verified.

The third results area referred to (3) mechanical properties—Charpy impact test. In this case, the fiber reinforcement was shown to affect the magnitude of work required to break the specimens. As in the tensile test, an increase in work of up to 2.7 times higher was recorded. An improper placement of the fiber in terms of its orientation was proven to reduce the strength, which was evident in the Type C sample, where the fibers were oriented at an angle of 0°, perpendicular to the direction of the force. In addition, it was also proven that, as with the tensile tests, a lower volume of material in the sample also led to a decrease in strength.

The results of this paper overall indicate a clear need of further addressing the improvement of mechanical properties of composite materials. It was proven in this research that by controlling the orientation of the reinforcing elements (in our case, carbon fibers), it was also possible to control the properties of the end products, which is especially important during the analysis of the loading forces that act on the part or specimen under study. This fact is even more relevant and evident in the case of parts of complex shapes. It was also demonstrated that although the porosity of the part increases by inserting fibers into the base material, a better strength is still achieved. Even when this fact may be seen as a paradox from a classic metal-based point of view, i.e., porosity is never good when referring to metals, in the case of plastics, it all proved to be a benefit if the proper fiber orientation is taken into account as it was investigated and presented in this paper. This is a finding that has received poor attention and has been, up until this point, poorly explored in the analyzed literature. At the same time, it was also verified that there are other restrictions on the production of composite parts, which mainly relate to a reduced interlaminar strength and susceptibility to delamination. These effects were investigated, and it was concluded that these can be controlled by correctly combining and setting print parameters and fiber orientation.

Through the research, the authors could also further explore and prove the advantages of using the Metrotom computed tomography for the purposes of monitoring the exact placement of the fibers in the material. It became evident that, based on the evaluation of the distribution of carbon fibers in the material, it is possible to analyze errors during the printing of the composite, which usually end up translating into material failure, as it was shown in Figure 11, where the samples were broken in the exact places where printing errors had been detected.

The authors' future research will continue to focus on the application of fibers to complex-shaped prototypes and components. The authors would also like to investigate the use of other fillers coming from waste (see, for example, an interesting take presented by [63]), making clear comparisons to the ones investigated within this paper and, thus, exploring more about the capabilities and drawbacks of polymeric composite materials. Further research could also incorporate other highly precise means such as the SEM analysis to complement the results obtained by computed tomography as well as to further study the differences, advantages, and disadvantages between both approaches in terms of unraveling the mechanism of fracture after performing the mechanical tests.

Author Contributions: Conceptualization, P.P., Š.V., D.R.D.S. and J.P.; Methodology, P.P., D.R.D.S., J.P. and R.G.; Validation, P.P. and D.R.D.S.; Formal Analysis, P.P., Š.V., R.G. and D.R.D.S.; Investigation, P.P., D.R.D.S. and J.P.; Resources, P.P. and Š.V.; Writing—Original Draft Preparation, P.P. and D.R.D.S.; Writing—Review and Editing: P.P., D.R.D.S., R.G. and J.P.; Visualization, P.P. and D.R.D.S.; Supervision, P.P. and J.P.; Project Administration, P.P.; Funding acquisition, J.P. and R.G. All authors have read and agreed to the published version of the manuscript.

Funding: This paper was completed in association with the project Innovative and additive manufacturing technology—new technological solutions for 3D printing of metals and composite materials, reg. no. CZ.02.1.01/0.0/0.0/17_049/0008407 financed by Structural Funds of the European Union and project.

Institutional Review Board Statement: Not applicable.

Informed Consent Statement: Not applicable.

Data Availability Statement: Not applicable.

Acknowledgments: The authors would like to thank in the first place Tomáš Ižvolt, as well as Roman Moravčík, and Michaela Kritikos for their cooperation in the realization of experimental measurements.

Conflicts of Interest: The authors declare no conflict of interest.

References

1. When Is FDM/CFF the Most Suitable? Available online: https://www.h3d.sk/en/fdm-ccf (accessed on 13 December 2021).
2. Zhang, H.; Wu, Y.; Wang, K.; Peng, Y.; Danqi, Y.P.; Wang, D.; Yao, S.; Wang, J. Materials selection of 3D-printed continuous carbon fiber reinforced composites considering multiple criteria. *Mater. Des.* **2020**, *196*, 109140. [CrossRef]
3. Singh, B.; Raman Kumar, R.; Chohan, J.S. Polymer matrix composites in 3D printing: A state of art review. *Mater. Today Proc.* **2020**, *33*, 1562–1567. [CrossRef]
4. Popan, A.; Popan, I.A.; Cosma, C.; Ceclan, V.; Balc, N. Experimental study on 3d printed parts made of continuous fiberglass reinforced polymer. *Acta Tech. Napoc. Ser.-Appl. Math. Mech. Eng.* **2021**, *64*, 81–86.
5. Pervaiz, S.; Qureshi, T.A.; Kashwani, G.; Kannan, S. 3D Printing of Fiber-Reinforced Plastic Composites Using Fused Deposition Modeling: A Status Review. *Materials* **2021**, *14*, 4520. [CrossRef]
6. Prüß, H.; Vietor, T. Design for Fiber-Reinforced Additive Manufacturing. *ASME J. Mech. Des.* **2015**, *137*, 111409. [CrossRef]
7. Mohan, N.; Senthil, P.; Vinodh, S.; Jayanth, N. A review on composite materials and process parameters optimisation for the fused deposition modelling process. *Virtual Phys. Prototyp.* **2017**, *12*, 47–59. [CrossRef]
8. Heidari-Rarani, M.; Rafiee-Afarani, M.; Zahedi, A.M. Mechanical characterization of FDM 3D printing of continuous carbon fiber reinforced PLA composites. *Compos. Part B Eng.* **2019**, *175*, 107147. [CrossRef]
9. Yang, C.; Tian, X.; Liu, T.; Cao, Y.; Li, D. 3D printing for continuous fiber reinforced thermoplastic composites: Mechanism and performance. *Rapid Prototyp. J.* **2017**, *23*, 209–215. [CrossRef]
10. Shuting, L.; Yingguang, L.; Nanya, L. A novel free-hanging 3D printing method for continuous carbon fiber reinforced thermoplastic lattice truss core structures. *Mater. Des.* **2018**, *137*, 235–244. [CrossRef]

11. Goh, G.D.; Dikshit, V.; Nagalingam, A.P.; Goh, G.L.; Agarwala, S.; Sing, S.L.; Wei, J.; Yeong, W.Y. Characterization of mechanical properties and fracture mode of additively manufactured carbon fiber and glass fiber reinforced thermoplastics. *Mater. Des.* **2018**, *137*, 79–89. [CrossRef]
12. Egorov, S.; Tarasova, T.; Terekhina, S. Production technology for polymeric composite materials by additive manufacturing methods. *IOP Conf. Ser. Mater. Sci. Eng.* **2020**, *971*, 022006. [CrossRef]
13. Baumann, F.; Scholz, J.; Fleischer, J. Investigation of a New Approach for Additively Manufactured Continuous Fiber-reinforced Polymers. *Procedia CIRP* **2017**, *66*, 323–328. [CrossRef]
14. Scrocco, M.; Chamberlain, T.; Chow, C.; Weinreber, L.; Ellks, E.; Halford, C.; Cortes, P.; Conner, B. Impact testing of 3D printed Kevlar-reinforced onyx material. In Proceedings of the 29th Annual International Solid Freeform Fabrication Symposium, Austin, TX, USA, 13–15 December 2018.
15. Ghebretinsae, F.; Mikkelsen, O.; Akessa, A.D. Strength analysis of 3D printed carbon fibre reinforced thermoplastic using experimental and numerical methods. *IOP Conf. Ser. Mater. Sci. Eng.* **2019**, *700*, 012024. Available online: https://iopscience.iop.org/article/10.1088/1757-899X/700/1/012024 (accessed on 15 June 2022). [CrossRef]
16. Hu, C.; Sun, Z.; Xiao, Y.; Qin, Q. Recent Patents in Additive Manufacturing of Continuous Fiber Reinforced Composites. *Recent. Patents Mech. Eng.* **2019**, *12*, 25–36. [CrossRef]
17. Sanei, S.H.R.; Popescu, D. 3D-Printed Carbon Fiber Reinforced Polymer Composites: A Systematic Review. *J. Compos. Sci.* **2020**, *4*, 98. [CrossRef]
18. Garcia, J.; Harper, R.; Lu, Y.C. Anisotropic Material Behaviors of Three-Dimensional Printed Carbon-Fiber Polymer Composites with Open-Source Printers. *ASME J. Manuf. Sci. Eng.* **2022**, *144*, 031014. [CrossRef]
19. Saeed, K.; McIlhagger, A.; Harkin-Jones, E.; McGarrigle, C.; Dixon, D.; Shar, M.A.; McMillan, A.; Archer, E. Characterization of continuous carbon fibre reinforced 3D printed polymer composites with varying fibre volume fractions. *Compos. Struct.* **2022**, *282*, 115033. [CrossRef]
20. Krzikalla, D.; Měsíček, J.; Halama, R.; Hajnyš, J.; Pagáč, M.; Čegan, T.; Petrů, J. On flexural properties of additive manufactured composites: Experimental, and numerical study. *Compos. Sci. Technol.* **2022**, *218*, 109182. [CrossRef]
21. Rajak, D.K.; Pagar, D.D.; Kumar, R.; Pruncu, C.I. Recent progress of reinforcement materials: A comprehensive overview of composite materials. *J. Mater. Res. Technol.* **2019**, *8*, 6354–6374. [CrossRef]
22. Yang, Y.; Boom, R.; Irion, B.; Heerden, D.J.; Kuiper, P.; Wit, H. Recycling of composite materials. *Chem. Eng. Process Process Intensif.* **2012**, *51*, 53–68. [CrossRef]
23. Otani, L.; Alves, H.; Melo, J.; Amico, S. Elastic moduli characterization of composites using the impulse excitation technique. *ATCP Phys. Eng.* **2014**. Available online: https://scholar.google.com/scholar_lookup?title=Elastic%20moduli%20characterization%20of%20composites%20using%20the%20impulse%20excitation%20technique&publication_year=2014&author=L.%20Otani&author=H.%20Alves&author=J.%20Melo&author=S.%20Amico (accessed on 15 June 2022). [CrossRef]
24. Ibrahim, I.; Jamiru, T.; Sadiku, R.; Kupolati, W. The use of polypropylene in bamboo fibre composite and their mechanical properties—A review. *J. Reinf. Plast. Compos.* **2015**, *34*, 1347–1356. [CrossRef]
25. Oliveira, M.S.; Pereira, A.C.; da Costa, F.; da Luz, F.S.; de Oliveira Braga, F.; Nascimento, L.F.C.; Lima, P.; Demosthenes, L.C.D.C.; Monteiro, S.N. Fique fiber-reinforced epoxy composite for ballistic armor against 7.62 mm ammunition. In *Green Materials Engineering*; Springer: Cham, Switzerland, 2019; pp. 193–199.
26. Fiore, V.; Di Bella, G.; Valenza, A. Effect of sheep wool fibers on thermal insulation and mechanical properties of cement-based composites. *J. Nat. Fibers* **2019**, *17*, 1532–1543. [CrossRef]
27. Kurdi, A.; Kan, W.H.; Chang, L. Tribological behaviour of high performance polymers and polymer composites at elevated temperature. *Tribol. Int.* **2018**, *130*, 94–105. [CrossRef]
28. Nimanpure, S.; Hashmi, S.A.R.; Kumar, R.; Bhargaw, H.N.; Kumar, R.; Nair, P.; Naik, A. Mechanical, electrical, and thermal analysis of sisal fibril/kenaf fiber hybrid polyester composites. *Polym. Compos.* **2017**, *40*, 664–676. [CrossRef]
29. Jamwal, A.; Prakash, P.; Kumar, D.; Singh, N.; Sadasivuni, K.K.; Harshit, K.; Gupta, S.; Gupta, P. Microstructure, wear and corrosion characteristics of Cu matrix reinforced SiC–graphite hybrid composites. *J. Compos. Mater.* **2019**, *53*, 2545–2553. [CrossRef]
30. Goriparthi, B.K.; Suman, K.N.S.; Rao, N.M. Effect of fiber surface treatments on mechanical and abrasive wear performance of polylactide/jute composites. *Compos. Part. A. Appl. Sci. Manuf.* **2012**, *43*, 1800–1808. [CrossRef]
31. Bhat, T.; Chevali, V.; Liu, X.; Feih, S.; Mouritz, A.P. Fire structural resistance of basalt fibre composite. *Compos. Part A Appl. Sci. Manuf.* **2015**, *71*, 107–115. [CrossRef]
32. Zegaoui, A.; Derradji, M.; Ma, R.; Cai, W.; Medjahed, A.; Liu, W.; Qadeer Dayo, A.; Wang, J. Silane-modified carbon fibers reinforced cyanate ester/benzoxazineresin composites: Morphological, mechanical and thermal degradation properties. *Vacuum* **2018**, *150*, 12–23. [CrossRef]
33. Ghouti, H.A.; Zegaoui, A.; Derradji, M.; Cai, W.-A.; Wang, J.; Liu, W.; Dayo, A.Q. Multifunctional Hybrid Composites with Enhanced Mechanical and Thermal Properties Based on Polybenzoxazine and Chopped Kevlar/Carbon Hybrid Fibers. *Polymers* **2018**, *10*, 1308. [CrossRef]
34. Daham, A.; Zegaoui, A.; Ghouti, H.A.; Derradji, M.; Cai, W.; Wang, J.; Liu, W.; Wang, J.; Moussa, Z. Structural, morphological and mechanical properties of hyperbranched polymers coated carbon fibers reinforced DCBA/BA-a composites. *Compos. Interfaces* **2019**, *27*, 905–919. [CrossRef]

35. Zhang, Q.; Yang, Q.-C.; Li, W.-J.; Gu, X.-L.; Dai, H.-H. Study on model of flexure response of carbon fiber textile reinforced concrete (CTRC) sheets with short AR-glass fibers. *Case Stud. Constr. Mater.* **2023**, *18*, e01791. [CrossRef]
36. de Azevedo, A.R.G.; Amin, M.; Nyarko, M.H.; Saad Agwa, I.; Zeyad, A.M.; Tayeh, B.A.; Adesina, A. Possibilities for the application of agro-industrial wastes in cementitious materials: A brief review of the Brazilian perspective. *Clean. Mater.* **2022**, *3*, 100040. [CrossRef]
37. Pokorný, P.; Václav, Š.; Petru, J.; Kritikos, M. Porosity Analysis of Additive Manufactured Parts Using CAQ Technology. *Materials* **2021**, *14*, 1142. [CrossRef]
38. Perepelkin, K.E. *Armiruyuschie Volokna i Voloknistye Polimernye Kompozity. Monografiya [Reinforcing Fibers and Fibrous Polymeric Composites. Monograph]*; Scientific Basics and Technologies Publ.: St. Petersburg, Russia, 2009; 380p.
39. Skornyakov, I.A.; Tarasova, T.V.; Terekhina, S.M. Investigation of the strength characteristics of samples made of nylon by FFF technology. *ARPN J. Eng. Appl. Sci.* **2019**, *14*, 2427–2432.
40. Onyx—Micro Carbon Fiber Filled Nylon that Forms the Foundation of Markforged Composite parts. Available online: https://markforged.com/materials/plastics/onyx (accessed on 1 February 2022).
41. MATERIAL DATASHEET, Composites. Available online: https://www-objects.markforged.com/craft/materials/CompositesV5.2.pdf (accessed on 1 February 2022).
42. Kritikos, M. Porosity Measurement by X—Ray Computed Tomography: Different Porosity Analysis Application. In: Digital Conversion on the Way to Industry 4.0. ISPR 2020. In *Lecture Notes in Mechanical Engineering*; Springer: Cham, Switzerland, 2021. [CrossRef]
43. Gómez, H.V.; Lee, C.H.; Smith, S.T. Dimensional metrology with X-ray CT: A comparison with CMM measurements on internal features and compliant structures. *Precis. Eng.* **2018**, *51*, 291–307. [CrossRef]
44. Gómez, H.V.; Morse, E.P.; Hocken, R.J.; Smith, S.T. Dimensional metrology of internal features with X-ray computed tomography. In Proceedings of the 29th ASPE Annual meeting, Boston, MA, USA, 12 October 2014; pp. 684–689.
45. Gómez, H.V.; Thousand, J.D.; Morse, E.P.; Smith, S.T. CT measurements and their estimated uncertainty: The significance of temperature and bias determination. In Proceedings of the 15th International Conference on Metrology and Properties of Engineering Surfaces, Charlotte, NC, USA, 2–5 March 2015; pp. 1–8.
46. Moravčík, R.; Hudáková, M.; Černičková, I.; Bošák, O. *Náuka o Materiáloch I. Návody na Cvičenia*, 1st ed.; Trnava AlumniPress: Trnava, Slovakia, 2017; p. 198. ISBN 978-80-8096-242-5.
47. Somireddy, M.; Singh, C.V.; Czekanski, A. Mechanical behaviour of 3D printed composite parts with short carbon fiber reinforcements. *Eng. Fail. Anal.* **2020**, *107*, 104232. [CrossRef]
48. Somireddy, M.; Czekanski, A. Anisotropic material behavior of 3D printed composite structures—Material extrusion additive manufacturing. *Mater. Des.* **2020**, *195*, 108953. [CrossRef]
49. Handwerker, M.; Wellnitz, J.; Marzbani, H. Review of mechanical properties of and optimisation methods for continuous fibre-reinforced thermoplastic parts manufactured by fused deposition modelling. *Prog. Addit. Manuf.* **2021**, *6*, 663–677. [CrossRef]
50. Blok, L.G.; Longana, M.L.; Yu, H.; Woods, B.K.S. An investigation into 3D printing of fibre reinforced thermoplastic composites. *Addit. Manuf.* **2018**, *22*, 176–186. [CrossRef]
51. Dickson, A.N.; Barry, J.N.; McDonnell, K.A.; Dowling, D.P. Fabrication of continuous carbon, glass and Kevlar fibre reinforced polymer composites using additive manufacturing. *Addit. Manuf.* **2017**, *16*, 146–152. [CrossRef]
52. Yasa, E.; Ersoy, K. Additive Manufacturing of Polymer Matrix Composites. *Aircr. Technol.* **2018**, *7*, 147–169. [CrossRef]
53. Ning, F.; Cong, W.; Qui, J.; Wei, J.; Wang, S. Additive manufacturing of carbon fiber reinforced thermoplastic composites using fused deposition modeling. *Compos. Part B* **2015**, *80*, 369–378. [CrossRef]
54. Tekinalp, H.L.; Kunc, V.; Velez-Garcia, G.M.; Duty, C.E.; Love, L.J.; Naskar, A.K.; Blue, C.A.; Ozcan, S. Highly oriented carbon fiber-polymer composites via additive manufacturing. *Compos. Sci. Technol.* **2014**, *105*, 144–150. [CrossRef]
55. Love, L.J.; Kunc, V.; Rios, O.; Duty, C.E.; Elliott, A.M.; Post, B.K.; Smith, R.J.; Blue, C.A. The importance of carbon fiber to polymer additive manufacturing. *J. Mater. Res.* **2014**, *29*, 1893–1898. [CrossRef]
56. Tian, X.; Liu, T.; Yang, C.; Wang, Q.; Li, D. Interface and performance of 3D printed continuous carbon fiber reinforced PLA composites. *Compos. Part A* **2016**, *88*, 198–205. [CrossRef]
57. Matsuzaki, R.; Ueda, M.; Namiki, M.; Jeong, T.K.; Asahara, H.; Horiguchi, K.; Nakamura, T.; Todoroki, A.; Hirano, Y. Three-dimensional printing of continuous-fiber composites by in-nozzle impregnation. *Sci. Rep.* **2016**, *6*, 23058. [CrossRef]
58. Ramesh, M.; Rajeshkumar, L.; Balaji, D. Influence of Process Parameters on the Properties of Additively Manufactured Fiber-Reinforced Polymer Composite Materials. A Review. *J. Materi. Eng. Perform.* **2021**, *30*, 4792–4807. [CrossRef]
59. Rossana, R.; Fernandes, A.Y.; Tamijani, M. Al-Haik. Mechanical characterization of additively manufactured fiber-reinforced composites. *Aerosp. Sci. Technol.* **2021**, *113*, 106653. [CrossRef]
60. Mohammadizadeh, A.; Imeri, I.; Fidan, M.; Elkelany, M. 3D printed fiber reinforced polymer composites—Structural analysis. *Compos. Part B Eng.* **2019**, *175*, 107112. [CrossRef]
61. Todoroki, A.; Oasada, T.; Mizutani, Y.; Suzuki, Y.; Ueda, M.; Matsuzaki, R.; Hirano, Y. Tensile property evaluations of 3D printed continuous carbon fiber reinforced thermoplastic composites. *Adv. Compos. Mater.* **2019**, *29*, 147–162. [CrossRef]

62. Araya-Calvo, M.; López-Gómez, I.; Chamberlain-Simon, N.; León-Salazar, J.L.; Guillén-Girón, T.; Corrales-Cordero, J.S.; Sánchez-Brenes, O. Evaluation of compressive and flexural properties of continuous fiber fabrication additive manufacturing technology. *Addit. Manuf.* **2018**, *22*, 157–164. [CrossRef]
63. Gnatowski, A.; Gołebski, R.; Petru, J.; Pagac, M. Analysis of Thermomechanical Properties and the Influence of Machining Process on the Surface Structure of Composites Manufactured from Metal Chips with a Polymer Matrix. *Polymers* **2022**, *14*, 3501. [CrossRef]

materials

MDPI

Article

Assessment of the Impact of Superficial Contamination and Thermo-Oxidative Degradation on the Properties of Post-Consumer Recycled Polypropylene

Laura Prior [1], Mónica S. A. Oliveira [1,2] and Tatiana Zhiltsova [1,2,*]

1 Department of Mechanical Engineering, Centre for Mechanical Technology and Automation (TEMA), University of Aveiro, 3810-193 Aveiro, Portugal
2 LASI—Intelligent Systems Associate Laboratory, 4800-058 Guimaraes, Portugal
* Correspondence: tvzhiltsova@ua.pt

Abstract: Single-use plastics are a matter of convenience in everyday life, with the majority allocated to packaging production. However, it comes with a high environmental price as its mass recycling is challenging due to the heterogeneity of composition, contaminations of different kinds, and degradation caused by service and processing. This study aims to ascertain the impact of removing contaminants from post-consumer recycled polypropylene (rPP) on its degradation and properties by implementing a systematic approach for decontamination by washing. Four lots of recycled plastics with different degrees of contamination were evaluated via Fourier transform infrared, melt flow indexer, and differential scanning calorimetry and tested for tensile strength. Degradation of the rPP was manifested by the deterioration in ductility, resulting in 14.58% elongation at break (unwashed rPP) compared with 191.41% (virgin PP)) and a significant reduction in oxidation induction time. In the unwashed rPP sample, a wave intensity peak at 1730 cm^{-1}, assigned to the saturated C = 0 stretch of the carbonyl functional group, was detected. This peak was gradually disappearing with an increase in the cleaning efficiency of rPP, highlighting the role of contaminants as degradation catalysts. The cold-washing method showed similar processing and mechanical performance improvement results compared with the other washing methods, while being more environmentally friendly and energy efficient.

Keywords: polymers; recycling; degradation; decontamination by washing; post-consumer; mechanical properties

Citation: Prior, L.; Oliveira, M.S.A.; Zhiltsova, T. Assessment of the Impact of Superficial Contamination and Thermo-Oxidative Degradation on the Properties of Post-Consumer Recycled Polypropylene. *Materials* **2023**, *16*, 1198. https://doi.org/10.3390/ma16031198

Academic Editors: Agnieszka Kijo-Kleczkowska and Adam Gnatowski

Received: 9 December 2022
Revised: 28 January 2023
Accepted: 28 January 2023
Published: 31 January 2023

1. Introduction

Polymers became an indispensable part of modern society, revolutionizing, simplifying, and improving the way of living and being. They combine low price and weight with high versatility of applications and processability, thus, becoming preferred materials for plastic food and non-food packaging, enabling better and safer conditioning of goods and, subsequently, reducing waste [1–3]. However, despite a steady increase in post-consumer plastic collection over recent years, only a third of this amount is effectively recycled, and the remainder is used either for energy recovery or disposed of in landfills [4]. Such a low rate of recycling may be explained by a quality issue inherent to post-consumer recycled plastics, which mainly originated from two sources: contamination either by immiscible polymer inclusions, by primary use, and by waste collection methods [5–8], and thermo-oxidative degradation occurring during synthesis, transformation, service life as a product, and reprocessing after recycling [9]. Thermo and photo-oxidative degradation is known to be a significant factor in the deterioration of recycled PP properties due to the mechanisms of molecular scission and the formation of oxidized moieties [10–12]. Determination of the degradation degree is essential to understanding recycled plastic products' future performance. A decrease in the melting temperature of photodegraded PP was linked to progressively lower molecular weight, lower rates of nucleation in the crystalline structures, and a larger number of chemical irregularities in the form of carbonyl groups [13–15]. An increase in the degradation level of PP due to intensified molecular scission and, therefore, shorter molecular chains, which promote increased mobility at a molecular level, is reflected as a rise in the crystallization temperature [15,16]. Higher crystallization temperatures were associated with a greater number of crystalline particles with the γ form developed during the crystallization of molecules with low molecular weight due to degradation through molecular scission [17].

Alteration of crystallinity degree is another degradation signature of widely researched polymers. The decrease in PP crystallinity, according to Ojeda et al. [12], was attributed to the increased concentration of chemical impurities (i.e., carbonyl and hyperoxide groups) due to antioxidant consumption in PP. However, Rabello and White [13] show that the crystallization degree of PP may vary either way, increasing after short-term exposure to photo-oxidative degradation and then decreasing after prolonged exposure. Short-term exposure to light and oxygen reduces the molecular mass and increases polarity, leading to greater polymer chain mobility and promoting reorganization into more orderly structures. However, with more prolonged exposure, the crystallinity degree becomes limited due to chemical irregularities like carbonyl and hydroperoxides due to continuous photooxidation. Similar results are reported by Wu et al. [14] and Elvira et al. [16] that associate the decrease of crystallinity with the increase of carbonyl groups that form during the oxidation of PP. It should be stressed that one of the most notorious effects of PP degradation is a deterioration of the mechanical properties. The oxidative degradation in semicrystalline plastics, in general, and in PP, in particular, initiates in the crystalline interphase and advances through the amorphous regions. Therefore, when these tie molecules linking polymer crystallites undergo scission, their bearing capacity to withstand the applied stress is significantly diminished, leading to rapid embrittlement even at low oxidation levels [18,19]. These observations were supported by various authors [10,12,20,21], who reported a decrease in the mechanical properties of recycled polyolefins, especially the abrupt reduction in the elongation at break.

Along with the assessment of recycled post-consumer plastics degradation state, significant efforts were dedicated to the investigation of the influence of contaminants removal by washing on their properties [7,10,22–27]. Most of these studies were focused on odor removal by different washing processes, which is a well-known problem of post-consumer plastic originating from food and nonfood packaging caused when organic compounds adhere to the surface and are embedded into the polymer matrix [22–24]. Inks, partially consumed additives, and possible degradation products were identified as origins of odor [22]. The applied washing methods have proven to be effective in partially removing odor contaminants, as demonstrated by Demets et al. [23], where a reduction of odor constituents in contaminated plastic films (rinsed and washed in a friction washer) was detected. However, after extrusion and pelletizing, these contaminants' content increased due to their release during melting and thermal degradation, being, nevertheless, lower than in unwashed plastic waste. Strangl et al. [28] reached a similar conclusion for extruded and pelletized post-consumer mixed packaging polyolefins. The presence of dirt in copolymer PP requires lower activation energy for 10% thermogravimetric weight loss compared with uncontaminated plastic, confirming the accelerating effect of degradation due to contaminants [29]. This conclusion was corroborated in the recent study of Veroneze et al. for artificially contaminated PP. They have reported that the residual contaminants decrease the molar mass and exacerbate the degradation reaction [27]. However, in the studies mentioned above, the impact of recyclates' purity on the mechanical properties of post-consumer recycled plastics was not investigated. Very few studies systematically tackle this important topic. Garofalo et al. [7] reported an improvement in the flexural modulus of post-consumer polyethylene (PE) contaminated with PP due to the removal of low molecular compounds, especially evident when hot-water washing with caustic soda was implemented.

Despite the considerable amount of research dedicated to the post-consumer degradation of PP, it is still essential to carry out structured research mainly focused on evaluating the influence of superficially adhered contaminants responsible for its post-consumer degradation. Therefore, this study aims to investigate the impact of the contamination of PP, recovered from post-consumer plastic packaging waste, on its degradation by assessing the rheological, thermal properties and mechanical performance. To this end, three different washing procedures were implemented, resulting in four batches of recycled plastics with different degrees of contamination. These four batches and the reference virgin PP were investigated with FTIR, MFI, and DSC and tested for tensile strength to identify the influence of contamination on post-consumer PP degradation. The present work's main contributions were identifying the most efficient washing method for the contaminants' removal and clarifying the potential contribution of the contaminants to the degradation of the reprocessed post-consumer PP.

2. Materials and Methods

2.1. Materials

The material used in this study was a recycled PP collected from several distributed recycling locations via the yellow bin system. First, the retrieved material, mainly food and nonfood packaging, was sorted according to the plastic recycling code, and only PP waste was retained. After the sorting stage, it was coarsely cut by hand to simplify the grinding step, which was accomplished using a plastic grinder from PLASMAQ MRU 18-20 (Plasmaq Cm, Lda, Barrosa, Leiria, Portugal). The plastic was shredded three times until the flakes reached a granulometry lower than 4 mm, achieved by sieving through the mesh strainer (mesh size 4 mm). Recycled PP (rPP) flakes were divided into four batches during the third material preparation step. One of the batches was not washed, designated further as (rPPu). In post-consumer plastic waste, the most frequently encountered superficially adhered contaminants are dirt and organic residues, removable by washing. It was accomplished by subjecting three other batches to different washing procedures: washed with cold water (rPPcw), washed with hot water (rPPhw), and hot-washed with cleaning agents (rPPhwca), i.e., caustic soda (NaOH) and surfactant Triton X-100. Washing was accomplished with the mechanical stirrer Yellow Line OST 20 digital (IKA®-Werke GmbH & Co. KG, Staufen, Germany) with an impeller speed of 1300 rpm agitated for 10 min. After washing and separation via sink float method, drying was accomplished in the Meemert Universal Oven UF30 (Memmert GmbH + CO.KG, Schwabach, Germany) at 60 °C for 24 h.

The washing and drying conditions for each recycled PP batch are reported in Table 1. A more detailed description of washing, sink float separation, and drying procedures may be consulted elsewhere [30]. The schematic of the transformation of the post-consumer PP waste into secondary raw material is demonstrated in Figure 1. Virgin PP Moplen HP500N (isotactic homopolymer) from LyondellBasell was used as a reference.

Table 1. Washing and drying conditions of recycled PP.

PP Batch	Washing Temperature (°C)	Drying Temperature (°C)	Cleaning Agents
rPPcw	20	60	none
rPPhw	75	60	none
rPhwca	75	60	Triton X-100 0.3% [1] and NaOH 0.5% [1]

[1] weight content in water.

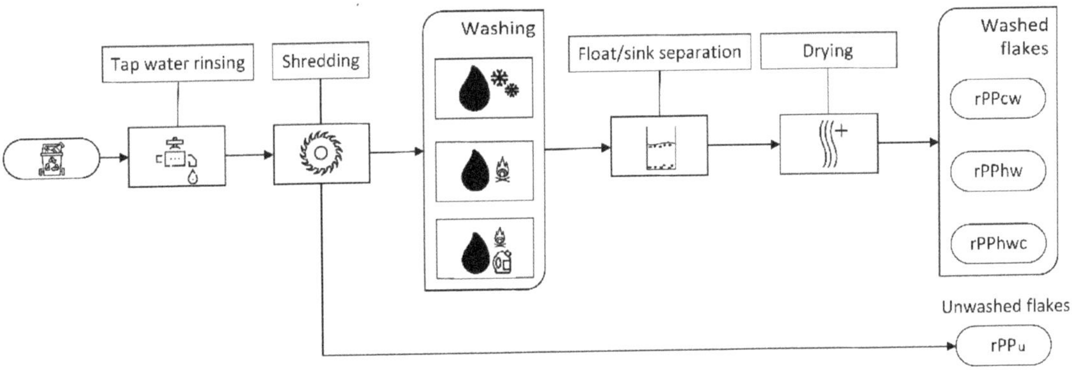

Figure 1. rPP flakes preparation procedure.

2.2. Methods

2.2.1. Melt Flow Index

The measurements of the melt flow index were performed on the machine Göttfert MI-3 (GÖTTFERT Werkstoff-Prüfmaschinen GmbH, Buchen, Germany) according to the standard ASTM D1238-04 [31] with a temperature of 230 °C and a load of 2.16 kg. Ten samples were analyzed for each recycled PP batch and virgin PP.

2.2.2. Differential Scanning Calorimetry

Melting (T_m) and crystallization (T_c) temperatures, and melting (ΔH_m) and cold crystallization (ΔH_c) enthalpies were obtained by differential scanning calorimetry (DSC) from samples with an approximate weight between 5 and 10 mg. DSC tests were conducted, based on the guidelines given by the standard ASTM D3418-03 [32], on DSC Discovery 250 (TA Instruments, New Castle, DE, USA), and data were treated with TRIOS software (proprietary software of TA Instruments). For each PP batch, three samples were analyzed. Each sample was exposed to two heating and cooling cycles, where the first one aimed to "erase" the thermal history of the material. Only the data from the second cycle was collected and analyzed. The samples were stabilized at 20 °C, then heated up to 200 °C, and cooled down to 20 °C at a heating and cooling rate of 10 °C/min.

Each sample's degree of crystallinity (χ) was calculated by equation 1 [33].

$$\chi(\%) = (\Delta H_m/(\Delta H^0{}_m) \times 100 \quad (1)$$

where ΔH_m (J/g) is the melting enthalpy of the polymer under analysis, and the melting enthalpy of 100% crystalline PP ($\Delta H^0{}_m$) is 207 J/g [33].

Oxidation induction time was obtained according to the detailed procedure in ASTM D3895-14 [34]. PP samples were films covering the bottom of the aluminium crucible without a lid, with a mass between 5 and 10 mg. They were heated under a nitrogen atmosphere from ambient temperature up to 200 °C at a heating rate of 20 °C/min and maintained isothermally for 5 min. Next, N_2 atmosphere was changed to an oxygen atmosphere and maintained until the exothermic peak was registered. OIT was determined as an intercept between the baseline of the oxidative reactive exotherm and its steepest linear slope. Three tests were performed for each PP batch.

2.2.3. Infrared Spectroscopy (FTIR)

The FTIR spectra in transmittance mode were acquired within a wavenumber range from 4000 to 350 cm^{-1}, using a resolution of 4 cm^{-1}. A total of 256 spectra were acquired and averaged for a sample (i.e., individual injection molded specimen) tested for each PP batch. The ATR ZnSe crystal was cleaned after each scan with ethanol, and a new background calibration was performed. The measurements were carried out using a Bruker Tensor 27 FT-IR Spectrometer (Bruker Optik GmbH, Billerica, MA, USA).

2.2.4. Mechanical Properties

For the tensile testing, ISO 527-2 [35] type 5 A specimens were obtained by injection molding with the mini injection machine HAAKE MiniJet II (Thermo Fisher Scientific, Waltham, MA, USA). The injection molding processing conditions of the virgin and recycled PP specimens are shown in Table 2.

Table 2. Injection molding processing conditions.

Parameter	Value
Plasticization chamber temperature (°C)	230
Mold temperature (°C)	40
Injection time (s)	3
Injection pressure (bar)	300
Packing time (s)	17
Packing pressure (bar)	240
Plasticization time (s)	180

The tensile tests were conducted on the machine Shimadzu AGS-X-10kN (Shimadzu Scientific Instruments (SSI), Columbia, MD, USA) following the standard ISO 527-1 [36]. These tests were executed at ambient temperature in two steps. First, the specimens were pulled with a tensile rate of 1 mm/min to obtain values for calculating the Young modulus. In the second stage, a tensile rate of 50 mm/min was applied and maintained until the specimens ruptured. The data from this second test was used to determine the yield stress (σ_y) and strain (ε_y), and tensile strength (σ_u) and strain at break (ε_b). It should be noted that the latter is especially relevant for the polymer degradation assessment because of this property's extraordinary sensitivity to any structural change [37]. For each PP batch, five specimens were tested.

3. Results and Discussion

3.1. Melt Flow Index

The melt flow index measurements for virgin and recycled PP with different levels of contamination are shown in Figure 2. The virgin PP has a melt flow index of 12.82 g/10 min, differing slightly from the reference (12 g/10 min) stated in the manufacturer's datasheet. All recycled PP samples show higher MFI than the reference material (vPP). It should be mentioned that the recycled PP samples comprise a wide variety of PP grades for packaging applications of unknown origin, and their direct comparison in terms of MFI with virgin PP is not viable. MFI does not seem to be an adequate technique to ascertain the possibility of polymer degradation in this case. However, some conclusions can be drawn regarding the processability of recycled PP batches as their MFI varies from about 31.56 to 42.14 g/10 min, representative of medium and low viscosity grades, typical grades for food packaging, and, therefore, suitable for reprocessing by injection molding. The contaminated rPPu has a lower MFI of 31.6 g/10 min with a standard deviation of 13.04, i.e., higher melt flow index dispersity, which may generate difficulties during processing. It is worth mentioning that, according to the appearance of the wash water (dark grey) and to the presence of the sediments (dirt, paper particles, heavier plastic flakes) after float sink separation, superficially adhered contaminants present in post-consumer PP are mainly particulate contaminants. Two factors may explain the high standard deviation and lower MFI of unwashed samples: the uneven concentration of the particulate matter contaminants on the surface of the flakes and the fact that the particulate matter contaminants have a similar effect to that of a filler. After removing contaminants by washing, rPPcw, rPPhw, and rPPhwca samples had higher MFI and significantly lower dispersity of measurements than the rPPu. The mean MFI values of the three applied washing methods are quite similar, varying between 40.4 and 42.1 g/10 min, which is representative of reground of injection molding and blow molding packaging.

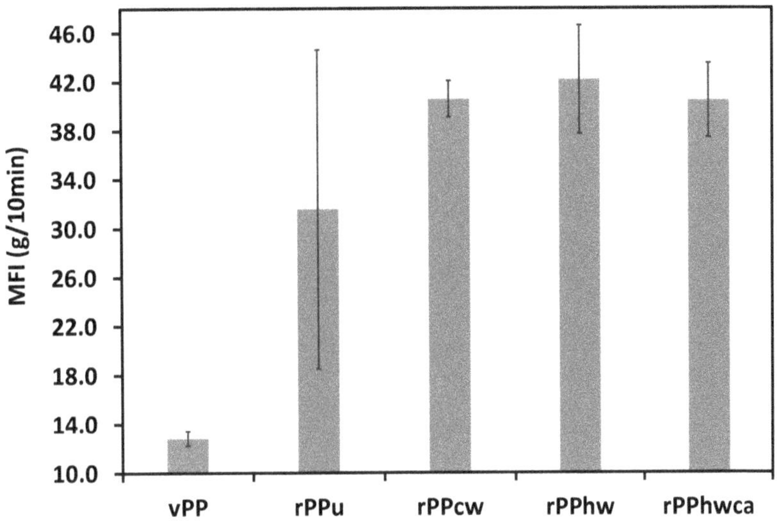

Figure 2. Melt flow index of virgin and recycled PP.

It should be stressed that rPPcw, rPPhw, and rPPhwca have larger MFI dispersity than virgin PP, which is also expected and may be explained by the heterogeneity of recycled PP composition from sample to sample, leading to variations in the respective melt flow indexes. The material batch with the less standard deviation of 1.49 was rPPcw, probably indicating that the cold-water washing method, consuming less energy and with low environmental impact, is the most efficient for recycled PP rheological stability.

3.2. *Differential Scanning Calorimetry*

3.2.1. Thermal Transitions and Degree of Crystallinity

The relevant data concerning melting peak temperature T_m, melting enthalpy ΔH_m, crystallization peak temperature T_c, and degree of crystallinity, obtained from DSC analysis, are compiled in Table 3.

Table 3. Relevant data from DSC thermograms.

Sample	T_m (°C)	T_c (°C)	ΔH_m (J/g)	χ (%)
vPP	165 (±0.31)	113 (±0.38)	109 (±2.41)	53
rPPu	163 (±0.30)	123 (±0.43)	98 (±0.66)	48
rPPcw	162 (±0.21)	122 (±0.16)	99 (±0.54)	48
rPPhw	163 (±0.76)	122 (±0.60)	97 (±2.69)	47
rPhwca	163 (±0.24)	123 (±072)	100 (±1.32)	49

Numbers in parentheses stand for standard deviation.

There is a slight decrease in the melting temperature in all recycled PP samples. The melting temperature between the different batches of recycled PP fluctuates within a range of 1.2 °C. Therefore, it is impossible to draw any definite conclusion about the impact of washing on the melting temperature. A considerable increase in the crystallization temperature was registered in all recycled PP samples compared to virgin PP (Figure 3). PP is a polymorphic material and can crystallize in different crystalline forms, the most common and stable being the α form, but the γ form is also likely to be found in PP and can even coexist with the α [38,39]. Therefore, this increase in crystallization temperature may be attributed to the degradation of recycled PP, which provokes molecular scission, resulting in shorter chains with lower molecular weight and, therefore, crystallization in γ form likely to occur at higher temperatures, as advocated by De Rosa et al. [17]. The difference in the peak crystallization temperature is barely perceptible between the different batches of washed and contaminated recycled PP. Therefore, no definite conclusions can be formulated regarding the influence of washing on the crystallization temperature.

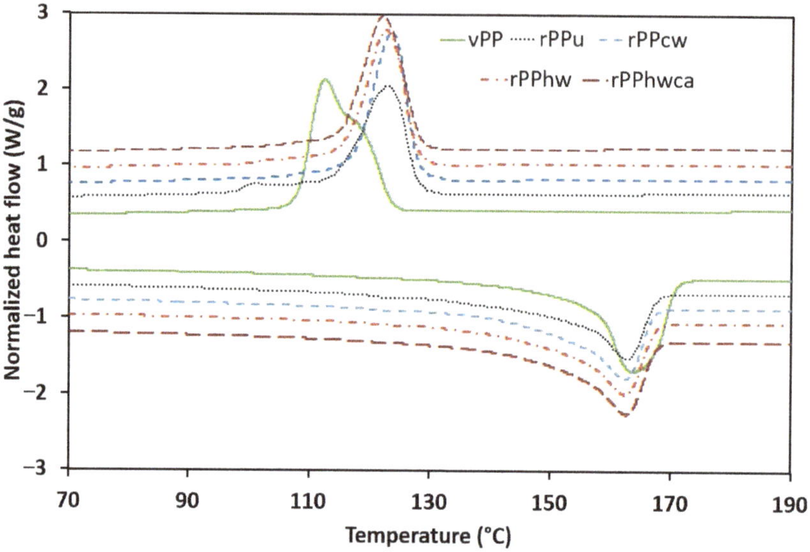

Figure 3. DSC curves for virgin and recycled PP (Upper curves from the second cooling scan and lower curves from the second heating scan have been arbitrarily shifted to improve comparability. Exothermal direction is up).

The degree of crystallinity was reduced in all recycled PP samples, varying between 47 and 48.5% compared with vPP (52.5%), as shown in Table 3. These results are corroborated by the data reported by Ojeda et al. [12] and Rabello and White [13], who stated that degraded PP, after prolonged exposure to photo-oxidation, has lower crystallinity content. Regarding the influence of the removal of contaminants on the crystallinity, the highest value of 48.5% was obtained for rPPhwca, while the lowest crystallinity content was obtained for rPPhw. The latter results seem to be in contradiction with the data reported in the literature, as an increase in the crystallinity is to be expected with the removal of the contaminants. A decrease in crystallinity due to contaminants, acting as the hinders of chain packaging, has been reported by Fitaroni et al. [40] for PET samples. Veroneze et al. [27] related a decrease in the enthalpy of fusion, influenced by the diffusivity of the polymer chains to the presence of the contaminants. Nevertheless, only the values for rPPhw are out of the expected trend, while the crystallinity content of the other recycled PP batches slightly increases with decontamination. This out-of-trend value of rPPhw may be explained by the high standard deviation of the correspondent melting enthalpy compared with the other batches (Table 3), essentially due to a lack of homogeneity between the samples.

3.2.2. Oxidation Induction Time

The oxidation induction time test allows accessing the level of stabilization of the material by determining the time of oxidative decomposition. As shown in Figure 4., all the recycled PP samples show considerably lower OIT when compared with vPP. The latter samples oxidize in pure oxygen in about 300 s. All the recycled PP samples show considerably lower OIT than vPP, indicating a significant reduction of the antioxidants' content due to degradation processes induced primary by processing and service life of the constituent's different PP grades. As a result, the polymer matrix of these materials is unprotected against oxidation, which leads to the deterioration of their properties and may compromise future applications. This result highlights the necessity of adding antioxidants to compensate, to some extent, for the deterioration of the post-consumer recycled plastics properties. As can be seen from Table 4, the oxidation induction times of all the recycled PP samples vary between 41.6 and 61.4 s. The latter time is the best performance shown by rPPhwca samples with the most thorough hot-washing treatment with cleaning agents, corroborating various researchers' conclusions [27,29,41] about the accelerating effect of thermo-oxidative degradation due to the presence of contaminants. Besides the antioxidant's destruction and influence of the particulate contaminants, which cannot be entirely removed by washing, the concentration, and composition of residual polymerization catalysts and metal additives may contribute to the formation of degradation products and, thus, decrease the thermo-oxidative stability of PP [42]. The increasing concentration of metals associated with polymerization catalysts in post-consumer PP was demonstrated by Curtzwiler et al. [43].

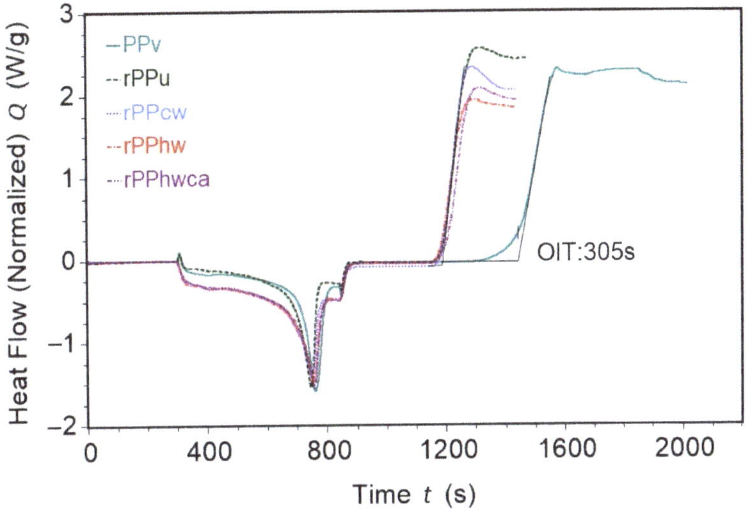

Figure 4. OIT curves of virgin-and recycled PP.

Table 4. Relevant data from DSC thermograms.

Sample	OIT(s)	Standard Deviation
	Mean	±
vPP	274.2	51.98
rPPu	53.0	9.62
rPPcw	41.6	5.54
rPPhw	41.6	1.39
rPhwca	61.4	4.42

It also should be noted that high variability of OIT for virgin PP was detected, which could originate from the storage conditions, leading in turn to its uneven oxidation, reflected as the measurements' dispersity.

3.3. FTIR Spectroscopy Analysis

The FTIR spectra of virgin and recycled PP are shown in Figure 5a, and their detailed presentation in the different wavelength ranges in Figure 5b–d. All spectra are quite similar, displaying the characteristic peaks of PP [44,45] described in more detail in Table 5.

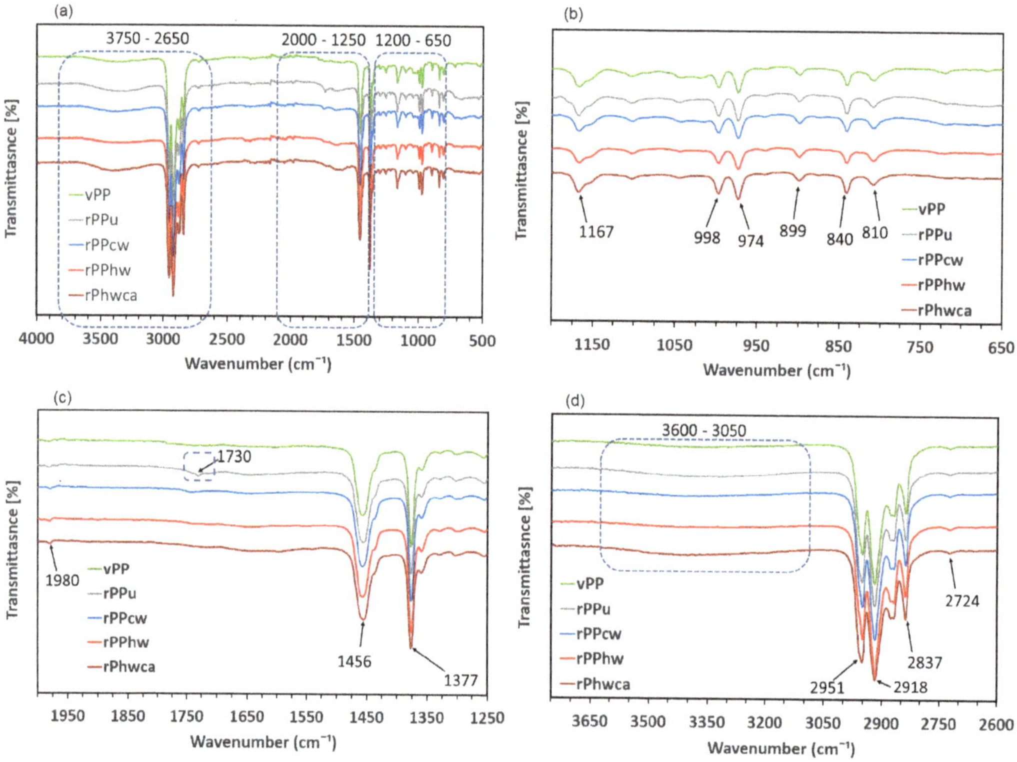

Figure 5. FTIR transmittance spectra for virgin and recycled PP: (**a**) wavelength range: 4000–500 cm^{-1} (**b**) wavelength range: 1200–650 cm^{-1}, (**c**) wavelength range: 2000–1250 cm^{-1}, (**d**) wavelength range: 3750–2650 cm^{-1}.

Table 5. FTIR peak maxima.

Wavenumber (cm^{-1})	Group Vibrations
810	CH2 r, CC s
840	CH2 r
899	CH3 r, CH2 r, CH b
974	CH3 r, CC s
998	CH3 r, CH b, CH2 w
1167	CC s, CH3 r, CH b
1257	CH b, CH2 t, CH3 r
1377	CH3 umbrella bending mode
1456	CH3 b assym., CH2 b
1730	C = O s (carboxylic acid group)
2837	CH2 s sym.
2918	CH2 s assym.
2951	CH3 s assym.
3050–3600	O=H s or N H s

b—bending; r—rocking; t—twisting; s—stretching; w—wagging.

However, some additional intensities of the vibrational bands were detected in some of the samples of the recycled PP under different washing procedures. As shown in Figure 5c, there is a peak at 1730 cm^{-1}, assigned to the saturated C = 0 stretch of the carbonyl functional group in rPPu (unwashed sample), indicating polymer degradation. This peak is not detected in vPP; its intensity decreases drastically in rPPcw and disappears in the rPPhw and rPPhwca. These findings are in line with the conclusions of Day et al. [29], who pointed out an acceleration of polymer degradation in the presence of impurities. Moreover, it corroborates the conclusions of Garofalo et al. [7], who reported an attenuation of intensity at 1730 cm^{-1} wavelength for washed mixed polyolefin-based recyclates. As for the broad water vapor band between 3050 cm^{-1} and 3600 cm^{-1}, highlighted in Figure 5d with the corresponding stretching vibrations of O=H or N-H [42], it is present in all samples, including vPP, its intensity being higher for PPu and rPPhwca. The slightly higher spectral intensity of the latter samples may be attributed to increased water sorption of the unwashed rPP and, thus, not subjected to drying and thoroughly washed rPPhwca, requiring more exposure to water for effective removal of the cleaning agents.

3.4. Mechanical Properties

The mechanical properties, accessed by tensile testing (E—Young's modulus; σ_Y—stress at yield; σ_u—ultimate tensile strength; ε_b—elongation at break) are summarized in Table 6. As shown in Figure 6, the recycled PP samples underwent some reduction in the yield stress and a severe decrease in ductility, rupturing almost instantly. The latter may be attributed to the thermo-oxidative degradation of the recycled PP samples, confirmed by the DSC tests, due to scission and disentanglement of the molecular chains in the amorphous phase and reduction of their interconnections with the crystalline regions [46]. A slight increase in the yield stress was observed for rPPcw, rPPhw, and rPPhwca compared with the unwashed rPP, corroborating the thermal, rheological, and FTIR spectroscopy results, which is also in agreement with the findings of Garofalo et al. [7]. The higher stiffness of all recycled PP samples is reflected in the higher ultimate tensile strength (Figure 7a), indicating the significant deterioration of their ductility, as demonstrated in Figure 7b. The elongation values at break for vPP samples comply with the data (>50%) presented in the Moplen HP500N data sheet. As expected, the vPP samples showed the highest elongation (191.41%) until rupture. The worst resistance to rupture among the recycled PP batches was obtained for rPPu, which broke at 14.58% of elongation without any necking. Thus, Figure 7b confirms the influence of contaminants on the reduction of ductility in recycled PP. It was impossible to draw any conclusions about the influence of the degradation and the implemented washing methods on the elastic properties of recycled PP, as the Young modulus varied between the batches from 1379.73 to 1619.88 MPa. The heterogeneity of recycled PP composition may explain this variation and the high dispersity of the results from sample to sample.

Table 6. Mechanical properties of virgin and recycled PP.

Batch	E (MPa)	σ_y (MPa)	σ_u (MPa)	ε_b (%)
vPP	1545.69 (±58.73)	38.33 (±1.73)	16.48 (±2.31)	191.41 (±31.27)
rPPu	1459.48 (±170.26)	33.20 (±0.79)	22.78 (±5.52)	14.58 (±4.42)
rPPcw	1379.72 (±59.13)	33.90 (±0.70)	23.32 (±4.45)	18.75 (±4.17)
rPPhw	1619.88 (±160.29)	33.90 (±077)	23.51 (±2.98)	16.97 (±1.79)
rPhwca	1516.02 (±190.92)	33.52 (±0.06)	19.68 (±5.07)	20.76 (±4.92)

Numbers in parentheses stand for standard deviation.

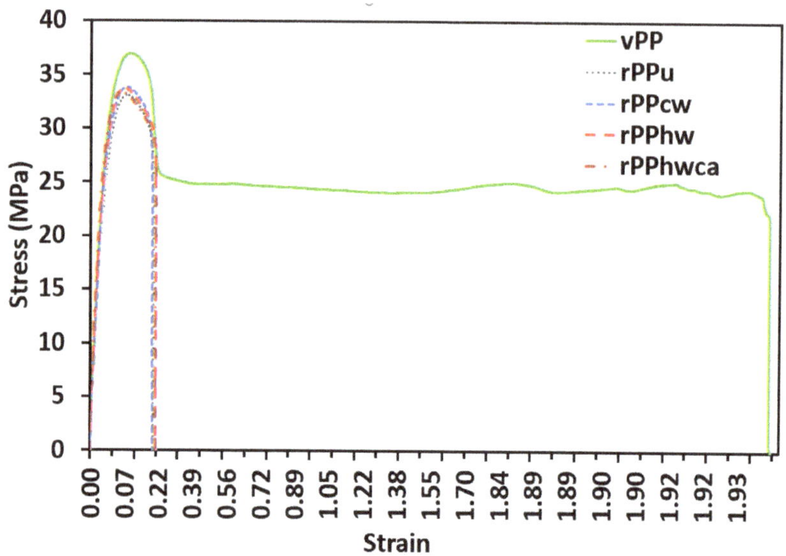

Figure 6. Stress vs. strain response of virgin and recycled PP.

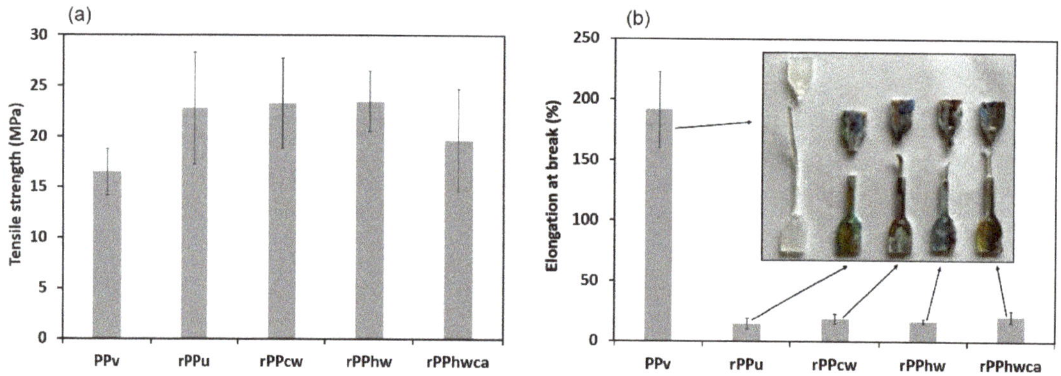

Figure 7. Mechanical properties of virgin and recycled PP: (**a**) tensile strength; (**b**) elongation at break.

4. Conclusions

Post-consumer PP waste contamination, heterogeneity, and degradation can complicate reprocessing and the quality of recycled components. Melt flow index measurements unambiguously pointed out the heterogeneity of the recycled PP composition and their constituent materials' origin (packaging), with MFI varying between 32 to 42 g/10 min, falling within the range of the medium and low viscosity grades. The melt flow index standard deviation was significantly minimized in all washed samples, varying between 1.49 and 4.42 g/10 min, compared with the standard deviation of 13.04 g/10 min of rPPu, indicating an improvement in their processability. Thermal characterization by DSC revealed that the recycled PP, independently of the washing method, required less energy to melt but crystallized under higher temperatures. Both occurrences can be attributed to the degradation processes by accumulating chemical impurities in the form of carbonyl groups resulting from the oxidative processes and reduction of molecular mass due to polymer chain scission. The latter is likely responsible for the lower crystalline content (47–49%) of recycled PP compared with virgin PP (53%). OIT tests demonstrated that for all the recycled PP samples, significantly less time was required to induce oxidation than in the reference vPP sample, indicating some depletion of the initially present antioxidants that protected the polymeric chains from the oxidation processes by preventing their scission. The weakened molecular structure resulted in the drastic reduction of ductility in all samples of recycled PP and, therefore, a significant decrease in the elongation at break (15–21%) compared with virgin PP (191%).

Nevertheless, it should be stressed that a slight ductility improvement was observed in the samples decontaminated by washing. The significance of this observation was confirmed by the FTIR spectroscopy data, where the wave intensity peak at 1730 cm^{-1}, assigned to the saturated C = 0 stretch of the carbonyl functional group, was detected in rPPu. The intensity of this peak decreased drastically in rPPcw and disappeared in the rPPhw and rPPhwca, highlighting the efficiency of washing for attenuation of the thermo-oxidative degradation intensity. These results demonstrate that contaminants act as degradation catalysts, and their removal improves the processability and mechanical properties. It can be concluded that the most promising washing procedure for improving the recycled PP quality is cold washing (rPPcw), as it showed very similar results to rPPhw and rPPhwca methods, with the advantages of energy savings and environmental safety.

Author Contributions: Conceptualization, L.P., T.Z. and M.S.A.O.; Data curation, L.P. and T.Z.; Formal analysis, L.P., T.Z. and M.S.A.O.; Funding acquisition, T.Z. and M.S.A.O.; Investigation, L.P., T.Z. and M.S.A.O.; Methodology, L.P., T.Z. and M.S.A.O.; Project administration, T.Z. and M.S.A.O.; Resources, T.Z. and M.S.A.O.; Supervision, T.Z. and M.S.A.O.; Validation, L.P., T.Z. and M.S.A.O.; Writing—original draft, L.P. and T.Z.; Writing—review & editing, T.Z. and M.S.A.O. All authors have read and agreed to the published version of the manuscript.

Funding: This research was funded by Fundação para a Ciência e Tecnologia (UIDB/00481/2020) and (UIDP/00481/2020) and by Programa Operacional Regional do Centro (CENTRO-01-0145-FEDER-022083) under the PORTUGAL 2020 Partnership Agreement, through the European Regional Development Fund.

Institutional Review Board Statement: Not applicable.

Informed Consent Statement: Not applicable.

Data Availability Statement: Not applicable.

Acknowledgments: T. Zhiltsova is grateful to Portuguese national funds (OE), through FCT, I.P., in the scope of the framework contract foreseen in the numbers 4, 5, and 6 of Article 23, of the Decree-Law 57/2016, of August 29, changed by Law 57/2017, of July 19).

Conflicts of Interest: The authors declare no conflict of interest.

References

1. Carraher, C.E. *Introduction to Polymer Chemistry;* CRC Press: Boca Raton, FL, USA, 2013; ISBN 9781466554955.
2. *Recycling of Polymers: Methods, Characterization and Applications;* Francis, R. (Ed.) John Wiley & Sons: Weinheim, Alemanha, 2017; ISBN 9783527338481.
3. European Commission A European Strategy for Plastics in a Circular Economy 2018. Available online: https://www.europarc.org/wp-content/uploads/2018/01/Eu-plastics-strategy-brochure.pdf (accessed on 12 July 2022).
4. Plastics Europe Plastics—the Facts 2021 An Analysis of European Plastics Production, Demand and Waste Data. Available online: https://plasticseurope.org/knowledge-hub/plastics-the-facts-2021/ (accessed on 24 September 2022).
5. Serranti, S.; Bonifazi, G. Post-Consumer Polyolefins (PP-PE) Recognition by Combined Spectroscopic Sensing Techniques. *Open Waste Manag. J.* **2010**, *3*, 35–45. [CrossRef]
6. Karaagac, E.; Jones, M.P.; Koch, T.; Archodoulaki, V.-M. Polypropylene Contamination in Post-Consumer Polyolefin Waste: Characterisation, Consequences and Compatibilisation. *Polymers* **2021**, *13*, 2618. [CrossRef]
7. Garofalo, E.; Di Maio, L.; Scarfato, P.; Pietrosanto, A.; Protopapa, A.; Incarnato, L. Study on Improving the Processability and Properties of Mixed Polyolefin Post-Consumer Plastics for Piping Applications. *Polymers* **2021**, *13*, 71. [CrossRef]
8. Kazemi, Y.; Ramezani Kakroodi, A.; Rodrigue, D. Compatibilization Efficiency in Post-Consumer Recycled Polyethylene/Polypropylene Blends: Effect of Contamination. Polym. *Eng. Sci.* **2015**, *55*, 2368–2376. [CrossRef]
9. Cruz, S.A.; Zanin, M. Evaluation and Identification of Degradative Processes in Post-Consumer Recycled High-Density Polyethylene. Polym. *Degrad. Stab.* **2003**, *80*, 31–37. [CrossRef]

10. Eriksen, M.K.; Christiansen, J.D.; Daugaard, A.E.; Astrup, T.F. Closing the Loop for PET, PE and PP Waste from Households: Influence of Material Properties and Product Design for Plastic Recycling. *Waste Manag.* **2019**, *96*, 75–85. [CrossRef] [PubMed]
11. Khoironi, A.; Hadiyanto, H.; Anggoro, S.; Sudarno, S. Evaluation of Polypropylene Plastic Degradation and Microplastic Identi Fi Cation in Sediments at Tambak Lorok Coastal Area, Semarang, Indonesia. *Mar. Pollut. Bull.* **2020**, *151*, 110868. [CrossRef]
12. Ojeda, T.; Freitas, A.; Birck, K.; Dalmolin, E.; Jacques, R.; Bento, F.; Camargo, F. Degradability of Linear Polyole Fi Ns under Natural Weathering. *Polym. Degrad. Stab.* **2011**, *96*, 703–707. [CrossRef]
13. Rabello, M.S.; White, J.R. Crystallisation and Melting Behaviour of Photodegraded Polypropylene—II. Re-Crystallisation of Degraded Molecules. *Polymer (Guildf)* **1997**, *38*, 6389–6399. [CrossRef]
14. Wu, H.; Zhao, Y.; Dong, X.; Su, L.; Wang, K.; Wang, D. Probing into the Microstructural Evolution of Isotactic Polypropylene during Photo-Oxidation Degradation. *Polym. Degrad. Stab.* **2021**, *183*, 12. [CrossRef]
15. Delva, L.; Ragaert, K.; Degrieck, J.; Cardon, L. The Effect of Multiple Extrusions on the Properties of Montmorillonite Filled Polypropylene. *Polymers* **2014**, *6*, 2912–2927. [CrossRef]
16. Elvira, M.; Tiemblo, P.; Gómez-Elvira, J.M. Changes in the Crystalline Phase during the Thermo-Oxidation of a Metallocene Isotactic Polypropylene. A DSC Study. *Polym. Degrad. Stab.* **2004**, *83*, 509–518. [CrossRef]
17. De Rosa, C.; Auriemma, F.; Circelli, T.; Waymouth, R.M. Crystallisation of the α and γ Forms of Isotactic Polypropylene as a Tool to Test the Degree of Segregation of Defects in the Polymer Chains. *Macromolecules* **2002**, *35*, 3622–3629. [CrossRef]
18. Scott, G. *Mechanisms of Polymer Degradation and Stabilisation*; Springer: Berlin/Heidelberg, Germany, 2012; ISBN 9789401138383.
19. Billingham, N.C. Localization of Oxidation in Polypropylene. *Makromol. Chemie. Macromol. Symp.* **1989**, *28*, 145–163. [CrossRef]
20. Ragaert, K.; Delva, L.; Geem, K. Van Mechanical and Chemical Recycling of Solid Plastic Waste. *Waste Manag.* **2017**, *69*, 24–58. [CrossRef]
21. Saikrishnan, S.; Jubinville, D.; Tzoganakis, C.; Mekonnen, T.H. Thermo-Mechanical Degradation of Polypropylene (PP) and Low-Density Polyethylene (LDPE) Blends Exposed to Simulated Recycling. *Polym. Degrad. Stab.* **2020**, *182*, 109390. [CrossRef]
22. Bledzki, A.K.; Kessler, A.; Lieser, J. Odour Reduction on Plastics and Its Measurement. *Polym. Test.* **1999**, *18*, 63–71. [CrossRef]
23. Demets, R.; Roosen, M.; Vandermeersch, L.; Ragaert, K.; Walgraeve, C.; De Meester, S. Development and Application of an Analytical Method to Quantify Odour Removal in Plastic Waste Recycling Processes. *Resour. Conserv. Recycl.* **2020**, *161*, 104907. [CrossRef]
24. Strangl, M.; Fell, T.; Schlummer, M.; Maeurer, A.; Buettner, A. Characterization of Odorous Contaminants in Post-Consumer Plastic Packaging Waste Using Multidimensional Gas Chromatographic Separation Coupled with Olfactometric Resolution. *J. Sep. Sci.* **2017**, *40*, 1500–1507. [CrossRef] [PubMed]
25. Welle, F. Post-Consumer Contamination in High-Density Polyethylene (HDPE) Milk Bottles and the Design of a Bottle-to-Bottle Recycling Process. *Food Addit. Contam.* **2005**, *22*, 999–1011. [CrossRef]
26. Garcia, P.S.; Scuracchio, C.H.; Cruz, S.A. Effect of Residual Contaminants and of Different Types of Extrusion Processes on the Rheological Properties of the Post-Consumer Polypropylene. *Polym. Test.* **2013**, *32*, 1237–1243. [CrossRef]
27. Veroneze, I.B.; Onoue, L.A.; Cruz, S.A. Thermal Stability and Crystallisation Behavior of Contaminated Recycled Polypropylene for Food Contact. *J. Polym. Environ.* **2022**, *30*, 3474–3482. [CrossRef] [PubMed]
28. Strangl, M.; Ortner, E.; Fell, T.; Ginzinger, T.; Buettner, A. Odor Characterization along the Recycling Process of Post-Consumer Plastic Film Fractions. *J. Clean. Prod.* **2020**, *260*, 121104. [CrossRef]
29. Day, M.; Cooney, J.D.; MacKinnon, M. Degradation of Contaminated Plastics: A Kinetic Study. *Polym. Degrad. Stab.* **1995**, *48*, 341–349. [CrossRef]
30. The Association of Plastics Reciclers Polyolefin Standard Laboratory Processing Practices. Available online: https://plasticsrecycling.org/images/Design-Guidance-Tests/APR-O-P-00-olefin-practices.pdf (accessed on 4 July 2022).
31. *ASTM International D 1238-04*; Standard Test Method for Melt Flow Rates of Thermoplastics by Extrusion Plastometer. ASTM International: West Conshohocken, PA, USA, 2004; pp. 1–13.
32. *ASTM D 3418-03*; Standard Test Method for Transition Temperatures of Polymers by Differential Calorimetry. ASTM International: West Conshohocken, PA, USA, 2004; pp. 1–7.
33. Ehrenstein, G.W.; Riedel, G.; Trawiel, P. *Thermal Analysis of Plastics*; Hanser Publishers: Munchen, Germany, 2004; ISBN 156990362X.
34. *ASTM D3895-14*; Standard Test Method for Oxidative-Induction Time of Polyolefins by Differential Scanning Calorimetry. ASTM International: West Conshohocken, PA, USA, 2015; p. 8.
35. *ISO 527-2*; Plastics—Determination of Tensile Properties—Part 2: Test Conditions for Moulding and Extrusion Plastics. ISO: Geneva, Switzerland, 1996; pp. 1–14.
36. *ISO 527-1*; Plastics—Determination of Tensile Properties—Part 1: General Principles. ISO: Geneva, Switzerland, 2012; pp. 1–28.
37. Antunes, M.C.; Agnelli, J.A.M.; Babetto, A.S.; Bonse, B.C.; Bettini, S.H.P. Correlating Different Techniques in the Thermooxidative Degradation Monitoring of High-Density Polyethylene Containing pro-Degradant and Antioxidants. *Polym. Test.* **2018**, *69*, 182–187. [CrossRef]
38. Cerrada, M.L.; Pérez, E.; Benavente, R.; Ressia, J.; Sarmoria, C.; Vallés, E.M. Gamma Polymorph and Branching Formation as Inductors of Resistance to Electron Beam Irradiation in Metallocene Isotactic Polypropylene. *Polym. Degrad. Stab.* **2010**, *95*, 462–469. [CrossRef]

39. *Polypropylene Handbook-Morphology, Blends and Composites*; Karger-Kocsis, J.; Bárány, T. (Eds.) Springer: Cham, Suica, 2019; Volume 43, ISBN 9783030129026.
40. Fitaroni, L.B.; de Oliveira, É.C.; Marcomini, A.L.; Paranhos, C.M.; Freitas, F.L.; Cruz, S.A. Reprocessing and Solid State Polymerization on Contaminated Post-Consumer PET: Thermal and Crystallisation Behavior. *J. Polym. Environ.* **2020**, *28*, 91–99. [CrossRef]
41. Paiva, R.; Veroneze, I.B.; Wrona, M.; Nerín, C.; Cruz, S.A. The Role of Residual Contaminants and Recycling Steps on Rheological Properties of Recycled Polypropylene. *J. Polym. Environ.* **2022**, *30*, 494–503. [CrossRef]
42. Goss, B.G.S.; Nakatani, H.; George, G.A.; Terano, M. Catalyst Residue Effects on the Heterogeneous Oxidation of Polypropylene. *Polym. Degrad. Stab.* **2003**, *82*, 119–126. [CrossRef]
43. Curtzwiler, G.W.; Schweitzer, M.; Li, Y.; Jiang, S.; Vorst, K.L. Mixed Post-Consumer Recycled Polyolefins as a Property Tuning Material for Virgin Polypropylene. *J. Clean. Prod.* **2019**, *239*, 117978. [CrossRef]
44. Smith, B.C. *Infrared Spectral Interpretation: A Systematic Approach*; CRC Press: Boca Raton, FL, USA, 2018; ISBN 9781351438377.
45. Karger-Kocsis, J. *Polypropylene: An A-Z Reference*; Kluwer Academic Publishers: Dordrecht, The Netherlands, 1998; ISBN 9781591241416.
46. Wang, K.; Peng, Y.; Matadi Boumbimba, R.; Bahlouli, N.; Pessey, D.; Ahzi, S.; Addiego, F.; Rémond, Y. Constitutive Modeling of the Tensile Behavior of Recycled Polypropylene-Based Composites. *Materials* **2019**, *12*, 2419. [CrossRef] [PubMed]

materials

MDPI

Article

Influence of Air Humidity Level on the Structure and Mechanical Properties of Thermoplastic Starch-Montmorillonite Nanocomposite during Storage

Natália Šmídová [1], Hamed Peidayesh [2], Anton Baran [1], Oľga Fričová [1], Mária Kovaľaková [1], Ružena Králiková [3,*] and Ivan Chodák [2,*]

1 Department of Physics, Faculty of Electrical Engineering and Informatics, Technical University of Košice, Park Komenského 2, 042 00 Košice, Slovakia
2 Polymer Institute, Slovak Academy of Sciences, Dúbravská cesta 9, 845 41 Bratislava, Slovakia
3 Department of Environmental Engineering, Faculty of Mechanical Engineering, Technical University of Košice, Park Komenského 5, 042 00 Košice, Slovakia
* Correspondence: ruzena.kralikova@tuke.sk (R.K.); ivan.chodak@savba.sk (I.C.)

Abstract: Thermoplastic starch (TPS) consisting of corn starch and glycerol as a plasticizer, and TPS-montmorillonite (MMT) nanocomposite were stored at room temperature in the air with relative humidities (RH) of 11, 55 and 85% for seven weeks. Mechanical testing and dynamic mechanical thermal analysis (DMTA) were performed to detect changes in their mechanical properties. Solid-state NMR spectroscopy monitoring the changes in molecular mobility in the samples provided an insight into relations between mechanical properties and local structure. The results of mechanical testing indicated that the addition of MMT results in the increase in the tensile strength and Young's modulus while elongation at break decreased, indicating the reinforcing effect of MMT. DMTA experiments revealed a decrease in glass transition temperature of starch-rich phase below room temperature for samples stored at higher RH (55 and 85%). This indicates that absorbed water molecules had additional plasticizing effect on starch resulting in higher mobility of starch chain segments. Recrystallization in these samples was deduced from the shape of cross-polarization magic-angle spinning ^{13}C NMR spectra. The shape of broad-line ^{1}H NMR spectra reflected changes in molecular mobility in the studied samples during seven weeks of storage and revealed that a high amount of water molecules impacts the starch intermolecular hydrogen bond density.

Keywords: thermoplastic starch nanocomposite; montmorillonite; relative humidity; NMR; mechanical properties

Citation: Šmídová, N.; Peidayesh, H.; Baran, A.; Fričová, O.; Kovaľaková, M.; Králiková, R.; Chodák, I. Influence of Air Humidity Level on the Structure and Mechanical Properties of Thermoplastic Starch-Montmorillonite Nanocomposite during Storage. *Materials* **2023**, *16*, 900. https://doi.org/10.3390/ma16030900

Academic Editors: Agnieszka Kijo-Kleczkowska and Adam Gnatowski

Received: 25 December 2022
Revised: 10 January 2023
Accepted: 11 January 2023
Published: 17 January 2023

1. Introduction

Thermoplastic starch (TPS) is generally prepared by the addition of plasticizers into the starch matrix using a thermomechanical process. An organic low molecular weight compound called plasticizer penetrates the starch molecules and creates hydrogen bonds with the hydroxyl groups of starch, resulting in an increase in free volume and consequently an increase in molecular mobility [1,2]. TPS-based materials have been considered in the biopolymer industry due to their biodegradability, availability from renewable resources, and low cost [3–5]. However, TPS suffers from recrystallization (retrogradation) and cocrystallization of amylose chains caused by humidity and its hydrophilic character. This evolution leads to substantial negative changes in mechanical properties during storage [6,7]. From this point of view, in many cases the addition of a small amount of nanofillers improves substantially the physico-chemical properties of TPS-based materials [8,9].

As the main groups of nanofillers, the layered silicates are frequently used for the preparation of polymer nanocomposites. Clay is considered to be an important mineral among the layered silicates, with a high amount of water molecules in the silicate layer. Most of the clays are structurally and chemically unique and may absorb various amounts of water that may affect the process of cations replacement [10–12]. Recently, a large number of studies are focused on montmorillonite (MMT) due to its well-controlled chemical properties, availability, and versatility towards the environment and health [13,14]. The surface chemistry of silicate through ion-exchange reactions and their ability to disperse in layers are the most important MMT properties that significantly affect the mechanical properties, thermal stability, and water resistance of the nanocomposite product. Hence, the hydration of Na^+ and K^+ ions occurring in the clays allows the clays to swell and provides the ability for polymer chains to intercalate between the layers [15] or under suitable conditions even exfoliation of the MMT may occur [16,17]. The ultimate properties of TPS-MMT nanocomposites are closely related to the arrangement of starch chains, molecular dynamics, and the dispersion of the clay fillers within the TPS matrix [18].

In designing TPS for a given purpose, mutual correlations between the source of starch, method of its modification, processing technique, physical and material parameters, as well as performance in various conditions should be considered. It is well known that TPS-based materials are sensitive to the surrounding relative humidity [19] and storage time, which substantially affect the polymer structure and molecular mobility and subsequently their mechanical properties. Investigating the structure and molecular mobility of TPS-based materials during aging is described in the literature [6,20–22]. However, to the best of our knowledge, no investigation has been done to evaluate the influence of aging under exactly defined relative humidities on the molecular mobility of TPS-MMT nanocomposite.

The main aim of the present work was to find the correlation between the key parameters of the structure of the TPS-MMT nanocomposite and performance properties according to the determined application. In this regard, changes in the starch chain arrangements and nanofiller particles in TPS-MMT nanocomposite as well as their mechanical properties were evaluated using solid-state ^1H and ^{13}C NMR spectroscopy, tensile testing and dynamic mechanical thermal analysis (DMTA).

2. Materials and Methods

2.1. Materials

Native corn starch Meritena® 100 was provided by Brenntag (Bratislava, Slovakia). Its water content was around 12 wt% as determined by drying in the oven at 105 °C for 5 h. Natrified montmorillonite (Cloisite® Na^+) with a cationic exchange capacity range of 80–95 mequiv/100 g was purchased from Southern Clay Products (Texas, Gonzales, USA). Glycerol was obtained from Centralchem, Ltd. (Bratislava, Slovakia). Double distilled water was used for the preparation of all solutions.

2.2. Preparation of TPS and TPS-MMT Nanocomposite Samples

The composition of the mixture consisted of 1 part of starch (dried), 0.5 parts of glycerol, and 2.3 parts of distilled water with or without dispersed Cloisite Na (MMT). To prepare nanocomposite samples, 0.02 parts (parts based on dry weight of starch) of MMT nanoparticles were dispersed in water and glycerol mixture for 5 min by mechanical mixing. Afterward, the starch was added to the prepared mixture and the suspension was annealed at 80 °C for 10 min under continuous mixing for complete gelatinization. The obtained mixtures were heated in an oven at 100 °C for 5 h followed by maintaining overnight at 60 °C to prevent moisture absorption. Then, the dried material was mixed in a laboratory mixer Plastograph Brabender PLE331 for 10 min at 130 °C and 100 rpm. Slabs 1 mm thick were prepared by compression molding (laboratory press, LabEcon 300) at 130 °C applying 2 min preheating without pressure and an additional 2 min at a pressure of 2.65 MPa.

The TPS and TPS-MMT samples were stored in dessicators at different relative humidities (RH) namely 11, 55 and 85% using saturated salt solutions at 25 °C with lithium chloride (LiCl), magnesium nitrate ($Mg(No_3)_2$), and potassium chloride (KCl) [23], respectively. In the next text, storage at particular RH is indicated by the number, e.g., nanocomposite stored at 55% RH is denoted as TPS-MMT55.

2.3. Tensile Testing

The compression-molded slabs were punched by pneumatic toggle press equipment to prepare dumbbell-shaped test specimens with a 3.5 mm × 30 mm dimension of the deformed area during the tensile test (thickness of each specimen of approximately 1 mm was exactly measured before testing to be used for strength and modulus calculations). The tensile properties were measured by using an Instron 3365 (Instron, Norwood, MA, USA) universal testing machine in uniaxial deformation at a cross-head speed of 1 mm/min up to 1% deformation (to determine Young's modulus) followed by a speed of 50 mm/min at higher deformations. The average values and standard deviations were calculated from seven specimens for all parameters.

2.4. Dynamic Mechanical Thermal Analysis (DMTA)

Dynamic-mechanical measurements were performed using the instrument DMA Q800 (TA Instruments, Hüllhorst, Germany). The specimens (ca. 10 mm × 7 mm × 1 mm) were measured in tensile mode at a frequency of 10 Hz and amplitude of dynamic deformation of 20 µm. The experiments were performed in the temperature range from −70 °C to 140 °C at a heating rate of 2 °C/min.

2.5. Water Content

First, the weight of TPS and TPS-MMT specimens (W_1) was measured after drying at 80 °C for 4 h and was taken as the weight of dry specimen. Then, the samples in triplicate were placed in desiccators with different relative humidities and weighed (W_2) after various periods. The total water content, W, in percent was calculated according to the equation below:

$$W\ (\%) = \frac{W_2 - W_1}{W_2} \times 100 \quad (1)$$

Dependence of the water content of the particular samples on the storage time will be discussed in Section 3.4.

2.6. ^{1}H and ^{13}C NMR Measurements

All NMR experiments were performed on a Varian solid-state NMR spectrometer (Palo Alto, CA, USA) equipped with a probe head with 4 mm ZrO_2 rotors at ambient temperature. ^{13}C and ^{1}H resonance frequencies were approximately 100 and 400 MHz, respectively.

The high-resolution ^{13}C NMR spectra were recorded using the cross-polarization technique (CP) and magic angle spinning (MAS) rate of 10 kHz. The measurements were performed with the radio frequency field strength of 63 kHz in accordance with Hartmann–Hahn condition. CP contact time 1 ms, acquisition time 40 ms, and relaxation delay between two consecutive scans 7 s were used. SPINAL pulse high-power proton decoupling of 63 kHz was applied. The number of scans was 10,000.

Broad-line (BL) ^{1}H NMR spectra were acquired using Chen sequence suppressing the broad probehead background, the ^{1}H MAS NMR spectra were accumulated at the spinning rate of 10 kHz using single pulse sequence. The same experimental conditions were applied during accumulation of both BL and MAS ^{1}H NMR spectra: $\pi/2$ pulse with duration 3.8 µs and 10 s recycle delay. Acquisition times 60 and 150 ms were used at measurement of BL and MAS ^{1}H NMR spectra, respectively.

The chemical shifts of all spectra were referenced to tetramethyl silane using adamantane as an external standard.

3. Results

3.1. ^{13}C CP MAS NMR

The ^{13}C CP MAS NMR spectra for both dried TPS and TPS-MMT samples are almost identical (Figure 1). The resonances in the spectra are related to the carbons of starch and glycerol (chemical structures shown in Figure 1) as follows: 103 ppm—starch C1 carbon, 82 ppm—starch C4 carbon in amorphous structure, around 73 ppm—starch C2, C3, C5 carbons and C4 carbon in ordered structure, 63 ppm—starch C6 carbon, signals 73 and 64 ppm—glycerol carbons. The shape of C1 carbon resonance indicates that the structure of both samples is fully amorphous [24–26].

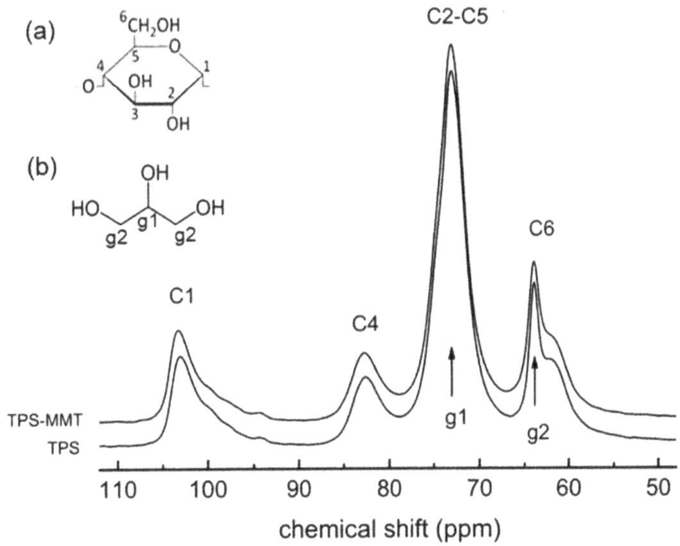

Figure 1. ^{13}C CP MAS NMR spectra measured for dried TPS and TPS-MMT samples. The C1–C6 and g1, g2 resonances relate to the carbons of starch monomer unit and glycerol, respectively. Chemical structures of (**a**) thermoplastic starch monomer unit (α-D-glucose) and (**b**) glycerol are depicted in the inset.

The spectra for TPS and TPS-MMT samples measured after one and seven weeks of storage at 11% RH are similar to the spectra of the dried samples (Figure 2). Some differences can be observed in the intensities of the glycerol signal at 62 ppm and the signal at 73 ppm which is a superposition of starch C2–C5 and glycerol COH resonances. Since the polarization transfer between ^{1}H and ^{13}C nuclei is restricted by their enhanced relative motion, a small decrease in the intensities of 62 and 73 ppm signals could be caused by increased mobility of glycerol molecules due to physical ageing of the samples during storage. Low water content does not significantly affect the starch chains mobility; it prevents the starch recrystallization and thus the samples remain fully amorphous during storage similarly as dried samples.

The spectra measured for TPS and TPS-MMT samples stored at 55 and 85% RH differ significantly from the spectra of the samples, both dried as well as stored at relative humidity of 11% (Figure 2). The shape of C1 signal indicates that ordered (crystalline) starch structure of B-type [24–26] was formed in these samples during the first week of storage. This change is accompanied by the decrease in the intensity of C4 signal and by narrowing of the C6 signal due to more enhanced polarization transfer between ^{1}H and ^{13}C nuclei present in ordered structure. These differences are the most pronounced in the spectra for TPS85 and TPS-MMT85 samples with the highest water content. In these spectra the superimposed signal of C2–C5 starch carbons and COH glycerol carbons is split and the intensity of the glycerol signal at 63 ppm decreased due to release of some glycerol molecules from the starch structure making them more mobile.

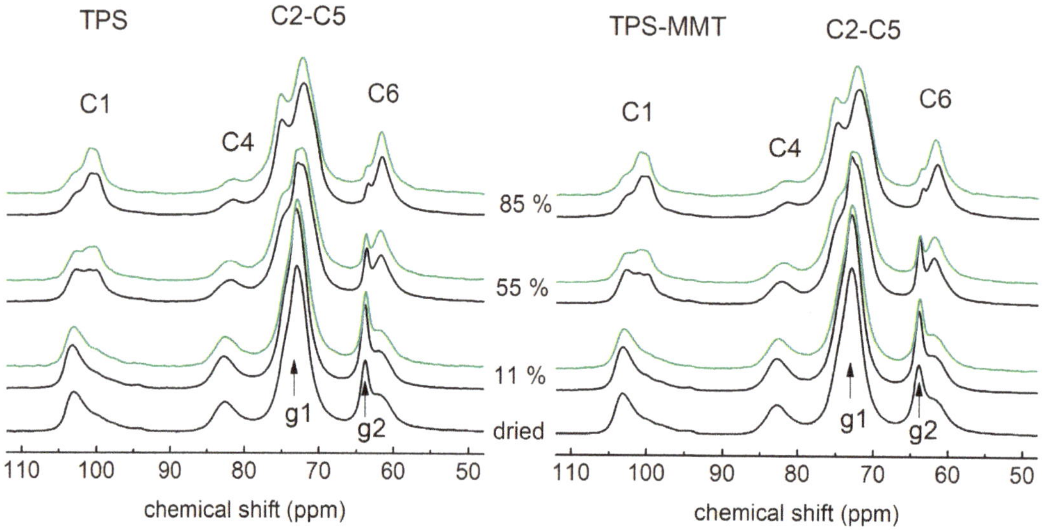

Figure 2. ^{13}C CP MAS NMR spectra measured for TPS (**left**) and TPS-MMT (**right**) samples dried and stored at indicated RHs for one week (black) and seven weeks (green). The C1–C6 and g1, g2 resonances relate to the carbons of starch monomer unit and glycerol, respectively.

Closer inspection on C1 and C4 resonances in the spectra for TPS55 and TPS-MMT55 samples reveals a change in the shape of C1 resonance and decrease in C4 resonance intensity after seven weeks of storage. This can be explained by ongoing recrystallization resulting in an increase in crystalline phase in the samples during storage at 55% RH. On the other hand, only minor changes were identified in the spectra for TPS85 and TPS-MMT85 samples stored for one and seven weeks, which indicates the structural stabilization of crystalline structure already after one week of storage at 85% RH. It can be suggested that an additional plasticizing effect of water resulting in enhanced starch chains mobility supports their arrangement in ordered structure, and thus it is responsible for rapid recrystallization during the first week.

3.2. ^{1}H BL NMR

TPS immediately after preparation is a complex system consisting of starch chains in amorphous phase, plasticizer and water molecules, which may interact through hydrogen bonding with each other as well as with starch chains, or the chains can stay relatively free. Moreover, the distribution of glycerol and water molecules in the sample volume is inhomogeneous, so plasticizer-rich and starch-rich domains are formed resulting in different mobility of starch chains [27,28] as indicated by measured temperature dependences of DMTA loss factor tan δ discussed later. If TPS contains a small amount of nanofiller, the components of TPS can also interact with nanofiller particles. In the case of the investigated samples, glycerol, water and starch chains can be intercalated in the galleries of MMT particles or they may interact with the surface of exfoliated MMT.

BL ^{1}H NMR spectra recorded on static polymer samples usually consist of one broad and one or several narrow lines whose widths reflect different mobility of hydrogen nuclei. The dominant mechanism of line broadening in ^{1}H NMR spectra of starch-based materials are ^{1}H-^{1}H dipolar interactions which can be averaged by molecular motion. For this reason, broad and narrow lines in the spectra are assigned to hydrogen nuclei in rigid and more mobile domains, respectively. Samples stored in desiccators absorb different amounts of water from ambient atmosphere acting as an additional plasticizer [29,30] that influences overall mobility of ^{1}H nuclei in the samples.

The BL ^{1}H NMR spectra for the studied TPS and TPS-MMT samples reflect the structural characteristics described above. Figure 3 shows their spectra measured after one week of storage at 11, 55 and 85% RH. The broad line relates to rigid starch chains located in both amorphous and crystalline regions and to a small amount of glycerol and water molecules entrapped in TPS structure if present. On the broad line, one or several narrow lines are superimposed; they are associated with water and glycerol molecules, and also with highly mobile starch chains if present in the sample.

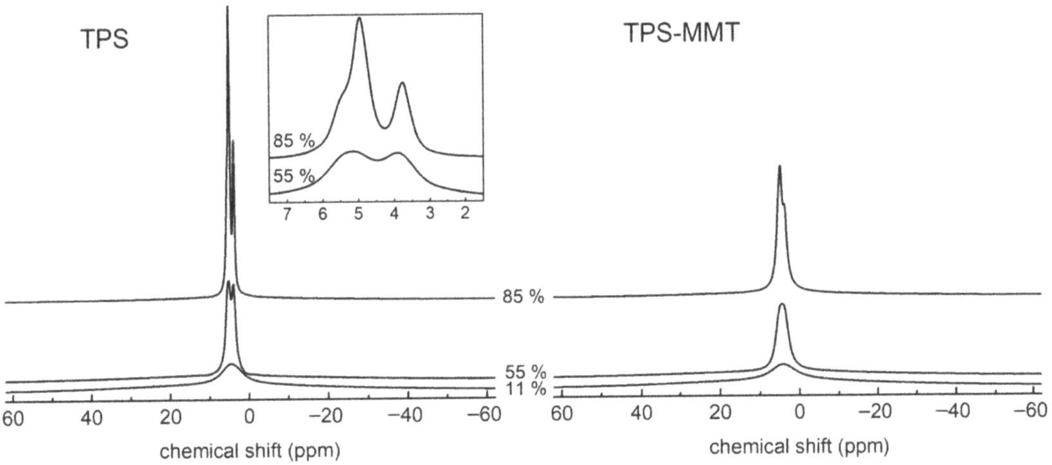

Figure 3. ^{1}H BL NMR spectra measured for TPS (**left**) and TPS-MMT (**right**) samples stored at indicated RHs for one week (all spectra are normalized according to the areas). The W, G(OH) and G(CH) signals relate to the hydrogens of water and glycerol OH and CH/CH_2 groups, respectively.

In the spectrum for TPS11 sample only one structureless narrow line is observed pointing to very low mobility of TPS components. The narrow signal is superimposed on the broad line with full width at half maximum (FWHM) of ~25 kHz. Water content in this sample is very low—approximately 0.8 wt% as determined from weight measurements (discussed later)—and it is suggested that water contributes to the overall intensity in the spectrum only to a very small extent. One can expect that entrapped glycerol molecules in TPS structure are present in the sample contributing to broad signal in the spectrum. This fact was confirmed by deconvolution discussed later.

Spectra measured for the TPS samples stored at higher RH (Figure 3 left) differ significantly from the TPS11 spectrum, since two narrow signals appear in the spectra which can be assigned to overlapping signals produced by hydrogens of water (4.8 ppm) and glycerol (5.3 and 3.7 ppm). FWHM of the broad line is about 22 and 20 kHz in the TPS55 and TPS85 spectrum, respectively. As all samples originate from the same batch they can differ only in their water content. TPS55 and TPS85 samples absorbed significant amounts of water so that they contained ~11 and 28 wt% of water after one week, respectively (discussed later). Absorbed water had an additional plasticizing effect on TPS structure [29,30] leading to the increased mobility of all TPS components resulting in averaging of ^{1}H-^{1}H dipolar interactions. Therefore, TPS55 and TPS85 samples provide significantly narrower lines in the BL ^{1}H NMR spectra compared to the TPS stored at 11% RH. The plasticizing effect of water manifested itself also by lowering the glass transition temperature (T_g) of the starch-rich phase (DMTA results). It is seen that spectral lines measured for TPS-MMT samples are broadened when compared to relevant lines in TPS spectra (Figure 3 left). This fact can be explained by interactions of glycerol and water molecules and starch chains with MMT particles occurring in nanocomposite samples which hinder molecular motion of TPS components. This results in the decreased mobility and line-broadening; thus, the narrow line splitting is not observed.

To assess the influence of long-term storage at various RHs, BL ^{1}H NMR experiments of our samples were acquired throughout seven weeks (Figure 4). Only central parts of the spectra are presented for better clarity.

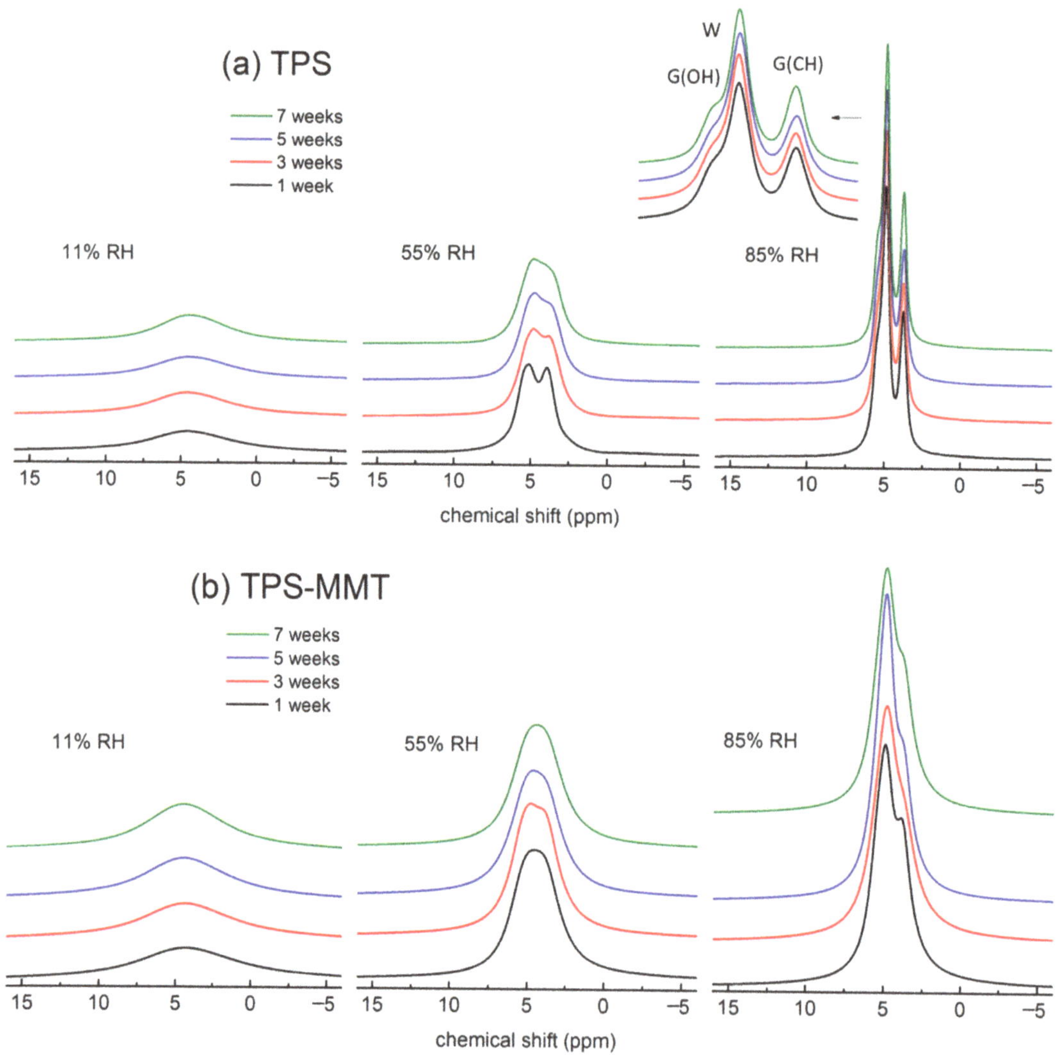

Figure 4. Central parts of the BL ^{1}H NMR spectra for (**a**) TPS and (**b**) TPS-MMT samples stored at 11 (**left**), 55 (**middle**) and 85% (**right**) RHs measured after indicated storage time (spectra are normalized according to the areas; spectra (**b**) are enlarged twice with respect to spectra in (**a**)). The W, G(OH) and G(CH) signals correspond to the hydrogens of water and glycerol OH and CH/CH_2 groups, respectively.

The shapes of the spectra for TPS11 and TPS-MMT11 samples (Figure 4a,b left) do not change significantly with time; they consist of one broad and one narrow line which keeps a structureless character. The width of the broad line remains approximately the same in all spectra pointing to the fact that mobility of rigid domains did not change during storage of the samples in the desiccator. On the contrary, the width of the narrow line decreased with storage time. This can be explained by physical ageing of the samples occurring at temperatures below T_g in starch-rich regions [31,32], during which the starch structure approaches thermodynamic equilibrium, free volume increases and glycerol and water molecules acquire higher mobility. DMTA measurements for both TPS and TPS-MMT samples stored at 11% RH showed that T_g of starch-rich phase for these samples is ~70 °C confirming they occur under T_g and physical ageing could take place. Crystalline structure does not form under T_g, which is in accordance with the results obtained from ^{13}C CP MAS NMR spectra (Figure 1).

The BL ^{1}H NMR spectra for TPS stored at 55% RH are shown in Figure 4a, middle. The spectrum resolved in two lines was obtained after one week of storage when the sample already underwent recrystallization as observed from ^{13}C NMR spectra and also reached equilibrium water content (discussed in Section 3.4). Recrystallization during the first week caused free volume increase in amorphous TPS domains and thus higher mobility of water and glycerol molecules. With longer storage time narrow lines broadened, which resulted in less resolved spectra. This fact can be explained by migration of the water and glycerol molecules in amorphous TPS domains making possible interactions between each other and/or with starch chains which decreased their mobility and enlarged their chemical shifts distribution. Formation of these hydrogen bonds makes the structure stiffer, and both tensile stress and Young's modulus increased while tensile strain decreased after seven weeks of storage (as discussed later). Similar features are observed also for the spectra measured for nanocomposite stored at 55% RH (Figure 4b middle).

The best resolved ^{1}H BL NMR spectra were detected for the TPS85 sample (Figure 4a right) due to the highest amount of absorbed water causing very high mobility of all TPS components. Concerning structural behavior in time, the widths of narrow lines slightly decreased after seven weeks of storage which is in contrast with TPS55 sample (the narrow lines broadened). The samples stored at 85% RH absorb water up to five weeks of storage (discussed in Section 3.4). During this period water molecules penetrate into an amorphous phase of the TPS structure that leads to a reduction in starch intermolecular hydrogen bond density. As a result, certain deterioration of mechanical properties was observed; tensile stress and Young's modulus decreased. Moreover, during penetration of water molecules into TPS structure, regions rich in glycerol or water could be formed and thus these molecules along with starch chains acquired higher mobility. This explains the decrease in narrow lines' widths for the TPS85 sample. Similar structural changes as took place in the TPS85 sample can be expected to occur in TPS-MMT85 sample as well; however, phase separation might be hindered by additional interactions of water and glycerol molecules and starch chains with nanofiller. This is confirmed by the fact that narrow lines widths did not change with time as deduced from BL ^{1}H NMR spectra (Figure 4b right).

In summary, we can say that high water content in the samples stored at 85% RH caused reduction in starch intermolecular H-bond density after seven weeks of storage. A similar effect was reported by Leroy et al. [33] who analyzed thermal and structural (WAXS) behavior of thermoplastic potato starch stored under 58 and 89% RH.

3.3. ^{1}H MAS NMR

Additional information concerning molecular mobility in the studied samples can be obtained using magic angle spinning (MAS) technique. This technique averages dipolar interactions which results in high-resolution solid-state NMR spectra. However, relatively narrow signals accompanied by spinning sidebands at multiples of the MAS rate can be still superimposed on a broad signal [26].

Figure 5 shows ^1H MAS NMR spectra measured for TPS and TPS-MMT samples stored for 1 week at 11, 55 and 85% RHs; one can notice that the spectra are significantly better resolved than corresponding BL ^1H NMR spectra (Figure 3). The peaks in these spectra are related to hydrogens of water, glycerol and starch chains. Despite using the MAS technique, the spectra for TPS11 and TPS-MMT11 samples show only signs of splitting in two lines with positions at 5.1 and 3.8 ppm. Better resolution was obtained in the TPS55 and TPS-MMT55 spectra indicating increased mobility of TPS components when the water content is higher. The spectral line at 5.1 ppm is split into two well-resolved lines with positions at ~5.3 and 4.8 ppm related to glycerol OH groups and water, respectively. The line at 3.8 ppm is related to glycerol CH and CH_2 groups [20]. The best resolution was obtained in the TPS85 spectrum indicating the highest mobility of TPS components due to the highest water content. Moreover, signal at 3.8 ppm shows splitting in several lines associated with starch hydrogen nuclei [20,34]. In the TPS-MMT85 spectrum this line is prevented from splitting due to additional interactions between starch chains and nanofiller. Highly mobile starch hydrogen nuclei form a signal at position 5.3 ppm in addition to signals between 3 and 4 ppm; however, the former signal is overlapped with the signal related to glycerol OH groups. It is possible that also in the case of TPS55 and TPS-MMT55 samples, highly mobile starch chains are present within the structure contributing to 5.3 and 3.8 ppm signals, but they are not resolved in the ^1H MAS NMR spectra. Their presence will be confirmed by deconvolution of the BL ^1H NMR spectra discussed later.

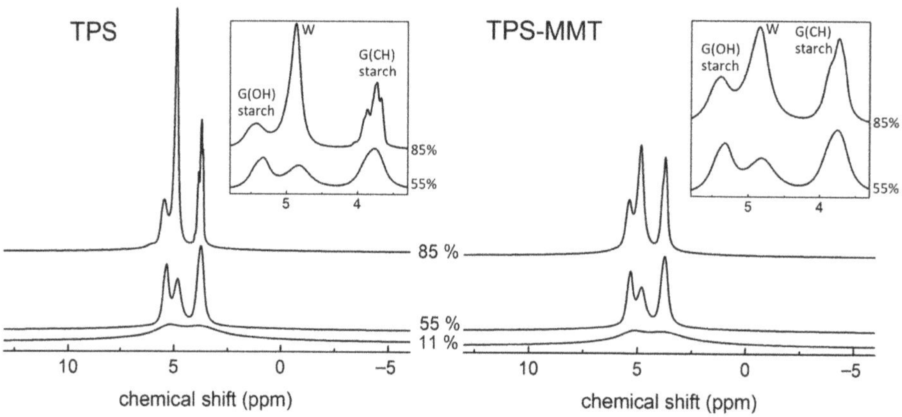

Figure 5. MAS ^1H NMR spectra measured for TPS (**left**) and TPS-MMT (**right**) samples stored one week at indicated RHs (all spectra are normalized according to the areas). The W, G(OH) and G(CH) signals correspond to the hydrogens of water and glycerol OH and CH/CH_2 groups, respectively; starch hydrogens signals overlap with G(OH) and G(CH) signals.

The linewidths of the signals in the spectra for TPS11 and TPS-MMT11 samples decreased with time, which could be explained by physical ageing as it was discussed in BL section. ^1H MAS NMR spectra for the TPS55, TPS-MMT55 and TPS85 samples show only negligible differences when measured after various storage times in contrast to BL spectra due to the effect of MAS averaging dipolar interactions. On the other hand, in the case of the TPS-MMT85 sample the signals at 5.3 and 3.8 ppm slightly broadened after seven weeks as a result of newly emerging interactions between glycerol and starch hydrogens with MMT particles. This is in accordance with results obtained from BL spectra.

3.4. *Water Content Estimation by Means of BL ^1H NMR*

The starch-based materials are very sensitive to storage conditions, especially to relative humidity, because starch can easily uptake water from ambient [35]. The water content then strongly influences the structure, recrystallization, and mechanical properties of these materials [33,36,37]. In TPS samples, mobile molecules of water denoted as free water and immobilised water molecules create strong bonds with starch chains [20,38] and present in MMT interlayer galleries termed bound water can be found. The well-defined ratios of starch, glycerol, water and montmorillonite (in the case of nanocomposite) used in the preparation process of TPS and nanocomposite samples directly offer mass ratio of starch and glycerol which remains unchanged during preparation, after drying or during storage. However, this does not apply to water, because its amount changes during the above-mentioned processes.

Deconvolutions of the BL ^{1}H NMR spectra provide intensities of deconvoluted lines which are proportional to the number of ^{1}H nuclei associated with particular lines. This fact allows the estimation of free water content in some cases.

The BL ^{1}H NMR spectra for TPS11 and TPS-MMT11 samples were deconvoluted in one broad and one narrow line at chemical shift of 5.5 and 4.4 ppm, respectively. It can be expected that due to the very low amount of water in these samples, the contribution of water hydrogens to the signals is negligible and thus only hydrogens of starch and glycerol form the signals. The deconvolutions provide broad to narrow lines intensity ratio of ~1.6, which is higher than theoretical starch/glycerol hydrogen ratio equal to 1.42 derived from composition of the sample. This fact can be explained by the presence of a certain portion of glycerol molecules immobilized in the TPS structure which contributes to broad signal. It was impossible to estimate water content from these deconvolutions.

The water content estimation was carried out for TPS and nanocomposite samples stored at 55 and 85% RHs. First, their BL ^{1}H NMR spectra were deconvoluted in one broad line related to hydrogen nuclei located in rigid starch chains and bound water molecules, and three narrow lines at chemical shifts of ~5.3, 4.8 and 3.8 ppm whose position was derived from MAS ^{1}H NMR spectra (Figure 5); free water hydrogens form signal at 4.8 ppm, glycerol OH groups at 5.3 ppm, glycerol CH and CH_2 groups at 3.8 ppm, and hydrogen nuclei in mobile starch chains at 5.3 and between 3 and 4 ppm [20,34]. Signal of the mobile starch chains might thus overlap with glycerol signal in our spectra. Indeed, the deconvolutions of the spectra provide the intensity ratio of broad line to narrow lines at 5.4 and 3.9 ppm approximately 1 which is lower than theoretical starch/glycerol hydrogen ratio of 1.42. Thus, we assume that at high RH (55 and 85%), the plasticizing effect of water on TPS structure resulted in high mobility of a certain fraction of starch chains comparable with mobility of water and glycerol molecules [28], since their signals appear in the spectra and overlap with glycerol signals.

The deconvolution for the TPS85 sample stored seven weeks is shown in Figure 6 as an example; intensities of 0.117, 0.380 and 0.192 related to narrow lines at 5.3, 4.8 and 3.8 ppm were obtained. The intensity of the broad line was 0.311. Based on this determination and knowing the theoretical starch/glycerol hydrogen ratio of 1.42, the relative amount of free water in the sample was calculated. The fact that a certain fraction of starch hydrogen nuclei contributes to narrow lines at 5.3 and 3.8 ppm was also considered. Estimated free water content by means of BL ^{1}H NMR for TPS and nanocomposite samples during storage at 55 and 85% RH is shown in Figure 7 along with water content determined from weighing of the samples. The most significant increase in water absorption was observed during the first week of storage for all samples. This was expected as the samples were dried before storage. During the following weeks the water content in TPS55 and TPS-MMT55 samples increased only slightly, but the increase in the samples stored at 85% RH was significant up to five weeks and then levelled off. Water content estimated by deconvolution of BL ^{1}H NMR spectra is systematically lower than that evaluated by weighing the samples, since only free water can be detected using the NMR technique in contrast to weighing, which allows estimation of total water content in the sample. It is in accordance with assumption that "free" and "bound" water is present in the structure. Moreover, a greater difference in estimation of water content by mentioned methods for nanocomposite samples reveals a lower amount of free water in these samples because water molecules can be intercalated in the galleries of MMT particles, or they interact with the surface of exfoliated MMT.

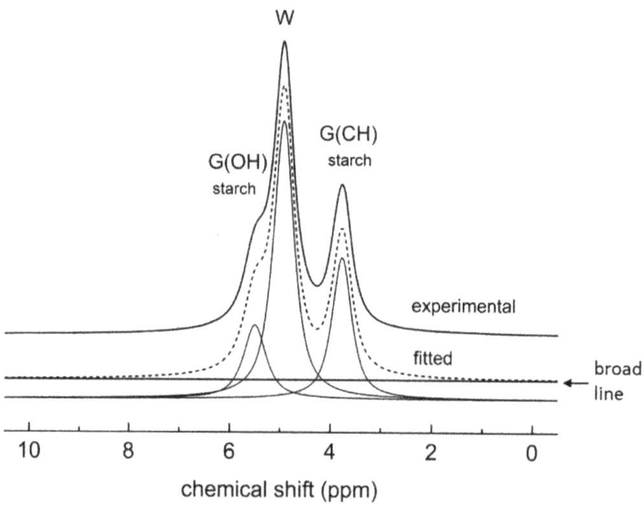

Figure 6. Deconvolution of the BL ^{1}H NMR spectrum for TPS85 sample stored seven weeks to one broad and three narrow lines. The W, G(OH) and G(CH) signals correspond to the hydrogens of water and glycerol OH and CH/CH_2 groups, respectively; starch hydrogens signals overlap with G(OH) and G(CH) signals.

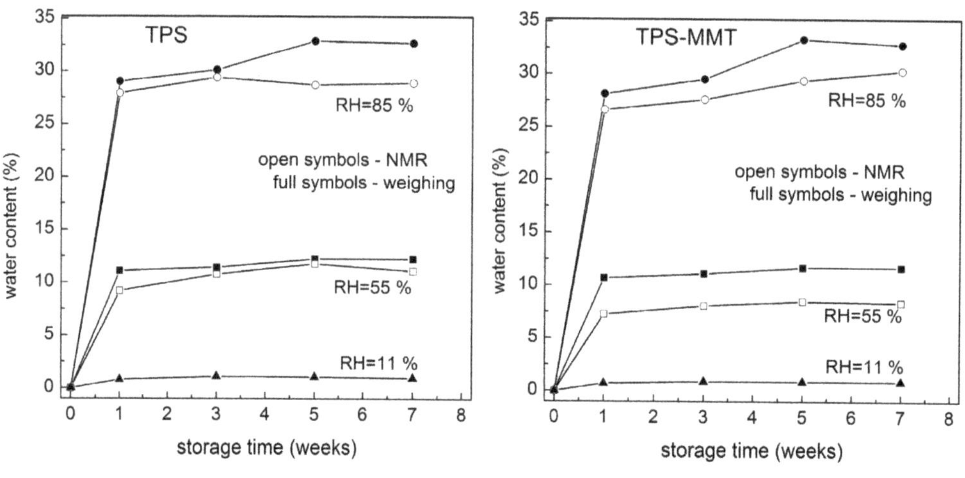

Figure 7. Dependence of water content on storage time at 11, 55 and 85% RHs for TPS and TPS-MMT samples based on their weighing (total water content) and estimated from BL ^{1}H NMR analysis (free water content).

3.5. Mechanical Properties

Mechanical properties are considered to be important parameters concerning TPS performance. These properties depend significantly on surrounding relative humidity and storage time. The values of ultimate tensile strength, elongation at break, and Young's modulus for TPS and TPS-MMT samples stored at various humidities for one and seven weeks are shown in Table 1. It is seen that the tensile strength and Young's modulus of the dried samples and the samples stored at 11% RH increase with the incorporation of MMT. This increase is ascribed to the formation of hydrogen bonds between starch chains, water, glycerol, and hydroxyl groups of MMT layers [3,39]. Concerning the relation of tensile strength changes with the humidity, it is seen that the tensile strength of both TPS and TPS-MMT samples is decreasing with the rising RH, while the tensile strength of the dried samples and the samples stored at 11% RH are identical. Moreover, the increase in RH results in a substantial decrease in Young's modulus compared to the dried samples. The higher amount of the absorbed water at higher RH has an additional plasticizing effect, leading to an increase in the mobility of starch chains. However, lower RH will not affect the structure. Therefore, these findings indicate the relationship between tensile strength, Young's modulus, and surrounding RH. As expected, the increase in RH and consequently increased chain mobility led to the increase in elongation at break. Similar behavior is observed for both TPS and TPS-MMT samples.

Table 1. Mechanical properties, namely tensile strength, elongation at break and Young's modulus of the TPS and TPS-MMT samples stored at various humidities for one and seven weeks.

	Sample Code	Dried	11% RH		55% RH		85% RH	
			1 Week	7 Weeks	1 Week	7 Weeks	1 Week	7 Weeks
Tensile Strength (MPa)	TPS	7.8 ± 2.7	9.3 ± 0.8	6.8 ± 0.5	1.3 ± 0.0	2.5 ± 0.1	1.3 ± 0.0	0.6 ± 0.1
	TPS-MMT	10.4 ± 1.1	11.5 ± 1.0	7.6 ± 0.2	1.4 ± 0.1	2.4 ± 0.4	1.3 ± 0.1	0.6 ± 0.1
Elongation at break (%)	TPS	1.5 ± 0.3	4.3 ± 1.6	27.1 ± 11.5	66.6 ± 2.1	34.4 ± 2.0	25.3 ± 1.2	8.2 ± 0.4
	TPS-MMT	1.7 ± 0.2	3.7 ± 0.8	31.4 ± 5.0	64.1± 8.9	27.8 ± 5.5	24.7 ± 1.1	7.6 ± 1.3
Young's Modulus (MPa)	TPS	970.2 ± 207.5	662.9 ± 74.7	297.4 ± 2.4	9.2 ± 0.5	25.2 ± 3.4	10.5 ± 0.4	9.4 ± 0.8
	TPS-MMT	1148.2 ± 47.4	703.0 ± 68.9	435.9 ± 4.3	10.0 ± 1.1	26.5 ± 1.3	10.9 ± 0.6	9.7 ± 0.5

Regarding the aging effect on mechanical properties, the samples stored at 11% RH showed a decrease in tensile strength and Young's modulus, while elongation at break increased which might be caused by physical ageing of the samples as discussed in ^{1}H BL NMR results when glycerol and water molecules acquire higher mobility. The TPS samples stored at 55% RH showed a slight increase in tensile strength and Young's modulus while the elongation at break decreased. This phenomenon is ascribed to a formation of new interactions between starch, glycerol, and water after reaching the equilibrium moisture content, resulting in the formation of hydrogen bonds and thus a slight increase in tensile strength and Young's modulus. On the other hand, the TPS samples stored at 85% RH presented a decrease in all mechanical properties. Penetration of a high amount of water into the TPS structure followed by reduction in hydrogen bond density can be considered as the main reason to explain the decrease in mechanical properties compared to 55% RH. All these results are supported by NMR spectra.

3.6. Dynamic Mechanical Thermal Analysis (DMTA)

DMTA measurements were performed to evaluate the viscoelastic response of the TPS samples, which is associated with the macromolecular motions in the structure. The storage modulus and tan δ of the TPS and TPS-MMT samples stored at various humidities for different times of storage are shown in Figures 8 and 9, respectively. As expected, the storage modulus of the TPS samples increases with the incorporation of MMT particles indicating reinforcement effects of the MMT layers. Moreover, it can be seen that the storage modulus of the samples decreases with the rising RH. This result is attributed to the plasticizing effect of water molecules. Tan δ as a function of temperature for TPS samples indicated two relaxation peaks. The low-temperature relaxation peak corresponds to the glass transition of glycerol-rich domains, while the high-temperature peak is ascribed to the starch-rich domains [1,39]. The temperatures of maxima tan δ determined from curves in Figure 9 are summarized in Table 2. As expected, T_g values for TPS-MMT nanocomposites are higher than for TPS samples due to the restrictions in starch chain mobility caused by a certain extent of intercalation of plasticizer molecules and starch chains into the MMT platelet galleries. Furthermore, as seen in Table 2, the first peak, which is controlled by the plasticizer molecular motion, showed a shift to lower temperature for samples stored at higher RH, indicating that regions rich in glycerol or water might be formed during water absorption. On the other hand, samples stored at 11% RH exhibited a slight decrease in T_g compared to the dried samples, whereas the T_g values for the samples stored at 55% RH decreased substantially. The presence of more water may decrease the intermolecular hydrogen bond density [33]. However, low water content could not significantly affect the starch chain mobility, leading to preventing recrystallization during storage. It should be noted that measuring the DMTA for the samples stored at 85% RH was possible only for the TPS-MMT samples stored for two weeks or less. After maintaining the TPS samples longer at high humidity, the material structure is too soft to be fastened in the instrument clamps since either the samples are broken during fastening before starting the measurement, or the sample is slipping inside the clamp so that the length of the measured sample is not constant.

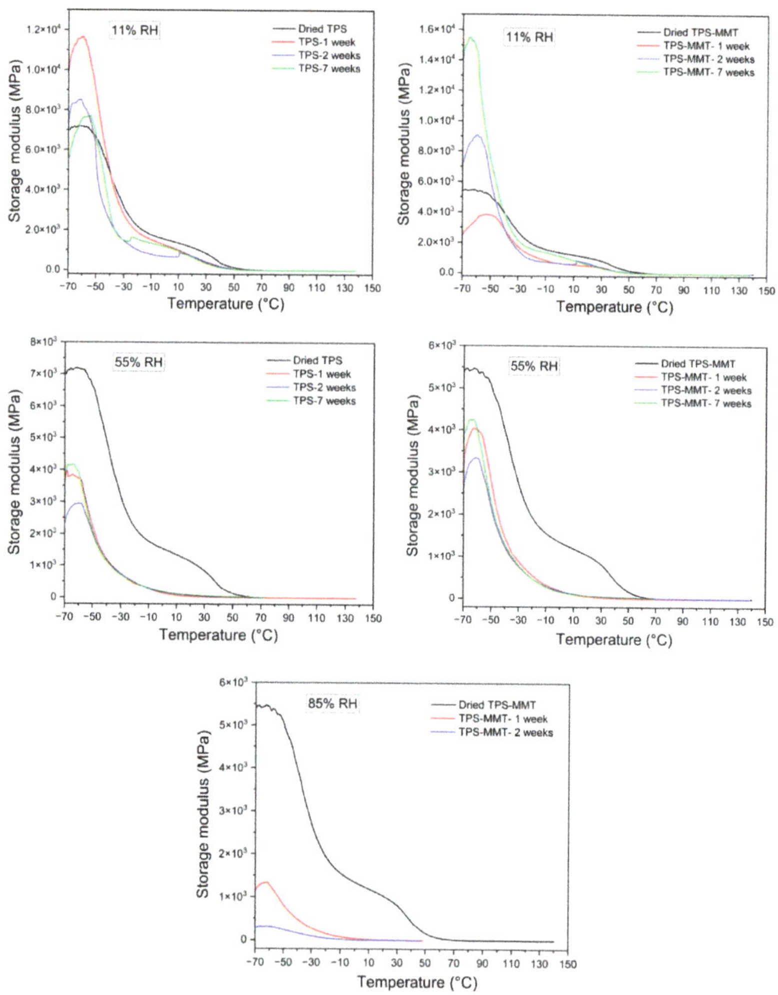

Figure 8. Storage modulus curves for TPS and TPS-MMT samples stored at 11, 55 and 85% RHs for one, two, and seven weeks of storage.

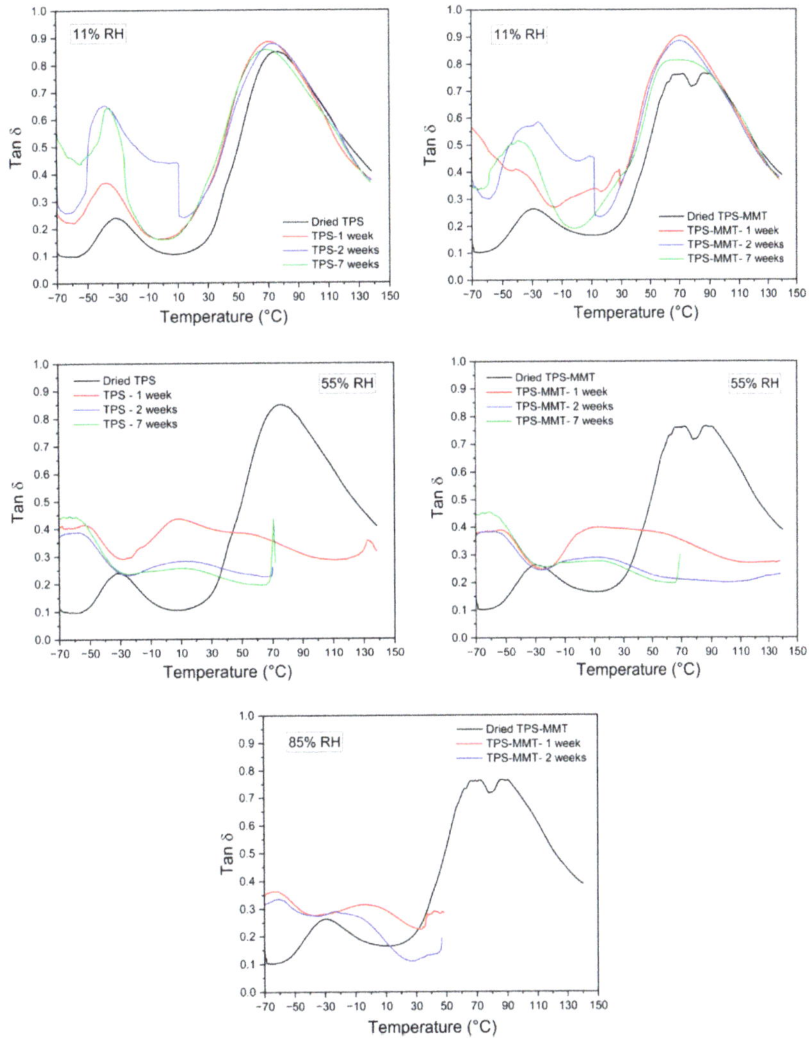

Figure 9. Tan δ curves for TPS and TPS-MMT samples stored at 11, 55 and 85% RHs for one, two, and seven weeks of storage.

Table 2. Temperatures of maximum values for tan δ appearance as extracted from Figure 9.

Sample	1st Peak T (°C)				2nd Peak T (°C)			
	Dried	11% RH	55% RH	85% RH	Dried	11% RH	55% RH	85% RH
TPS	−31.3				75.5			
TPS-MMT	−28.9				86.7			
TPS- 1 week		−37.1	−54.0	n/a		71.3	8.9	n/a
TPS- 2 weeks		−37.6	−56.6	n/a		74.0	14.1	n/a
TPS- 3 weeks		−37.0	−56.0	n/a		72.4	6.1	n/a
TPS- 5 weeks		−35.6	−53.2	n/a		71.2	20.2	n/a
TPS- 7 weeks		−35.2	−58.5	n/a		69.9	11.3	n/a
TPS-MMT- 1 week		−40.2	−52.9	−61.8		70.9	13.3	−4.2
TPS-MMT- 2 weeks		−32.5	−57.5	−60.6		70.7	11.6	−22.6
TPS-MMT- 3 weeks		−20.3	−57.7	n/a		64.0	10.2	n/a
TPS-MMT- 5 weeks		−33.3	−55.4	n/a		72.7	2.8	n/a
TPS-MMT- 7 weeks		−38.5	−60.6	n/a		69.8	13.6	n/a

4. Conclusions

The effect of storage at various relative humidities on the structure, molecular mobility and mechanical properties of TPS and its nanocomposite was studied. It was shown that the amount of water has a crucial influence on all measured properties. Water content of the samples was determined from the weight measurements and by means of deconvolution of BL ^{1}H NMR spectra, where the latter provided systematically lower values since only free water can be distinguished in the spectra.

Both TPS and nanocomposite samples stored at 11% RH at room temperature were brittle and preserved their glassy state. They did not recrystallize due to the small amount of water and for this reason only physical ageing could take place throughout the whole period of storage. On the contrary, the samples stored at higher RHs (55 and 85%) absorbed a significant amount of water acting as a plasticizer and lowering their T_g under room temperature which allowed the samples to recrystallize during the first week of storage. After seven weeks of storage the samples stored at 55% RH showed improvements in mechanical properties (tensile strength and Young's modulus increased) due to newly formed interactions between starch, glycerol, and water making the structure stiffer. On the other hand, during storage at the highest RH (85%) the water molecules significantly reduced starch intermolecular hydrogen bonding density leading to deterioration of all mechanical properties measured.

The reinforcing effect of MMT nanoparticles manifested itself by improving the mechanical properties of nanocomposite samples when compared to TPS samples and led to lower molecular mobility of the TPS components. From the point of view of the material final application in practice, we could conclude that the best performance of the studied TPS-based nanocomposite is reached when stored at 55% RH.

Author Contributions: Methodology, M.K. and I.C.; Validation, N.Š., H.P., A.B., O.F. and M.K.; Investigation, N.Š., H.P., A.B. and O.F.; Writing—original draft, N.Š., H.P. and A.B.; Writing—review & editing, O.F., M.K., R.K. and I.C.; Supervision, I.C. All authors have read and agreed to the published version of the manuscript.

Funding: Financial support from the Slovak Grant Agency (project no. 1/0751/21) and Slovak Research and Development Agency (project no. 21-0120) are acknowledged.

Data Availability Statement: The data presented in this study cannot be shared at this time as they form a part of an ongoing research.

Conflicts of Interest: The authors declare no conflict of interest.

References

1. Peidayesh, H.; Heydari, A.; Mosnáčková, K.; Chodák, I. In situ dual crosslinking strategy to improve the physico-chemical properties of thermoplastic starch. *Carbohyd. Polym.* **2021**, *269*, 118250. [CrossRef] [PubMed]
2. Baran, A.; Vrábel, P.; Koval'aková, M.; Hutníková, M.; Fričová, O.; Olčák, D. Effects of sorbitol and formamide plasticizers on molecular motion in corn starch studied using NMR and DMTA. *J. Appl. Polym. Sci.* **2020**, *137*, 48964. [CrossRef]
3. Peidayesh, H.; Ahmadi, Z.; Khonakdar, H.A.; Abdouss, M.; Chodák, I. Fabrication and properties of thermoplastic starch/montmorillonite composite using dialdehyde starch as a crosslinker. *Polym. Int.* **2020**, *69*, 317–327. [CrossRef]
4. Peidayesh, H.; Mosnáčková, K.; Špitalský, Z.; Heydari, A.; Opálková Šišková, A.; Chodák, I. Thermoplastic Starch–Based Composite Reinforced by Conductive Filler Networks: Physical Properties and Electrical Conductivity Changes during Cyclic Deformation. *Polymers* **2021**, *13*, 3819. [CrossRef] [PubMed]
5. Peidayesh, H.; Ahmadi, Z.; Khonakdar, H.A.; Abdouss, M.; Chodák, I. Baked hydrogel from corn starch and chitosan blends cross-linked by citric acid: Preparation and properties. *Polym. Adv. Technol.* **2020**, *31*, 1256–1269. [CrossRef]
6. Fričová, O.; Hutníková, M. Changes in molecular mobility of sorbitol plasticized starch during aging. *J. Appl. Polym. Sci.* **2022**, *139*, e52948. [CrossRef]
7. Baran, A.; Fričová, O.; Vrábel, P.; Popovič, Ľ.; Peidayesh, H.; Chodák, I.; Hutníková, M.; Koval'aková, M. Effects of urea and glycerol mixture on morphology and molecular mobility in thermoplastic starch/montmorillonite-type nanofiller composites studied using XRD and NMR. *J. Polym. Res.* **2022**, *29*, 257. [CrossRef]
8. Fekete, E.; Angyal, L.; Csiszár, E. The effect of surface characteristics of clays on the properties of starch nanocomposites. *Materials* **2022**, *15*, 7627. [CrossRef]

9. Karger-Kocsis, J.; Kmetty, Á.; Lendvai, L.; Drakopoulos, S.X.; Bárány, T. Water-assisted production of thermoplastic nanocomposites: A review. *Materials* **2015**, *8*, 72–95. [CrossRef]
10. Huang, M.; Yu, J.; Ma, X. High mechanical performance MMT-urea and formamide-plasticized thermoplastic cornstarch biodegradable nanocomposites. *Carbohyd. Polym.* **2006**, *63*, 393–399. [CrossRef]
11. Ikeo, Y.; Aoki, K.; Kishi, H.; Matsuda, S.; Murakami, A. Nano clay reinforced biodegradable plastics of PCL starch blends. *Polym. Adv. Technol.* **2006**, *17*, 940–944. [CrossRef]
12. Namazi, H.; Mosadegh, M.; Dadkhah, A. New intercalated layer silicate nanocomposites based on synthesized starch-g-PCL prepared via solution intercalation and in situ polymerization methods: As a comparative study. *Carbohyd. Polym.* **2009**, *75*, 665–669. [CrossRef]
13. Kalambur, S.; Rizvi, S.S. Biodegradable and functionally superior starch–polyester nanocomposites from reactive extrusion. *J. Appl. Polym. Sci.* **2005**, *96*, 1072–1082. [CrossRef]
14. Perotti, G.F.; Tronto, J.; Bizeto, M.A.; Izumi, C.M.S.; Temperini, M.L.A.; Lugão, A.B.; Parra, D.F.; Constantino, V.R.L. Biopolymer-clay nanocomposites: Cassava starch and synthetic clay cast films. *J. Braz. Chem. Soc.* **2014**, *25*, 320–330. [CrossRef]
15. Paul, D.R.; Robeson, L.M. Polymer nanotechnology: Nanocomposites. *Polymer* **2008**, *49*, 3187–3204. [CrossRef]
16. Ali, S.S.; Tang, X.; Alavi, S.; Faubion, J. Structure and physical properties of starch/poly vinyl alcohol/sodium montmorillonite nanocomposite films. *J. Agric. Food Chem.* **2011**, *59*, 12384–12395. [CrossRef]
17. Derungs, I.; Rico, M.; López, J.; Barral, L.; Montero, B.; Bouza, R. Influence of the hydrophilicity of montmorillonite on structure and properties of thermoplastic wheat starch/montmorillonite bionanocomposites. *Polym. Adv. Technol.* **2021**, *32*, 4479–4489. [CrossRef]
18. Coativy, G.; Chevigny, C.; Rolland-Sabaté, A.; Leroy, E.; Lourdin, D. Interphase vs confinement in starch-clay bionanocomposites. *Carbohyd. Polym.* **2015**, *117*, 746–752. [CrossRef]
19. Godbilot, L.; Dole, P.; Joly, C.; Rogé, B.; Mathlouthi, M. Analysis of water binding in starch plasticized films. *Food Chem.* **2006**, *96*, 380–386. [CrossRef]
20. Šoltýs, A.; Hronský, V.; Šmídová, N.; Olčák, D.; Ivanič, F.; Chodák, I. Solid-state ^{1}H and ^{13}C NMR of corn starch plasticized with glycerol and urea. *Eur. Polym. J.* **2019**, *117*, 19–27. [CrossRef]
21. Šmídová, N.; Šoltýs, A.; Hronský, V.; Olčák, D.; Popovič, L.; Chodák, I. Aging-induced structural relaxation in cornstarch plasticized with urea and glycerol. *J. Appl. Polym. Sci.* **2021**, *138*, 50218. [CrossRef]
22. Fričová, O.; Hutníková, M.; Kovaľaková, M.; Baran, A. Influence of aging on molecular motion in PBAT-thermoplastic starch blends studied using solid-state NMR. *Int. J. Polym. Anal. Charact.* **2020**, *25*, 275–282. [CrossRef]
23. Greenspan, L. Humidity fixed points of binary saturated aqueous solutions. *J. Res. Natl. Bur. Stand. Sect. A Phys. Chem.* **1977**, *81*, 89–96. [CrossRef]
24. Mutungi, C.; Passauer, L.; Onyango, C.; Jaros, D.; Rohm, H. Debranched cassava starch crystallinity determination by Raman spectroscopy: Correlation of features in Raman spectra with X-ray diffraction and ^{13}C CP/MAS NMR spectroscopy. *Carbohyd. Polym.* **2012**, *87*, 598–606. [CrossRef]
25. Thérien-Aubin, H.; Zhu, X.X. NMR spectroscopy and imaging studies of pharmaceutical tablets made of starch. *Carbohyd. Polym.* **2009**, *75*, 369–379. [CrossRef]
26. Spěváček, J.; Brus, J.; Divers, T.; Grohens, Y. Solid-state NMR study of biodegradable starch/polycaprolactone blends. *Eur. Polym. J.* **2007**, *43*, 1866–1875. [CrossRef]
27. Schmitt, H.; Guidez, A.; Prashantha, K.; Soulestin, J.; Lacrampe, M.F.; Krawczak, P. Studies on the effect of storage time and plasticizers on the structural variations in thermoplastic starch. *Carbohyd. Polym.* **2015**, *115*, 364–372. [CrossRef]
28. Kawai, K.; Hagura, Y. Discontinuous and heterogeneous glass transition behavior of carbohydrate polymer–plasticizer systems. *Carbohyd. Polym.* **2012**, *89*, 836–841. [CrossRef]
29. Hulleman, S.H.D.; Janssen, F.H.P.; Feil, H. The role of water during plasticization of native starches. *Polymer* **1998**, *39*, 2043–2048. [CrossRef]
30. Forssell, P.M.; Mikkilä, J.M.; Moates, G.K.; Parker, R. Phase and glass transition behaviour of concentrated barley starch-glycerol-water mixtures, a model for thermoplastic starch. *Carbohyd. Polym.* **1997**, *34*, 275–282. [CrossRef]
31. Shogren, R.L. Effect of moisture content on the melting and subsequent physical aging of cornstarch. *Carbohyd. Polym.* **1992**, *19*, 83–90. [CrossRef]
32. Shogren, L.R.; Jasberg, B.K. Aging properties of extruded high-amylose starch. *J. Environ. Polym. Degrad.* **1994**, *2*, 99–109. [CrossRef]
33. Leroy, L.; Stoclet, G.; Lefebvre, J.M.; Gaucher, V. Mechanical Behavior of Thermoplastic Starch: Rationale for the Temperature-Relative Humidity Equivalence. *Polymers* **2022**, *14*, 2531. [CrossRef] [PubMed]
34. Nowacka-Perrin, A.; Steglich, T.; Topgaard, D.; Bernin, D. In situ ^{13}C solid-state polarization transfer NMR to follow starch transformations in food. *Magn. Reson. Chem.* **2022**, *60*, 671–677. [CrossRef] [PubMed]
35. Johnson, K.A.; Mauer, L.J. Effects of controlled relative humidity storage on moisture sorption and amylopectin retrogradation in gelatinized starch lyophiles. *J. Food Sci.* **2019**, *84*, 507–523. [CrossRef] [PubMed]
36. Van Soest, J.J.G.; De Wit, D.; Vliegenthart, J.F.G. Mechanical properties of thermoplastic waxy maize starch. *J. Appl. Polym. Sci.* **1996**, *61*, 1927–1937. [CrossRef]

37. Thunwall, M.; Boldizar, A.; Rigdahl, M. Compression molding and tensile properties of thermoplastic potato starch materials. *Biomacromolecules* **2006**, *7*, 981–986. [CrossRef]
38. Brouillet-Fourmann, S.; Carrot, C.; Lacabanne, C.; Mignard, N.; Samouillan, V. Evolution of interactions between water and native corn starch as a function of moisture content. *J. Appl. Polym. Sci.* **2002**, *86*, 2860–2865. [CrossRef]
39. Fričová, O.; Hutníková, M.; Peidayesh, H. DMA study of thermoplastic starch/montmorillonite nanocomposites. *AIP Conf. Proc.* **2021**, *2411*, 050004.

materials

MDPI

Article

Characterization and under Water Action Behaviour of a New Plaster-Based Lightened Composites for Precast

Manuel Álvarez *, Daniel Ferrández, Patricia Guijarro-Miragaya and Carlos Morón

Departamento de Tecnología de la Edificación, Escuela Técnica Superior de Edificación, Universidad Politécnica de Madrid, 28040 Madrid, Spain
* Correspondence: manuel.alvarezd@upm.es

Abstract: Plaster is a construction material widely used for the production of prefabricated parts in building construction due to its high capacity for hygrothermal regulation, its good mechanical performance, and its fireproof nature, among other factors. Its historical use has been linked to ornamental elements, although more recent research is oriented towards the industrialisation of plaster composites and the design of prefabricated parts for false ceilings and interior partitions. In this work, the behaviour against water of four new plaster-based composite materials is studied, using additions of two types of super absorbent polymers (sodium polyacrylate and potassium polyacrylate) and a lightening material (vermiculite) in their manufacturing process. In addition, the transmission of water vapour through the samples was studied together with the water absorption capacity of the samples in order to check the suitability of the use of plaster-based materials exposed to these environments. The results of this study show that composites with the addition of super absorbent polymers as well as vermiculite significantly improve their water performance compared to traditional materials up to 7.3% water absorption with a minimal (13%) reduction in mechanical strength compared to current materials with similar additions. In this sense, a plaster material is obtained with wide possibilities of application in the construction sector that favours the development of sustainable and quality buildings, in line with Goal 9 for Sustainable Development included in the 2030 Agenda.

Keywords: super absorbent polymer; water performance; building precast; geomaterials

Citation: Álvarez, M.; Ferrández, D.; Guijarro-Miragaya, P.; Morón, C. Characterization and under Water Action Behaviour of a New Plaster-Based Lightened Composites for Precast. *Materials* **2023**, *16*, 872. https://doi.org/10.3390/ma16020872

Academic Editors: Agnieszka Kijo-Kleczkowska and Adam Gnatowski

Received: 25 November 2022
Revised: 10 January 2023
Accepted: 12 January 2023
Published: 16 January 2023

1. Introduction

Plaster is still today one of the most widely used building materials in construction [1]. Among the main advantages of this material are its versatility and ease of application. Plaster powder mixed with water generates a liquid paste that allows any type of shape to be formed and can be applied directly in the execution of surface finishes or for the production of prefabricated elements [2]. In this sense, as a result of its purity and fineness of grind, plaster becomes an optimal material for interior finishes in buildings [3].

Despite its great advantages, exposure to the action of water remains one of the major limitations of gypsum composite materials for use in the building sector [4,5]. The behaviour of these materials when in contact with water is still a health problem [6]. As this is a disadvantage of these materials, most of the studies in the existing literature do not carry out water resistance tests or trials, focusing the research on the mechanical, thermal, and acoustic characterisation aspects of the developed composites [7–12]. Although it is true that, on some occasions, the coefficient of water absorption by capillarity is determined, this information is usually treated in a complementary manner without going into the problem in depth [9]. In this sense, as water behaviour is one of the main weaknesses of gypsum composites, it becomes a great opportunity to carry out studies to understand and characterise these properties with the intention of introducing improvements in the manufacturing process of these materials for their use in wet chambers.

Among the possible solutions is the incorporation of plastic materials into the manufacturing process of gypsum and plaster composites. These polymeric materials are positioned as a technically valid and economically viable option to improve the performance of prefabricated plaster and plasterboard commonly used in buildings [13,14]. In this sense, some studies show how the incorporation of plastic waste in the plaster composite matrix improves the performance of these materials, reducing the total water absorption coefficient, and decreases their permeability to water vapour [15]. These additions also make it possible to introduce circular economy criteria into the building process, favouring waste management and reincorporating construction and demolition waste into the production system of new prefabricated products [16]. In addition, it has been possible to verify how plaster composites with the incorporation of polymeric materials in their composition make it possible to obtain mechanical strengths that meet the requirements of the UNE-EN 13279-2 standard for gypsum composite materials applied in construction [17,18], as well as to achieve a higher acoustic absorption coefficient compared to traditional plaster [19,20].

Another of the most studied polymers for addition to plaster composites is polyvinyl acetate (PVA). Used in conjunction with polypropylene fibre reinforcement (0–1.2% volume fraction), Zhu et al. developed a new composite with a faster curing process and improved both the flexural strength and toughness of the material by about 20% [21]. In other research, polymeric composites have been used to modify the matrix of plaster composites, favouring the occurrence of chemical reactions during the setting process that allow the introduction of occluded air into the material, thus reducing the density of the final composite [22].

Expanded polystyrene (EPS) and extruded polystyrene (XPS) residues represent some of the most widely used polymeric materials to reduce the density of gypsum and plaster composites, with a wide field of application in the production of prefabricated slabs and panels with low thermal conductivity and good acoustic behaviour, as reported in the literature [5,11,23]. Among the most recent research is that conducted by Bicer et al., who developed an experimental campaign in which, by adding small-diameter EPS and tragacanth, they managed to improve thermal conductivity by 20% and reduce water absorption by 30%, making this type of polymer a valid resource for improving comfort conditions in homes [24].

Finally, there are those known as super absorbent polymers (SAPs). These are polymers supplied in granular form that are capable of absorbing and retaining large amounts of water. These materials are highly resistant to attack by large chemical agents such as acids or chlorides [25,26]. They are mainly used in agricultural sectors so that they can be mixed with the soil to store water [27]. In building construction, they are often used as a sealant in underwater concrete joints [28], as well as sheathing for conductor cables to ensure their proper functioning. A study by Liu et al. showed another possible use for these materials, as when applied to cold-formed metal structures, they prevent fires by increasing their fire resistance [29]. However, despite these advantages, these polymers are among the least used for application in construction materials.

The main objective of this research is to analyse the behaviour against water of a new plaster composite material made with SAPs and lightened with vermiculite. Therefore, an experimental campaign is proposed in which different types of plaster with additions of sodium polyacrylate and potassium polyacrylate with and without the incorporation of vermiculite are produced, analysing their physical–mechanical properties, and studying the technical feasibility of this new material for use in prefabricated panels and slabs.

2. Materials and Methods

2.1. Materials

Figure 1 shows the raw materials used to develop the proposed research: plaster, water, sodium polyacrylate, potassium polyacrylate, and vermiculite.

Figure 1. Materials used: (**a**) gypsum plaster; (**b**) potassium polyacrylate; (**c**) sodium polyacrylate; (**d**) vermiculite.

2.1.1. Binder

Iberyola E-35 fast-setting plaster, supplied by Placo Saint-Gobain in accordance with the UNE-EN-13279-1 standard [30], was used as the matrix element. Its main characteristics include its thermal conductivity of 300 mW/m-K, its purity index of over 90%, and its fineness of grind, with a particle size of 0–0.2 mm [31]. In addition, this binder is classified as A1 in term of reaction to fire according to Spanish regulations [32].

2.1.2. Water

The water used for this study was tap water from Canal de Isabel II (Madrid, Spain). The chemical analysis conducted annually by the water service of the Madrid City Council accredits the absence of harmful chemical agents such as chlorides, sulphates, etc., in sufficient quantities to alter the properties of the plaster paste. The result is an inert water, of a soft nature, and classified as acceptable for human consumption [33].

2.1.3. Super Absorbent Polymers (SAPs)

Table 1 shows the main chemical characteristics of the two super absorbent polymers used in this research [26,34]. The incorporation of these polymers into the manufacturing process of the plaster composites allows for the generation of an internal pore network in the matrix of the hardened material. This effect is due to the evaporation of the water absorbed by these SAPs during the setting process of the plaster composites. Additionally, this evaporation is favoured by the oven drying process of the hardened samples after seven days according to UNE-EN-13279-2 [35].

Table 1. Super absorbent polymer (SAP) main characteristics [26,34].

	Appearance	Particle Size (μm)	Apparent Density (g/mL)	Humidity (%)	pH	Ignitability
$(C_3H_3NaO_2)_n$	Granulated	150–850	0.5–0.75	≤5	6.0 ± 0.5	Not flammable
$(C_3H_3KO_2)_n$	Crystals	1000–2000	0.56	≤5	6.8	Not flammable

2.1.4. Vermiculite

Vermiculite is a naturally occurring material of volcanic origin consisting of iron or magnesium silicates, belonging to the mica group [23], whose chemical formula (Formula (1)) is as follows:

$$(\mathrm{Mg, Ca})\ 0.7\ (\mathrm{Mg, Fe, Al})\ 6.0\ [((\mathrm{Al, Si})_8\mathrm{O}_{20})]\ (\mathrm{OH})\ 4.8\ \mathrm{H}_2 \quad (1)$$

The sheets that form the vermiculite reflect and disperse the energy it receives, resulting in good thermal insulation performance. In addition, these omnidirectional sheets reflect sound waves, while another part of them is absorbed by the air inside. This makes vermiculite also a good thermal insulator at many frequencies. As a volcanic material, its melting point is quite high at 1370 °C, making it a non-combustible and very stable mineral. This makes it a suitable material for fire retardation and fire insulation. When subjected to high temperatures, it expands, resulting in a lightweight, non-flammable material with a high capacity to absorb and retain water. Table 2 lists the physical and chemical characteristics of this product.

Table 2. Vermiculite properties.

Characteristics	Value
Appearance	Granulated
Particle size (mm)	1.4
Apparent density (kg/m^3)	120
pH	7
Electrical conductivity (ms/m)	10
Ignitability	Not flammable

2.2. Dosages and Samples Elaboration

For this study, two SAPs and a lightening compound were used, so a total of five dosages was prepared the manufacturing process was manual and in accordance with the current UNE-EN-13279-2:2014 standard [35]. Once demoulded, they were stored in the laboratory for one week at constant temperature and relative humidity (RH) (23 °C ± 2 °C and 50% ± 5%). Subsequently, the specimens were dried after seven days for 24 h prior to testing using a LabProcess [36] oven, model ARGO LAB TCN200, at a temperature of 45 °C ± 1 °C.

Table 3 shows the dosages and composition used. The nomenclature is as follows: E0.7-SAP-V, where E0.7 refers to the water/plaster ratio by mass used, SAP indicates the type of polymer added, which can be sodium polyacrylate (Na) or potassium polyacrylate (K), and finally V refers to those compounds that add the vermiculite lightening load.

Table 3. Dosages used and composition. Weight in grams (g).

Name	Plaster	Water	$(C_3H_3NaO_2)_n$	$(C_3H_3KO_2)_n$	Vermiculite
E0.7	1000	700	-	-	-
E0.7-Na	1000	700	15	-	
E0.7-Na-V	1000	700	15	-	30
E0.7-K	1000	700	-	15	
E0.7-K-V	1000	700	-	15	30

The amounts of SAP and vermiculite were incorporated as additions to the plaster paste, without there being any substitution of the raw materials that make up the matrix of the composite. In the same way, the lightening agents were added to the matrix as an admixture. In total, 15 g of SAP and 30 g of vermiculite were added.

To develop this study, tests described below were conducted. With this, 7 series of three specimens of each chosen dosage were prepared. Table 4 shows the dimensions and the tests conducted on each of these specimens.

Table 4. Samples elaborated and their uses.

Series	Dimensions	Tests
SERIES I	4 × 4 × 16 cm^3	- Shore C hardness - Flexural strength - Compressive strength
SERIES II	Ø14 cm	- Water retention
SERIES III	4 × 4 × 16 cm^3	- Capillarity water absorption
SERIES IV	4 × 4 × 16 cm^3	- Total water absorption ➢ Shore C hardness ➢ Flexural strength ➢ Compressive strength
SERIES V		- Water-stove cycles ➢ Shore C hardness ➢ Flexural strength ➢ Compressive strength
SERIES VI		- Wet chamber ➢ Shore C hardness ➢ Flexural strength ➢ Compressive strength
SERIES VII	Ø16.5 cm h: 2 cm	- Water vapour permeability

2.3. *Experimental Plan*

As the aim was to check composites' water reaction studied, a series of tests, some of which are not standardised, was conducted based on the existing literature for characterising gypsum and plaster composites [37–40]. All tests were conducted in the construction materials laboratory of the Escuela Técnica Superior de Edificación de Madrid. As superabsorbent polymers were used, which contribute to the water absorption of the materials used, the mechanical characterisation was meaningful, as well as the reaction to exposure of the different tests to which these materials were exposed.

2.3.1. Water Absorption

The aim of these tests was to characterise the water absorption of the materials. To this end, three different tests were conducted. On the one hand, the water retention test included in the annex of the UNE-EN-13279-2 standard [35], which expresses water retention as a percentage of the total mass, was conducted. On the other hand, the non-standardised capillary absorption test was conducted to determine the height that the rising water reaches through the plaster material, and finally, the total water absorption coefficient was determined.

The capillary water absorption test lasts 15 min and is conducted as follows. A container is filled with water to a height of one centimetre. A grid-like support is then placed at a height of one centimetre, so that the water is flush with the support. Once this is done, the 4 × 4 × 16 cm^3 samples are placed vertically, and the stopwatch is started. After the first five minutes, a line is drawn at the height the water has reached on one side. After this first measurement, a new line is drawn every minute. At the end of the test, the height in mm reached each minute is measured and recorded, and the result is expressed in mm/min.

On the other hand, for the water retention test, the materials needed for this test are the following: filter paper, non-woven gauze, Ø140 mm circular plastic ring, two square plastic plates at least 30 mm larger than the circular ring, and the mortar in question. The purpose of this test is to measure, during the setting process, the amount of water retained by the mass, by means of the difference in weights. The complete development is set out in the UNE-EN-459-2:2021 standard [41].

Finally, total water absorption was conducted. The purpose of this test is to check the total water absorption capacity of the material, its capacity to lose water, and how it affects its mechanical behaviour. This test is determined in the UNE-EN-520 standard [42]. It consists of completely submerging the samples for two hours in a container with water. Once this has been done, they are weighed to obtain the total amount of water absorbed. They are then stored in a laboratory environment (21 °C and 35% RH) for seven days. They are then weighed again to check the amount of water they have been able to remove. Once this has been done, they are mechanically characterised by testing surface hardness, flexural strength, and compressive strength.

2.3.2. Water Vapour Permeability

Water vapour permeability is another of the variables studied in this article. This test is included in the UNE-EN-ISO-12572 standard [43]. For this purpose, samples of each dosage were prepared with a thickness of 1 cm and a diameter of 16.5 cm, as shown in Table 4. These samples were placed in an airtight plastic container containing 200 mL of aqueous solution. To guarantee an environment of 98% relative humidity inside the containers, the dissolution was conducted until the water was saturated by the addition of potassium nitrate salts. In addition, to ensure that the water vapour diffused only through the plate, the edges of the plate were sealed to the vessel with silicone.

After the preparation process, twice a week for eight weeks, the different samples were weighed with the container to record the water vapour losses. With these weights, and using the procedure indicated in the standard, the water vapour resistance "R" was obtained.

2.3.3. Wet Chamber

The purpose of this test is to observe the variations in weight and mechanical behaviour of the dosages studied when subjected to a constant temperature and high relative humidity for a prolonged period of time. This test is not standardised but follows the test conducted by del Río Merino in his doctoral thesis [40]. It consists of reserving the samples for five days in a wet chamber with hygrothermal conditions of 21 °C and 75% relative humidity.

After this time, the specimens were weighed to calculate the percentage of water acquired during this time. The mechanical characterisation of the specimens was then conducted, consisting of shore C surface hardness, flexural strength, and compressive strength tests.

2.3.4. Water–Stoves Cycles

The purpose of the water–stove cycling test is to test the response to exposure to extreme contrasts of humidity and dryness. After the test specimens have been made and left in a laboratory environment for seven days (23 °C and 35% humidity), they are placed in a container with water so that they are completely submerged for two days. After this time, they are removed, weighed, and placed in an oven at 60 °C for a further two days. The test tubes are weighed again, and this cycle is repeated twice. After finishing, the mechanical characterisation is conducted, consisting of measuring the surface hardness and the flexural and compressive strengths. This test is not standardised; as in the previous test, the test conducted by del Río Merino in his doctoral thesis [40] was followed.

2.3.5. Mechanical Characterisation

With the exception of the water vapour permeability, capillary absorption, and water retention samples, all samples produced were tested for flexural and compressive strength and Shore C surface hardness in order to compare them with the reference samples as shown in Figure 2.

- Bending and compressive test. These tests were conducted according to UNE-EN-13279-2 [35] using a hydraulic press machine, model AUTOTEST 200-10SW, from IBERTEST (Madrid, Spain).
- Surface hardness. It was conducted with the help of a Shore C durometer.

Figure 2. Tests performed: (**a**) surface hardness; (**b**) flexural and compressive strength; (**c**) total water absorption; (**d**) water vapour permeability.

3. Results and Discussion

This section presents the results derived from the planned experimental campaign, as well as the discussion derived from their interpretation.

3.1. Water Retention and Absorption and Capillarity Test

This section shows the results obtained for the three tests related to water absorption in the plaster composites produced. Firstly, in Table 5, the values obtained for the water retention test are shown.

Table 5. Water retention test results.

Type	Water Retention (%)
E0.7	61.77%
E0.7-Na	80.33%
E0.7-Na-V	70.19%
E0.7-K	84.20%
E0.7-K-V	73.00%

The water retention test shown in Table 5 was used to measure the amount of water that the plaster mixture was able to release when setting. The results of this test showed a water retention of 61.77% for the E0.7 dosage. This means that it lost a considerable amount of water compared to that initially existing in the mixture. Those dosages with the addition of sodium (80.33%) and potassium (84.2%) polyacrylates showed higher water retention due to the nature of these superabsorbent polymers. Finally, it was observed that the incorporation of vermiculite in the preparation of the plaster composites reduced the amount of water retained in the samples prepared with SAP, although without reaching the values obtained for the reference plaster E0.7.

Figure 3 shows the results obtained for the total water absorption test and the height reached after the capillary water absorption test.

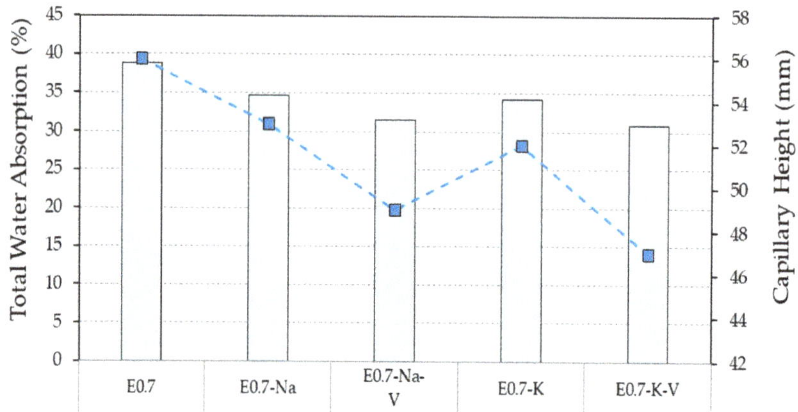

	E0.7	E0.7-Na	E0.7-Na-V	E0.7-K	E0.7-K-V
Total Water Absorption (%)	38.8	34.7	31.5	34.2	30.8
Capillarity Height (mm)	56	53	49	52	47

Figure 3. Capillarity height and total water absorption test results compared.

The capillary water absorption and total water absorption tests analyse material behaviour once it has hardened. Therefore, they are related to material response throughout its service life. Figure 3 shows how the dosage without any addition (E0.7) reached the highest height of all (56 mm) and retained the highest amount of water when submerged (38.8% of its initial weight). The addition of SAP to the mixture reduced the total absorption of both by 4.1% for sodium polyacrylate and 4.6% for potassium polyacrylate. Likewise, the capillary height also decreased in both cases, being lower in the samples with the addition of potassium polyacrylate, as with the total water absorption. This effect was more pronounced in the plaster composites incorporating vermiculite in their composition compared to the E0.7 reference, further reducing the values obtained in the samples containing only SAP.

3.2. Water Vapour Permeability

Figure 4 shows the relative water vapour permeability results. It can be seen that the behaviour of most of the compounds was quite similar. From week one to two, there was a high diffusion of water vapour through the samples. From week two to five, the value remained stable with small variations. Finally, from week five to eight, water vapour permeability decreased 50%.

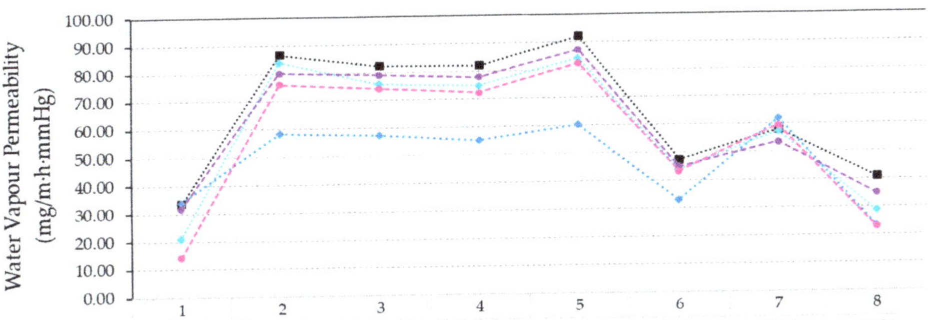

	1	2	3	4	5	6	7	8
E0.7	33.56	86.60	82.27	81.99	92.01	47.36	57.91	40.86
E0.7-Na	33.84	58.45	57.43	55.38	60.50	32.82	62.04	23.07
E0.7-Na-V	21.11	83.89	75.77	74.96	84.16	43.84	57.37	28.68
E0.7-K	31.93	80.10	79.02	77.83	87.24	44.92	53.58	34.91
E0.7-K-V	14.34	76.04	74.15	72.36	82.43	43.30	59.53	22.73

Figure 4. Relative water vapor permeability test results.

Although all the plaster composites studied followed the same pattern, the E0.7-Na sample presented a greater opposition to the diffusion of water vapour through its surface, obtaining lower transmission values than the rest, especially in the central weeks (2–5), where the difference was up to 25%. Week seven was the most stable and similar for all samples, as the maximum difference in transmission was 2%. It should be noted that all samples with polymer addition reduced the water vapour permeability compared to the reference. This is mainly due to the water absorption capacity of the polymers, which trapped water vapour while passing through the samples. This resulted in a clogging of the pores generated in the plaster matrix, thus reducing the surface area where water vapour could pass through.

In this test, in addition to determining the permeability to water vapour passing through the processed samples, the water vapour retention capacity inside each of the plaster composites could be observed. Table 6 shows the mass variation observed in the samples before and after completion of the water vapour permeability test, as well as the percentage difference between the two. As a first result, it should be noted that all the samples tested retained water. The sample E0.7-Na-V was the one that retained the highest amount of water, reaching 1%, which was twice the amount obtained in samples with the lowest amount of water retention, namely, E0.7-K and E0.7 (0.5%). Samples with the addition of vermiculite increased the amount of water retained after the test. Although not very significantly, this may be due to the greater specific surface area in these plaster matrices as a result of the expanded vermiculite. This allowed water vapour to condense on its surface when it passed through the material.

Table 6. Water retention from water vapour permeability.

Specimen	Pre-Test Weight (g)	Post-Test Weight (g)	Difference (%)
E0.7	220.6	221.6	0.5%
E0.7-Na	282	283.7	0.6%
E0.7-Na-V	325.7	329.1	1.0%
E0.7-K	231.9	233.1	0.5%
E0.7-K-V	242.5	244.7	0.9%

3.3. *Wet Chamber*

The wet chamber tests (21 °C and 75% relative humidity) yielded two different types of results. On the one hand, the percentage of water absorption by the specimens during the duration of the test, which can be observed in Table 7, was determined. On the other hand, the results derived from the mechanical characterisation, hardness, bending, and compression are shown in Section 3.5 and were compared and analysed with the other tests conducted in the mechanical characterisation section.

All tested samples retained a higher percentage of water compared to the reference plaster E0.7 (1.15%). However, contrary to the water vapour transmission test, those samples with added vermiculite retained less water than those containing only SAP. The uninterrupted exposure of the specimens to the wet environment for two days showed a higher water retention than water vapour transmission samples even though they were subjected to a saturated environment for a longer period of time.

Table 7. Wet chamber test results weight variation.

Specimen	Weight (Dry) (g)	Weight after Test (g)	Δ(W)
E0.7	270.0	273.1	1.15%
E0.7-Na	271.3	276.4	1.88%
E0.7-Na-V	269.0	273.2	1.56%
E0.7-K	271.3	275.8	1.66%
E0.7-K-V	258.6	262.3	1.43%

3.4. *Water–Stove Cycles*

The results of the water–stove cycle test was analysed by comparing the weight increase at the end of each of the cycle phases (first cycle C1 and second cycle C2). Table 8 shows, in grams and percentage ratio, the differences obtained.

Table 8. Water–stove cycles test weight variation results.

Specimen	Initial Weight (Dry)	Weight C1 (Wet)	Δ(W) *	Weight C1 (Dry)	Δ(W) *	Weight C2 (Wet)	Δ(W) *	Weight C2 (Dry)	Δ(W) *
E0.7	273	360	32%	271.9	−0.40%	359.5	32%	272.4	−0.22%
E0.7-Na	274.2	379.4	38%	272.3	−0.69%	377.2	39%	272.8	−0.51%
E0.7-Na-V	272	370.6	36%	269.8	−0.81%	368.8	37%	270.2	−0.66%
E0.7-K	273.4	378.6	38%	271.2	−0.80%	375.1	38%	274.4	0.37%
E0.7-K-V	259.5	352.7	36%	258.8	−0.27%	350.1	35%	259.8	0.12%

* Weight compared to initial weight (dry). Weight in grams (g).

Water absorption by the samples analysed was lower in the E0.7 reference (32%) than in all samples with additions. In water–stove cycles, samples with the highest percentage of water retention were those without vermiculite, namely, E0.7-Na and E0.7-K, with 38–39% and 38–38% in the respective cycles 1 and 2. The addition of vermiculite resulted in a 2–3% difference compared to samples with only SAP. It should be noted that after the drying cycles, the weight achieved was in all cases lower than the initial dry weight prior to the test. The difference was higher than 1% in all samples.

3.5. *Mechanical Characterization*

In this section, the mechanical characterisation results between the reference, the water and stove cycles, and the wet chamber are analysed and compared.

Figure 5 shows the results of the surface hardness tests in Shore C units. It can be observed that the results of all the tests were lower than the reference ones regardless of the type of sample tested. The maximum hardness value reached was 89 Shore C in the reference sample E.7-K-V, while the lowest value was obtained in the reference sample E0.7 subjected to water and stove cycles, with 73 Shore C.

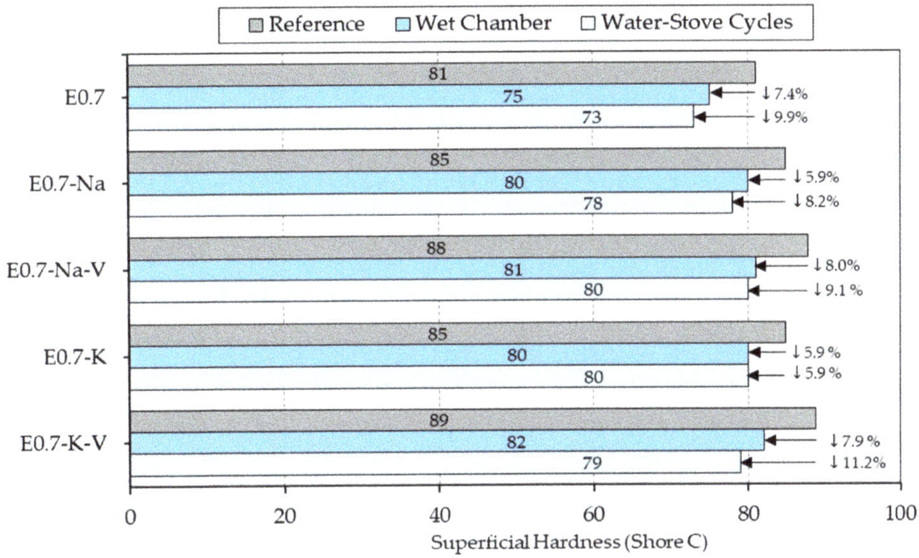

Figure 5. Shore C hardness test results compared.

Both in the reference samples and in the samples subjected to durability cycles, all the mixes with the addition of SAP and vermiculite exceeded the surface hardness values obtained by the reference plaster E0.7. This translates into a direct improvement of this material characteristic. The samples with the addition of potassium polyacrylate achieved the most extreme results for these tests in such a way that the E0.7-K-V mix subjected to water–stove durability cycles obtained a reduction in its hardness with a maximum value of 11.2% compared to reference, and the E0.7-K sample subjected to cycles reduced its surface hardness minimally (5.9% compared to reference).

The results derived from the flexural and compressive strength tests of the different plasters produced are shown in Figure 6.

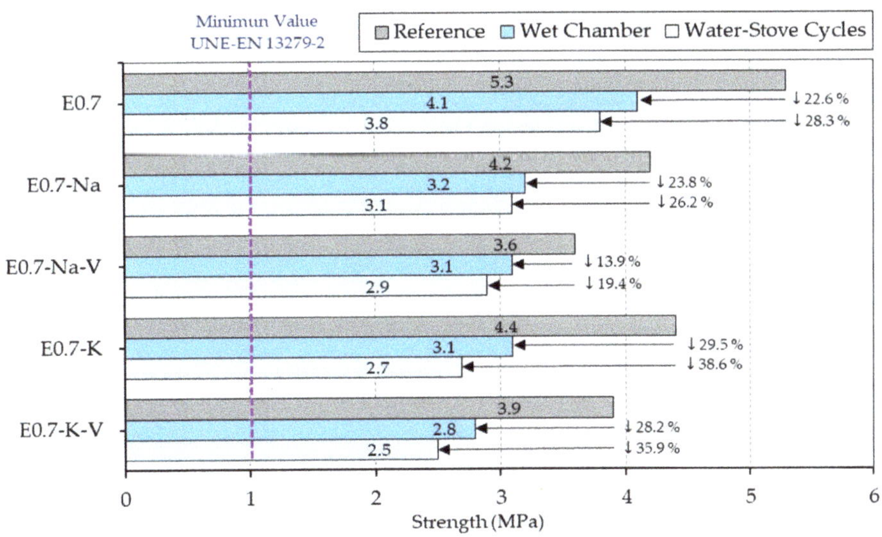

Figure 6. Flexural strength test results compared.

As can be seen, none of the samples with SAP and vermiculite addition showed flexural strength values higher than sample E0.7. Furthermore, in all the plaster composites studied, a decrease in flexural strength was observed when the samples were subjected to ageing cycles under the action of water. On the other hand, the wet chamber cycles were less aggressive than the water–stove cycles, since the latter had a greater effect on the mechanical behaviour of the lightened plaster composites produced in this research.

Figure 6 shows how all the plaster composites with potassium polyacrylate incorporation showed a higher flexural strength than those made with sodium polyacrylate addition. Likewise, in all the samples with the addition of SAP in which vermiculite was incorporated, there was a decrease in their mechanical strength. In all the mixes tested, flexural strength values higher than the minimum recommended in the UNE-EN-13279-2 standard [35] were obtained.

Finally, Figure 7 presents the compressive strength results of the samples tested. As in the flexural tests, no sample exceeded the strength obtained for reference E0.7 plaster (10.8 MPa). The lowest strength values were obtained by the E0.7-Na-V sample (5.9 MPa) subjected to water–stove cycles. Again, the samples tested using wet chamber cycles performed better than those exposed to water–stove cycles. This is because the changes produced inside the specimens in the water–stove cycles caused a weakening of the gypsum crystals, which had a direct impact on their internal bonding and thus on their mechanical strength [44].

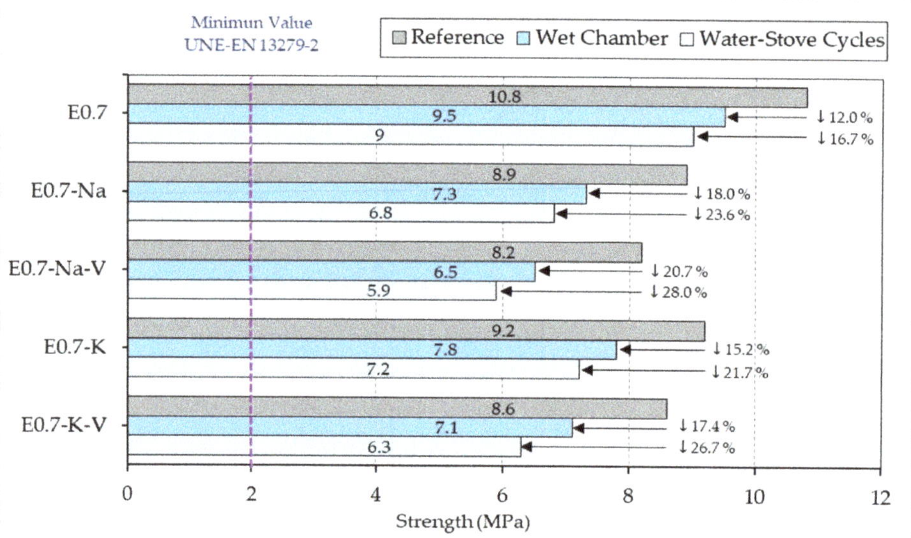

Figure 7. Compressive strength results compared (MPa).

In order to better compare the results and clearly establish the conclusions of the present study, the results obtained in the different tests are shown together in Figure 8. In addition to the water performance analysed, the average bulk density of the processed samples has been added. The figure also shows the minimum values required by the standards for the mechanical resistance to bending (1 MPa) and compression (2 MPa).

A direct relationship between the incorporation of SAP and vermiculite and a significant decrease in density can be observed in Figure 8, as occurred in other studies in the existing literature [44–46]. Compared to reference sample E0.7, the lightened plaster composites with SAP addition showed an average decrease in bulk density of 2–3%, and the samples with vermiculite addition further reduced the density to obtain a 6–8% lightened composite compared to traditional plasters. The density of the samples with potassium polyacrylate was lower than those with sodium polyacrylate, mainly due to the size and physicochemical characteristics of these additions. Potassium polyacrylate has a greater facility to release cations than sodium polyacrylate. Thus, interacting with water makes it more hydrophilic. In water behaviour tests, it was observed that in samples containing polyacrylates there was a loss of mass compared to the reference, which has a direct impact on bulk density. This difference was between 3% lower in those subjected to a wet chamber and 6% in those subjected to water and stove cycles. There was also a clear relationship between density and mechanical strength. These results obtained showed a whole new way to study these kinds of materials, as there are no similar studies in the literature. The main use of plaster-based materials for interior uses has led to a lack of standardized tests to study its behaviour under water action.

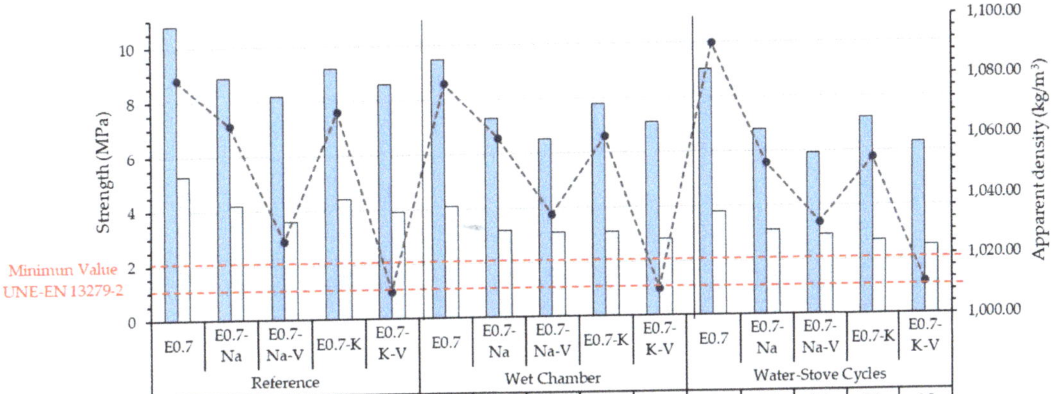

	E0.7	E0.7-Na	E0.7-Na-V	E0.7-K	E0.7-K-V	E0.7	E0.7-Na	E0.7-Na-V	E0.7-K	E0.7-K-V	E0.7	E0.7-Na	E0.7-Na-V	E0.7-K	E0.7-K-V
	Reference					Wet Chamber					Water-Stove Cycles				
Compressive	10.8	8.9	8.2	9.2	8.6	9.5	7.3	6.5	7.8	7.1	9	6.8	5.9	7.2	6.3
Flexural	5.3	4.2	3.6	4.4	3.9	4.1	3.2	3.1	3.1	2.8	3.8	3.1	2.9	2.7	2.5
Density	1,080.0	1,064.6	1,026.1	1,069.0	1,009.0	1,078.0	1,059.6	1,034.0	1,059.8	1,009.0	1,090.6	1,050.4	1,030.7	1,052.1	1,010.7

Figure 8. Results evaluation and comparison.

Generally, results of the different tests were similar. The mechanical strength values (bending and compression) achieved by the samples subjected to the wet chamber test showed that the exposure of this material to wet areas did not change its mechanical strength significantly. This means that composites with SAP behaviour are acceptable for use in the manufacture of precast products in wet environments, and the values obtained were well above the minimum required by the standards [31]. Water absorption by the material in these environments helps to regulate the relative humidity in damp rooms.

The flexural strength decreased with specimens subjected to the wet chamber and decreased even more for those specimens subjected to water–stove cycles. The flexural strength of the tested specimens ranged from 2.5 MPa (E0.7-K-V subjected to water–stove cycles) to 4.1 MPa (E0.7 subjected to the wet chamber). Despite this decrease, the minimum value, E0.7-K-V subjected to water–stove cycles (2.5 MPa), was only 35% lower than E0.7 subjected to the same test and 52% lower than the reference sample E0.7 (5.2 MPa). However, this value was still above the 2 MPa required by the standard [35].

The same situation occurs with compressive strength. The decrease was greater in lightened plaster composites subjected to water–stove cycles than in those subjected to the wet chamber. After the accelerated ageing cycles, the results ranged from 6.3 MPa (E-0.7-K-V subjected to water and oven cycles) to 9.5 MPa (E0.7 subjected to the wet chamber test). As can be seen, E0.7-K-V was once again the sample with the worst results, with a decrease of 41% compared to reference E0.7 and 33% compared to E0.7 subjected to the same test.

Despite of the good results obtained and behaviour under water action performance of this new material, there were some shortcomings that could not be avoided and considered. The non-standardized tests performed were based in previous works that are hardly endorsed. Thus, they are still non-standardized, which means that there may be some variables that were not considered and could be improved.

Finally, Figure 9 presents all results together to establish a selection of the samples with the best and worst behaviour under water action. In all the lightened plaster composites, a linear decrease in their mechanical properties was observed when subjected to accelerated water exposure cycles. It was observed that water action reduced the technical performance of the plaster materials designed for this research. It was observed that the lightened plaster composite with the best mechanical behaviour was 0.7-K subjected to wet chamber cycles, while the sample with the worst mechanical behaviour after the accelerated water exposure cycle tests was E0.7-Na-V. This combination was the one that offered the lowest guarantees for the production of prefabricated elements exposed to wet areas. The samples subjected to wet chamber testing showed more homogeneous bending behaviour than those subjected to water–stove cycles. There was a significant difference in behaviour between the reference samples and those subjected to water behaviour tests. While the reference samples progressively and linearly reduced their strength, the samples subjected to water reaction tests moved away from this distribution. In both tests, E0.7-Na and E-0.7-K-V obtained more similar values in compressive tests and E0.7-Na-V and E0.7-K in bending tests. Finally, it can be seen that the durability test water–stove cycles generated more harmful effects and a decrease in mechanical resistance compared to the wet chamber cycle test. In general, samples with the same composition performed the same way in mechanical strength tests after being subjected to accelerated water exposure cycles.

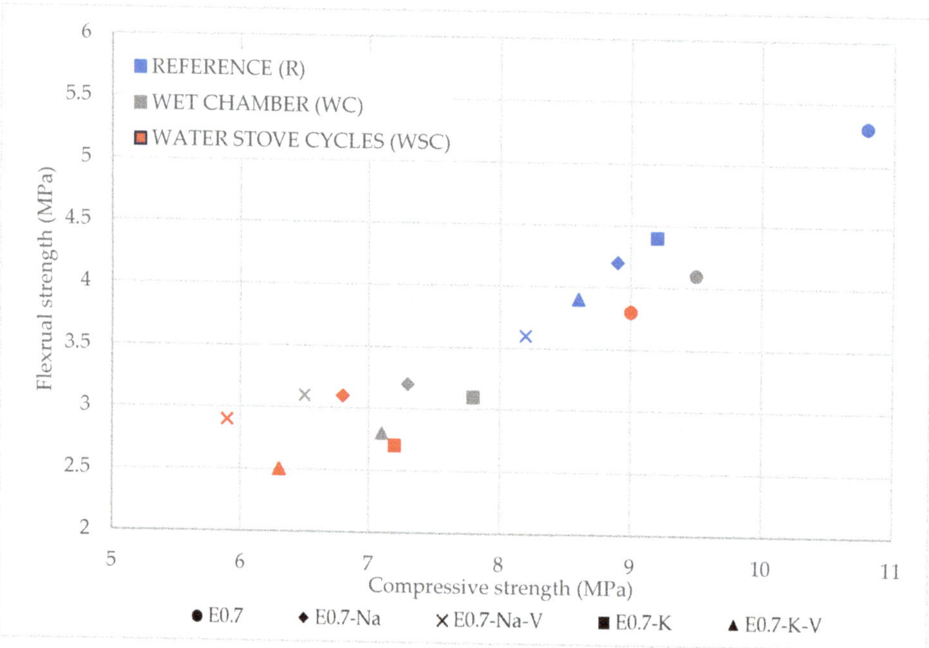

Figure 9. All mechanical performance results.

4. Conclusions

In this research, the water behaviour of a new type of lightened plaster composite was studied. These plaster composites have been registered under invention patent ES-2841130-B2 [47] and were specially designed for the production of prefabricated products. The results obtained in this study allow us to study their technical feasibility for use in the manufacture of slabs and panels. This also allows us to explore their application possibilities for the interior cladding of wet areas in both residential and service buildings. Finally, after reviewing and discussing the results, the following conclusions were reached.

- Firstly, related to these materials' physical properties, bulk density decreased as both SAP and vermiculite were added. Density was lower in samples with potassium polyacrylate addition than those with sodium polyacrylate addition. This density reduction resulted in lower building loads, as well as lower transport costs and ease of on-site execution.
- Water–stove cycle samples obtained a lower mechanical strength than wet chamber cycle samples. However, after both tests, the samples continued showing mechanical strength results suitable for development and exposure to wet environments.
- Wet environments exposure and water action on these materials have repercussions for their mechanical behaviour, although they do not prevent or limit their use in wet areas in buildings. All lightened plaster composites produced in this work exceeded the minimum flexural and compressive strength values set out in the UNE-EN 13279-2 standard. The worst mechanical behaviour in all the tests conducted was shown by the E0.7-Na-V composite.
- The lowest water vapour permeability was reached by E0.7-Na, with 27% less than the rest of the samples. However, all the lightened composites acted in a similar way to each other, improving the lightened plaster composites' hygroscopic regulation capacity. The addition of sodium and potassium polyacrylate increased plaster composites' water absorption capacity and reduced the capillary height reached by the water.

These conclusions have a number of implications for the application of these composites in the building sector. The good results obtained allow us to affirm that SAPs are a great opportunity for the development of prefabricated materials exposed to water action and wet environments. Materials with this addition maintain a mechanical behaviour well above the minimum required for this performance. As a main consequence, they could improve workability, as well as help to stabilise the hygrothermal conditions of the places where they are exposed. With this, pathologies resulting from condensation, both superficial and interstitial, would be avoided. Finally, the main limitation is the higher cost of the final lightened plaster product and the decrease in its mechanical properties.

Author Contributions: Conceptualization, D.F. and M.Á.; methodology, M.Á. and P.G.-M.; software, D.F. and M.Á.; validation, M.Á. and P.G.-M.; formal analysis, M.Á. and D.F.; investigation, M.Á. and P.G.-M.; resources, C.M.; data curation, M.A and D.F.; writing—original draft preparation, M.Á.; writing—review and editing, D.F.; visualization, M.Á. and D.F.; supervision, D.F. and C.M.; project administration, D.F.; funding acquisition, D.F. All authors have read and agreed to the published version of the manuscript.

Funding: This research received no external funding.

Institutional Review Board Statement: Not applicable.

Informed Consent Statement: Not applicable.

Data Availability Statement: Not applicable.

Acknowledgments: The authors would like to thank the Laboratorio de Materiales de la Escuela Técnica Superior de Edificación de Madrid for their collaboration and help in carrying out the tests.

Conflicts of Interest: The authors declare no conflict of interest.

References

1. Arroyo, F.; Christoforo, A.; Salvini, V.; Pelissari, P.; Pandolfelli, V.; Luz, A.; Cardoso, C. Development of plaster foam for thermal and acoustic applications. *Constr. Build. Mater.* **2020**, *262*, 120800. [CrossRef]
2. Freire, M.T.; Veiga, M.D.R.; Silva, A.S.; de Brito, J. Restoration of ancient gypsum-based plasters: Design of compatible materials. *Cem. Concr. Compos.* **2021**, *120*, 104014. [CrossRef]

3. Muntongkaw, S.; Pianklang, S.; Tangboriboon, N. Modifications to improve properties of gypsum ceiling composites as multifunctional construction by embedding Typha angustifolia fiber and natural rubber latex compound. *Case Stud. Constr. Mater.* **2021**, *15*, e00658. [CrossRef]
4. Barriguete, A.V.; del Río Merino, M.; Sanchez, E.A.; Ramírez, C.P.; Arrebola, C.V. Analysis of the feasibility of the use of CDW as a low-environmental-impact aggregate in conglomerates. *Constr. Build. Mater.* **2018**, *178*, 83–91. [CrossRef]
5. Doleželová, M.; Scheinherrová, L.; Krejsová, J.; Keppert, M.; Černý, R.; Vimmrová, A. Investigation of gypsum composites with different lightweight fillers. *Constr. Build. Mater.* **2021**, *297*, 123791. [CrossRef]
6. Bouzit, S.; Laasri, S.; Taha, M.; Laghzizil, A.; Hajjaji, A.; Merli, F.; Buratti, C. Characterization of Natural Gypsum Materials and Their Composites for Building Applications. *Appl. Sci.* **2019**, *9*, 2443. [CrossRef]
7. Aranda, B.; Guillou, O.; Lanos, C.; Daiguebonne, C.; Freslon, S.; Tessier, C.; Laurans, M.; Baux, C. Effect of multiphasic structure of binder particles on the mechanical properties of a gypsum-based material. *Constr. Build. Mater.* **2016**, *102*, 175–181. [CrossRef]
8. Freire, M.T.; Veiga, M.D.R.; Silva, A.S.; de Brito, J. Studies in ancient gypsum-based plasters towards their repair: Physical and mechanical properties. *Constr. Build. Mater.* **2019**, *202*, 319–331. [CrossRef]
9. Wu, H.-C.; Xia, Y.-M.; Hu, X.-Y.; Liu, X. Improvement on mechanical strength and water absorption of gypsum modeling material with synthetic polymers. *Ceram. Int.* **2014**, *40*, 14899–14906. [CrossRef]
10. Sadeghiamirshahidi, M.; Vitton, S.J. Analysis of drying and saturating natural gypsum samples for mechanical testing. *J. Rock Mech. Geotech. Eng.* **2019**, *11*, 219–227. [CrossRef]
11. Santa Cruz Astorqui, J.; Río Merino, M.D.; Villoria Sáez, P.; Porras Amores, C. Analysis of the viability of prefabricated elements for partitions manufactured with plaster and eps from waste recycling. *Dyna* **2019**, *94*, 415–420. [CrossRef]
12. Erbs, A.; Nagalli, A.; de Carvalho, K.Q.; Mymrin, V.; Passig, F.H.; Mazer, W. Properties of recycled gypsum from gypsum plasterboards and commercial gypsum throughout recycling cycles. *J. Clean. Prod.* **2018**, *183*, 1314–1322. [CrossRef]
13. Geraldo, R.H.; Pinheiro, S.M.; Silva, J.S.; Andrade, H.M.; Dweck, J.; Gonçalves, J.P.; Camarini, G. Gypsum plaster waste recycling: A potential environmental and industrial solution. *J. Clean. Prod.* **2017**, *164*, 288–300. [CrossRef]
14. Pedreño-Rojas, M.; Rodríguez-Liñán, C.; Flores-Colen, I.; De Brito, J. Use of polycarbonate waste as aggregate in recycled gypsum plasters. *Materials* **2020**, *13*, 3042. [CrossRef] [PubMed]
15. Erbs, A.; Nagalli, A.; de Carvalho, K.Q.; Mazer, W.; Erbs, M.D.M.; Paz, D.H.; Lafayette, K.P. Development of plasterboard sheets exclusively from waste. *J. Build. Eng.* **2021**, *44*, 102524. [CrossRef]
16. Dobrotă, D.; Dobrotă, G.; Dobrescu, T. Improvement of waste tyre recycling technology based on a new tyre marking. *J. Clean. Prod.* **2020**, *260*, 121141. [CrossRef]
17. Gencel, O.; Diaz, J.J.D.C.; Sütçü, M.; Koksal, F.; Rabanal, F.A.; Martinez-Barrera, G.; Brostow, W. Properties of gypsum composites containing vermiculite and polypropylene fibers: Numerical and experimental results. *Energy Build.* **2014**, *70*, 135–144. [CrossRef]
18. Romero-Gómez, M.; Silva, R.; Flores-Colen, I.; de Brito, J. Influence of polypropylene residues on the physico-mechanical and water-resistance properties of gypsum plasters. *J. Clean. Prod.* **2022**, *371*, 133674. [CrossRef]
19. Capasso, I.; Pappalardo, L.; Romano, R.A.; Iucolano, F. Foamed gypsum for multipurpose applications in building. *Constr. Build. Mater.* **2021**, *307*, 124948. [CrossRef]
20. Guna, V.; Yadav, C.; Maithri, B.; Ilangovan, M.; Touchaleaume, F.; Saulnier, B.; Grohens, Y.; Reddy, N. Wool and coir fiber reinforced gypsum ceiling tiles with enhanced stability and acoustic and thermal resistance. *J. Build. Eng.* **2021**, *41*, 102433. [CrossRef]
21. Zhu, C.; Zhang, J.; Peng, J.; Cao, W.; Liu, J. Physical and mechanical properties of gypsum-based composites reinforced with PVA and PP fibers. *Constr. Build. Mater.* **2018**, *163*, 695–705. [CrossRef]
22. Álvarez, M.; Ferrández, D.; Morón, C.; Atanes-Sánchez, E. Characterization of a new lightened gypsum-based material reinforced with fibers. *Materials* **2021**, *14*, 1203. [CrossRef] [PubMed]
23. Merino, M.D.R.; Sáez, P.V.; Longobardi, I.; Astorqui, J.S.C.; Porras-Amores, C. Redesigning lightweight gypsum with mixes of polystyrene waste from construction and demolition waste. *J. Clean. Prod.* **2019**, *220*, 144–151. [CrossRef]
24. Bicer, A.; Kar, F. Thermal and mechanical properties of gypsum plaster mixed with expanded polystyrene and tragacanth. *Therm. Sci. Eng. Prog.* **2017**, *1*, 59–65. [CrossRef]
25. Afridi, S.; Sikandar, M.A.; Waseem, M.; Nasir, H.; Naseer, A. Chemical durability of superabsorbent polymer (SAP) based geopolymer mortars (GPMs). *Constr. Build. Mater.* **2019**, *217*, 530–542. [CrossRef]
26. Alquera, S.L. Ficha de Seguridad Poliacrilato de Sodio. 2021. Available online: https://www.alquera.com/poliacrilato-de-sodio/ (accessed on 11 January 2023).
27. Qi, Z.; Hu, X. Water absorbency of super absorbent polymer based on flexible polymeric network. *Eur. Polym. J.* **2022**, *166*, 111045. [CrossRef]
28. Liu, L.; Fang, Z.; Huang, Z.; Wu, Y. Solving shrinkage problem of ultra-high-performance concrete by a combined use of expansive agent, super absorbent polymer, and shrinkage-reducing agent. *Compos. Part B Eng.* **2022**, *230*, 109503. [CrossRef]
29. Liu, K.; Chen, W.; Ye, J.; Bai, Y.; Jiang, J.; Tong, Y.; Chen, W. Improved fire resistance of cold-formed steel walls by using super absorbent polymers. *Thin-Walled Struct.* **2021**, *160*, 107355. [CrossRef]
30. *UNE-EN 13279-1*; Yesos de Construcción y Conglomerantes a Base de Yeso Para la Construcción. Parte 1: Definiciones y Especificaciones. UNE: Madrid, Spain, 2006.

31. Saint-Gobain, P. Iberyola e-30/e35 Technical Information. 2022. Available online: https://www.placo.es/Producto/iberyolar (accessed on 11 January 2023).
32. *UNE-EN-13501-1*; Clasificación en Función del Comportamiento Frente al Fuego de los Productos de Construcción y Elementos para la Edificación. Parte 1: Clasificación a Partir de Datos Obtenidos en Ensayos de Reacción al Fuego. UNE: Madrid, Spain, 2019.
33. de Calidad, S. Características Medias del Agua de Consumo en el Sistema de Distribución. Madrid. 2021. Available online: https://www.canaldeisabelsegunda.es/documents/20143/636811/Caracteristicas+Medias+2021.pdf/ (accessed on 6 May 2022).
34. Alquera, S.L. Ficha Técnica Poliacrilato de Potasio; 2020; pp. 1–6. Available online: https://www.alquera.com/poliacrilato-de-potasio/ (accessed on 11 January 2023).
35. *UNE-EN 13279-2:2014*; Yesos de Construcción y Conglomerantes a Bas se de Yeso Para la Construcción. Parte 2: Métodos de Ensayo. UNE: Madrid, Spain, 2014.
36. LabProcess. *Estufa de Convección Natural TCN 200 PLUS*; LabProcess: Barcelona, Spain, 2021.
37. Doleželová, M.; Krejsová, J.; Scheinherrová, L.; Keppert, M.; Vimmrová, A. Investigation of environmen-tally friendly gypsum-based composites with improved water resistance. *J. Clean. Prod.* **2022**, *370*, 133278. [CrossRef]
38. Durgun, M.Y. Effect of wetting-drying cycles on gypsum plasters containing ground basaltic pumice and polypropylene fibers. *J. Build. Eng.* **2020**, *32*, 101801. [CrossRef]
39. Vidales-Barriguete, A.; Atanes-Sánchez, E.; del Río-Merino, M.; Piña-Ramírez, C. Analysis of the im-proved water-resistant properties of plaster compounds with the addition of plastic waste. *Constr. Build. Mater.* **2020**, *230*, 116956. [CrossRef]
40. del Río Merino, M. *Elaboración y Aplicaciones Constructivas de Paneles Prefabrica-dos de Escayola Aligerada y Reforzada con Fibras de Vidrio e y Otros Aditivos*; Universidad Politécnica de Madrid: Madrid, Spain, 1999. [CrossRef]
41. *UNE-EN-459-2:2021*; Cales Para la Construcción. Parte 2. Métodos de Ensayo. UNE: Madrid, Spain, 2021.
42. *UNE-EN-520:2005+A1:2010*; Placas de Yeso Laminado. Definiciones, Especificaciones y Métodos de Ensayo. UNE: Madrid, Spain, 2010.
43. *UNE-EN ISO 12572*; Prestaciones Higrotérmicas de los Productos y Materiales para Edificación. Determinación de las Propiedades de Transmisión de Vapor de Agua. Método del Vaso. UNE: Madrid, Spain, 2018.
44. del Río-Merino, M.; Vidales-Barriguete, A.; Piña-Ramírez, C.; Vitiello, V.; Cruz-Astorqui, J.S.; Castelluccio, R. A review of the research about gypsum mortars with waste aggregates. *J. Build. Eng.* **2022**, *45*, 103338. [CrossRef]
45. Caselle, C.; Baud, P.; Kushnir, A.; Reuschlé, T.; Bonetto, S. Influence of water on deformation and failure of gypsum rock. *J. Struct. Geol.* **2022**, *163*, 104722. [CrossRef]
46. Martias, C.; Joliff, Y.; Favotto, C. Effects of the addition of glass fibers, mica and vermiculite on the mechanical properties of a gypsum-based composite at room temperature and during a fire test. *Compos. Part B Eng.* **2014**, *62*, 37–53. [CrossRef]
47. Vega, D.F.; Fernández, C.M.; Dorado, M.Á.; Miragaya, P.G. Material de Construcción Aligerado para la Elaboración de Prefabricados. ES Patent ES-2841130_B2, 28 February 2022.

materials

MDPI

Article

A Top-Down Approach and Thermal Characterization of Luminescent Hybrid BPA.DA-MMA@Ln_2L_3 Materials Based on Lanthanide(III) 1H-Pyrazole-3,5-Dicarboxylates

Renata Łyszczek [1,*], Dmytro Vlasyuk [1], Beata Podkościelna [2], Halina Głuchowska [1], Ryszard Piramidowicz [3] and Anna Jusza [3]

1 Department of General and Coordination Chemistry and Crystallography, Institute of Chemical Sciences, Faculty of Chemistry, Maria Curie–Skłodowska University, M. Curie-Skłodowskiej Sq. 2, 20-031 Lublin, Poland
2 Department of Polymer Chemistry, Institute of Chemical Sciences, Faculty of Chemistry, Maria Curie–Skłodowska University, Gliniana 33, 20-614 Lublin, Poland
3 Institute of Microelectronics and Optoelectronics, Warsaw University of Technology, Koszykowa 75, 00-662 Warsaw, Poland
* Correspondence: renata.lyszczek@mail.umcs.pl

Abstract: In this study, novel hybrid materials exhibiting luminescent properties were prepared and characterized. A top-down approach obtained a series of polymeric materials with incorporated different amounts (0.1; 0.2; 0.5; 1, and 2 wt.%) of dopants, i.e., europium(III) and terbium(III) 1H-pyrazole-3,5-dicarboxylates, as luminescent sources. Methyl methacrylate and bisphenol A diacrylate monomers were applied for matrix formation. The resulting materials were characterized using Fourier transform infrared spectroscopy (FTIR) and thermal analysis methods (TG-DTG-DSC, TG-FTIR) in air and nitrogen atmosphere, as well as by luminescence spectroscopy. The homogeneity of the resulting materials was investigated by means of optical microscopy. All obtained materials exhibited good thermal stability in both oxidizing and inert atmospheres. The addition of lanthanide(III) complexes slightly changed the thermal decomposition pathways. The main volatile products of materials pyrolysis are carbon oxides, water, methyl methacrylic acid and its derivatives, bisphenol A, 4-propylphenol, and methane. The luminescence properties of the lanthanide complexes and the prepared hybrid materials were investigated in detail.

Keywords: 1H-pyrazole-3,5-dicarboxylates; polymers; hybrid materials; luminescence; thermal analysis

Citation: Łyszczek, R.; Vlasyuk, D.; Podkościelna, B.; Głuchowska, H.; Piramidowicz, R.; Jusza, A. A Top-Down Approach and Thermal Characterization of Luminescent Hybrid BPA.DA-MMA@Ln_2L_3 Materials Based on Lanthanide(III) 1H-Pyrazole-3,5-Dicarboxylates. *Materials* **2022**, *15*, 8826. https://doi.org/10.3390/ma15248826

Academic Editors: Agnieszka Kijo-Kleczkowska and Adam Gnatowski

Received: 12 October 2022
Accepted: 6 December 2022
Published: 10 December 2022

1. Introduction

Intensive development of luminescent hybrid materials has been observed in recent years, due to their huge potential applications in many fields including optoelectronics, laser systems, chemical sensors, optical communication devices, protection coating, etc. [1–10]. Hybrid materials are typically composed of at least two different components connected at the nano-metric or molecular level, where one of the components plays the matrix role [3,11,12]. These components introduce different properties, thus the final advanced materials exhibit improved and desirable features. This approach allows the generation of structurally diverse materials designed for special applications. Many hybrid materials are based on polymer matrices, due to their easy processing, mechanical resistance, low density, good flexibility, and controllable cost. Polymers such as poly(methyl methacrylate), poly(vinyl alcohol), polyethylene, polycarboxylate, poly(divinylbenzene), and others have been well examined as matrices [13–17]. On the contrary, the polymerization of Bisphenol A diacrylate (BPA.DA) and methyl methacrylate (MMA) monomers generates a polymer that has been comparatively neglected as a matrix in the production of hybrid materials, despite its wide use in various fields of science and industry, including packaging and coating materials, due to its high transparency, heat-deflection temperature, and impact strength, good ultraviolet stability, fire resistance, and excellent mechanical properties [18–24]. These desirable features arise from the fact that aromatic BPA.DA plays a crucial crosslinking role during the polymerization of these monomers, making this polymer an excellent matrix for the formation of hybrid materials containing dopants insoluble in monomers.

The luminescent properties of hybrid materials can originate from different sources, such as the matrix, additives at small concentrations, or resulting from the synergistic effects of the properties of the parent components. Among numerous compounds, the selected lanthanide complexes are perfect candidates for luminescent dopants, owing to the unique optical properties of Ln(III) ions. Their emission spectra show narrow bands derived from the f-f transitions in the visible (VIS) and near-infrared (NIR) ranges, but their intensities are low because lanthanide ions suffer from weak light absorption [3,25,26]. Many lanthanide complexes with organic ligands exhibit intense luminescence, because of the effective intramolecular energy transfer from the coordinated ligands to the luminescent central lanthanide ions. In such cases, the lanthanide complexes exhibit sensitized luminescence known also as the "antenna effect" [3,25–30].

The earliest reported luminescent hybrid materials were doped with molecular lanthanide complexes constructed predominantly of chelating agents [3,31–33]. Despite the preferable luminescent properties of such complexes and their good solubility, many of them suffer from low thermal stability that strongly impacts their final material properties. One possible way of overcoming such a disadvantage is to use more sustainable compounds as dopants. Carboxylate lanthanide complexes are considered to be thermally stable due to the formation of strong Ln-O bonds. In many of these complexes, organic ligands act as a light-harvesting system participating in the indirect sensitization of Ln(III) ions, which has led to the frequent application of these compounds as luminescence origins in hybrid materials [8,34–36]. In previously reported luminescent hybrid materials based on polymeric matrices, lanthanide carboxylates of molecular and polymeric structures were used as dopants [35,36].

Thermal analysis methods play a significant role in the characterization of novel materials, including luminescent hybrid materials for a variety of possible applications. Studies of the thermal behavior of different classes of chemicals and organic, inorganic, and hybrid materials should be conducted to assess their thermal stability and to verify their potential to cause accidents, fire, or environmental pollution [37,38]. Therefore, these studies are essential for supporting theoretical models with practical purposes. Furthermore, many new or existing materials can undergo thermal degradation, combustion, or pyrolysis (depending on the reaction temperature and the degree of oxygen consumption), and the results obtained can be used to draw appropriate conclusions on a possible recycling procedure for energy recovery and and/or production of valuable oily chemical substances. In common practice, studies of the thermal behavior of materials are considered a preliminary test measure to discover the most appropriate temperature ranges at which pre-treatment can be carried out to elicit a given property for an required application [39,40]. Materials and composites including antibiotics, polyhedral oligomeric silsesquioxane nanocomposites polymers, and organic–inorganic hybrids are frequently tested by experimental thermogravimetric-differential scanning calorimetry (TG-DSC). Data are obtained for kinetic parameters related to thermal decomposition reactions, in order to evaluate their thermal stability [41].

This article is a continuation of our research on the synthesis and characterization of hybrid materials formed by the incorporation of newly prepared lanthanide(III) complexes bearing desired functional features, e.g., luminescence, into the polymeric matrices [35,36]. Such a procedure in the case of the formation of hybrid materials containing coordination polymers is regarded as a top-down approach. In contrast, generation of metal complexes along with target hybrid materials is referred to as a bottom-up procedure [16].

The purpose of this paper is to present the top-down approach to synthesizing new hybrid materials from BPA.DA-MMA, a rarely used polymer matrix, and lanthanide(III) complexes of 1H-pyrazole-3,5-dicarboxylic acid such as $[Eu_2(Hpdca)_3(H_2O)_6]$ and

$[Tb_2(Hpdca)_3(H_2O)_6]$. The selected organic ligand forms with lanthanides(III) coordination polymers (LnCPs) because of the diverse coordinating modes and variabilities of deprotonation of the carboxylic groups [42,43]. The most recent report on LnCPs constructed from H_3pdca acid focused mainly on their synthesis methods, crystal structures, and thermal properties [43]. However, investigations into the potential applications of such lanthanide complexes, including formation of luminescent hybrid materials, have not yet been reported. Therefore, we aimed to determine whether the above lanthanide complexes would function as luminescent additives in the prepared materials.

Incorporation of lanthanide(III) coordination polymers at different concentrations of 0.1, 0.2, 0.5, 1, and 2 wt.% into the BPA.DA-MMA matrix led to the formation of materials which possessed both the characteristic luminescence of lanthanide(III) ions and the attractive features of organic polymers, including mechanical strength, flexibility, transparency, and ease of processing. A principal feature that determines the potential application of the materials is their thermal stability. Therefore, particular attention was paid to the examination of the thermal behaviour of the obtained materials. The pathways of thermal decomposition in the reported materials were examined in air using the thermogravimetric (TG) and differential scanning calorimetry (DSC) methods. The thermal degradation of the materials was also evaluated by means of the evolved thermal analysis method incorporating the TG-FTIR technique, allowing the mass loss and FTIR spectra of the evolved compounds to be recorded during the samples' heating in an air and nitrogen atmosphere. The impact of the processing on the pathways of thermal decomposition in the polymeric matrix and hybrid materials in an oxidizing atmosphere was also evaluated. The luminescence properties of the $[Eu_2(Hpdca)_3(H_2O)_6]$ and $[Tb_2(Hpdca)_3(H_2O)_6]$ complexes and those of the obtained hybrid materials were furthermore investigated. The role of 1H-pyrazole-3,5-dicarboxylate ligand in the enhancement of lanthanide-centered emission ions was established.

2. Materials and Methods

Bisphenol A glycerolate (1 glycerol/phenol) diacrylate (BPA.DA), methyl methacrylate (MMA), and 2,2-dimethoxy-2-phenylacetophenone (Irgacure 651, IQ) were purchased from Sigma-Aldrich, (Darmstadt, Germany). All the chemical reagents and materials were obtained from commercial sources and used without further purification.

The lanthanide complexes $[Eu_2(Hpdca)_3(H_2O)_6]$ and $[Tb_2(Hpdca)_3(H_2O)_6]$ were prepared using procedures previously described in the reaction of lanthanide(III) salt with sodium salt of 1H-pyrazole-3,5-dicarboxylic acid under hydrothermal conditions [43].

2.1. Synthesis of Hybrid Materials

MMA and BPA.DA monomers were mixed in proportions of 70:30 wt.%, and then Irgacure 651 was added at 1 wt.% as the photoinitiator. Next, the solid $[Eu_2(Hpdca)_3(H_2O)_6]$ complex was added to the obtained mixture in proportions of 0.1, 0.2, 0.5, 1, and 2 wt.%. The prepared mixtures were heated at 80 °C to remove air bubbles. The well-homogenized mixtures were poured into glass molds (10 × 12 × 0.2 cm) with a Teflon spacer and exposed to UV radiation (8 lamps, each 40 W). The UV polymerization time was 25–30 min, then the resulting composites were heated at 80 °C for 2 h for final cross-linking. The free matrix and the hybrid materials were obtained in the form of blocks.

2.2. Instrumentation and Methods

The attenuated total reflectance Fourier transform infrared spectra (ATR-FTIR) of the free matrix and the prepared hybrid materials were recorded using a Nicolet 6700 spectrophotometer equipped with a Smart iTR accessory (diamond crystal) over the range 4000–600 cm^{-1}.

The transmittance FTIR spectra of pure matrix, BPA.DA-MMA@2%Eu_2L_3, and BPA.DA-MMA@2%Tb_2L_3 materials were collected using a Jasco FT/IR-4600LE spectrophotometer in the range 4000–400 cm^{-1}. Thermal analyses of the synthesized materials were carried out, applying the thermogravimetric (TG) and differential scanning calorimetry (DSC) methods using a SETSYS 16/18 analyzer (Setaram). The samples (mass about 5–9 mg) were heated in alumina crucibles up to 1000 °C at a heating rate of 10 °C min^{-1} in a dynamic air atmosphere (12.5 cm^3 min^{-1}). The TG curves and the FTIR spectra of the evolved gases from the investigated samples were recorded using a Q5000 thermal analyzer (TA Instruments) coupled with a Nicolet 6700 spectrophotometer. Samples of 20–30 mg were heated in a dynamic nitrogen and air atmosphere (25 cm^3 min^{-1}) at a heating rate of 10 and 20 °C min^{-1}. The samples were heated to 700 °C in open platinum crucibles. The transfer line was heated up to 250 °C and the gas cell of the spectrophotometer was heated to 240 °C.

The fragments of the solid composites were studied using a Morphologi G3 optical microscope (Malvern, UK). The scanning electron microscopy (SEM) images were recorded with a Quanta 3D FEG (FEI).

The excitation spectra, emission spectra, and profiles of fluorescence decay were recorded using a Horiba PTI QuantaMasterbased modular spectrofluorimetric system equipped with double monochromators in the excitation and emission paths, enabling both CW and pulsed excitations (Xe CW lamp of 75 W maximum power, Xe pulsed lamp of maximum frequency 300 Hz, pulse duration 1 µs) over a wide spectral range (200–2000 nm). All measurements were taken at room temperature, and all spectra were corrected for the spectral characteristics of the detector's response.

3. Results and Discussion

3.1. Morphological Analysis

A series of hybrid materials BPA.DA-MMA@Ln_2L_3 composed of the cross-linked polymeric matrix BPA.DA-MMA (BPA.DA-bisphenol A diacrylate and MMA-methyl methacrylate monomers) and europium(III) and terbium(III) complexes of 1H-pyrazole-3,5-dicarboxylic acid (Ln_2L_3: Eu_2L_3 = [$Eu_2(Hpdca)_3(H_2O)_6$] or Tb_2L_3 = [$Tb_2(Hpdca)_3(H_2O)_6$]) were synthesized and characterized. The polymeric matrix and hybrid materials containing different amounts of dopants, i.e., lanthanide complexes (Table S1), were prepared using the ultraviolet polymerization procedure. The scheme of the hybrid materials' synthesis and proposed structure of the materials is presented in Figure 1.

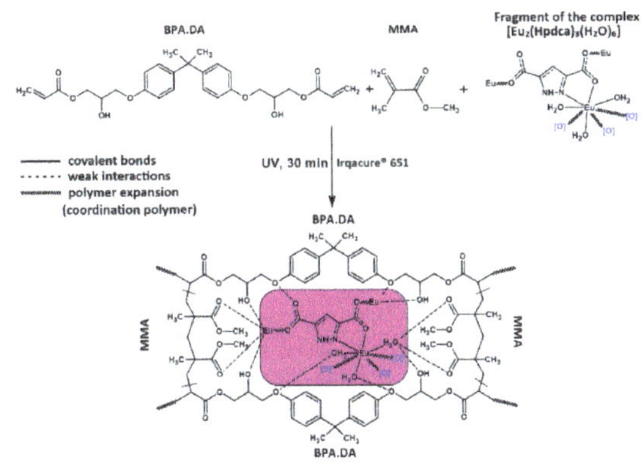

Figure 1. Scheme of the hybrid materials' synthesis and proposed structure of obtained materials based on the europim(III) complex.

The polymeric matrix was transparent, and the hybrid materials became cloudy with the increasing content of lanthanide dopants. Microscopic examination showed that in most cases the particles of the incorporated lanthanide complexes of irregular shapes had diameters in the range of 10–290 µm (Figure 2). The scanning electron microscope (SEM) images of the block pure matrix and hybrid materials are shown in Figure 3. The materials' surfaces are not smooth and irregular shaped particles can be distinguished.

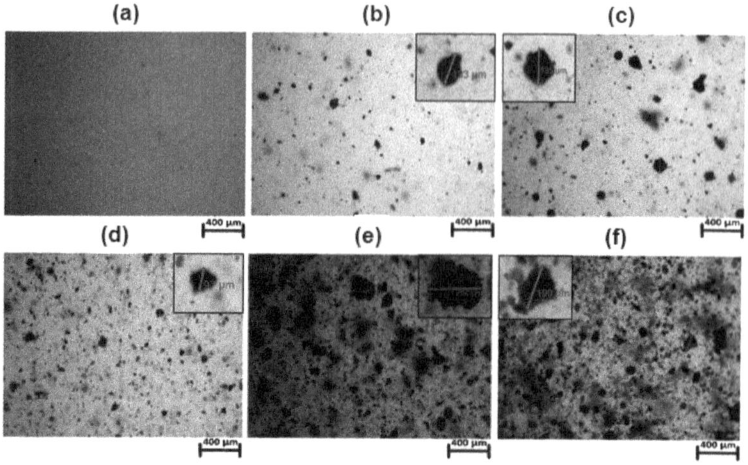

Figure 2. Optical microscope images: (**a**) BPA.DA-MMA; (**b**) BPA.DA-MMA@0.1%Eu_2L_3; (**c**) BPA.DA-MMA@0.2%Eu_2L_3; (**d**) BPA.DA-MMA@0.5%Eu_2L_3; (**e**) BPA.DA-MMA@1%Eu_2L_3; (**f**) BPA.DA-MMA@2%Eu_2L_3.

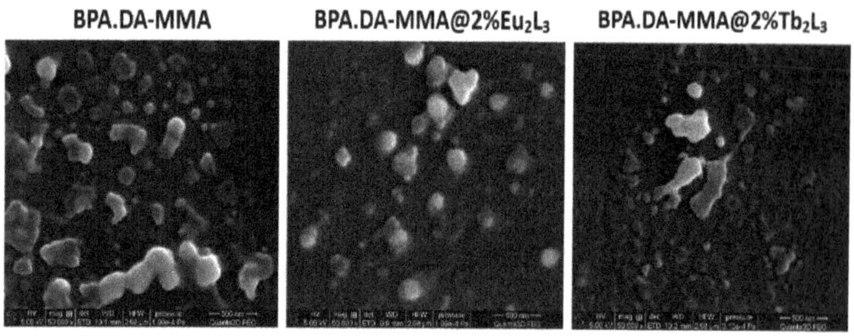

Figure 3. SEM images (1 µm magnification) of free matrix and hybrid materials doped with 2 wt.% of Eu_2L_3 and 2 wt.% of Tb_2L_3.

3.2. Infrared Spectroscopy Analysis

The ATR-FTIR spectra of the pure matrix (BPA.DA-MMA) and the hybrid materials were strongly dominated by the bands derived from the polymeric matrix BPA.DA-MMA (Figures S1 and S2). The stretching vibrations of the $C_{Ar}H$ and OH groups from bisphenol A moieties and the stretching vibrations of CH from the aliphatic chains appeared in the wavenumber range 3600–3100 cm^{-1}. The FTIR spectra exhibitedbands at 2950 and 2875 cm^{-1} originating from the stretching vibrations of the CH_2 and CH_3 groups (Figures S1 and S2). The bands of the deformation vibrations in those groups occurred at 1455 and 1385 cm^{-1}. The intense band at 1726 cm^{-1} is attributed to the stretching vibrations of the carbonyl groups ν(C=O) from the polymer ester parts. Several peaks at 1634, 1607, 1581 and 1508 cm^{-1} resulted from the stretching vibrations of ν($C_{Ar}C_{Ar}$) from the aromatic rings. The intense and broad bands at 1236 cm^{-1} and 1180 cm^{-1} were assigned to the stretching vibrations of the ν(C-O) and ν(C-O-C) groups from ester moieties. Several bands in the region 1060–600 cm^{-1} at wavenumbers 1039, 1011, 984, 828, 758, 737, and 727 cm^{-1} can be assigned to the skeletal stretching modes of ν(CC) of methyl(methacrylate) moieties, as well as $\gamma(C_{Ar}C_{Ar})$, $\gamma(C_{Ar}H)$ out-of-plane bending and the torsion motions of aromatic rings [24,44–47].

Incorporation of small amounts of dopants did not strongly affect the ATR-FTIR spectra of hybrid materials. The characteristic bands of lanthanide(III) complexes overlapped with those of the polymeric matrix. The only differences among the infrared spectra of the pure matrix and the obtained materials were observed on the transmittance FTIR spectra (Figure S3). The infrared spectrum of the pure matrix exhibited a very strong diagnostic band at 1732 cm^{-1} due to ν(C=O), which shifted slightly towards the lower wavenumbers 1728 cm^{-1} in the spectra of BPA.DA-MMA@2%Ln_2L_3 materials [18,35,36]. This observation can confirm the participation of functional groups in binding metal centers of the embedded lanthanide complexes, formation of hydrogen bonds, and/or the presence of other kinds of weak interactions between the components of the hybrid materials. The addition of a small amount of the lanthanide complexes did not cause visible changes in the infrared spectra of the obtained materials, which may be the result of the low intensity of the characteristic vibrations of the dopants. Therefore, the bands in the ranges 1597–1579 and 1359–1348 cm^{-1} derived from the asymmetric and symmetric stretching vibrations of carboxylate groups (-COO) characteristic for lanthanide complexes were obscured by the matrix bands in the ranges 1607–1581 cm^{-1} [43]. Furthermore, the bands from skeletal stretching vibrations $(C_{Ar}C_{Ar})$ of metal complexes overlapped with corresponding bands originated from the skeletal vibration of the polymeric matrix.

3.3. TG-DSC Analysis of BPA-MMA Matrix and Hybrid Materials

The TG curves of the BPA.DA-MMA matrix and the synthesized hybrid materials had similar shapes under the same measurement conditions (Figures 4 and S4–S7). However, a more careful analysis of the recorded TG curves revealed certain subtle differences due to the presence of lanthanide dopants in the investigated hybrid materials (Tables 1 and 2). The additives were found to affect the thermal stability of hybrid materials as well as the mechanism of their decomposition. The influencing effects of atmosphere temperature and materials' processing on their thermal stability and pathways of decomposition were also evaluated.

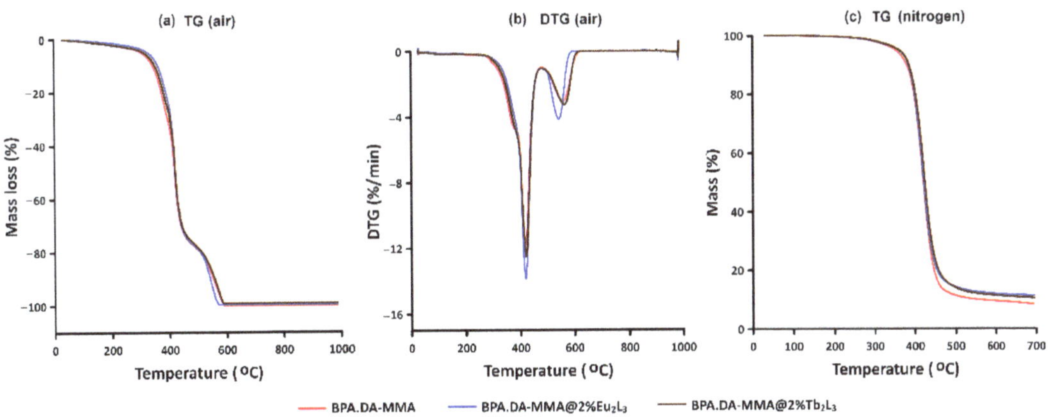

Figure 4. (**a**) TG and (**b**) DTG curves of the BPA.DA-MMA matrix, BPA.DA-MMA@2%Eu_2L_3, and BPA.DA-MMA@2%Tb_2L_3 materials (block form) in air atmosphere. (**c**) TG curves of materials (block form) in nitrogen atmosphere.

Table 1. Thermal characterization of matrix and materials (block) based on the TG-DSC curves in air.

	Stage I	Effect I		Effect II		Stage II	Effect III	
	ΔT_1 (°C) Δm_1 (%)	T_{onset}/T_{peak} (°C)	ΔH_1 (J/g)	T_{onset}/T_{peak} (°C)	ΔH_2 (J/g)	ΔT_2 (°C) Δm_2 (%)	T_{onset}/T_{peak} (°C)	ΔH_3 (J/g)
BPA.DA-MMA	229–448 74.8	395.4/401.2	11.3	443/448.4	129.8	449–592 25.2	- -	-
BPA.DA-MMA@Eu_2L_3	163–454 73.3	392.4/401.2	512.6	428.3/444.8	77.9	455–586 26.5	- -	-
BPA.DA-MMA@Tb_2L_3	119–454 74.1	385.2/389.7	220.3	404.2/414	343.8	455–607 25.6	431.6/440.6	130.1

Table 2. Comparison of thermal behavior of investigated materials in the form of block (B) and powder (P) in air atmosphere.

Mass Loss (%)	Temperature (°C) for Materials BPA.DA-MMA@Eu_2L_3 with Different Content of Metal Complex (B/P)				
	0.1%	0.2%	0.5%	1%	2%
1	198/164	203/165	187/159	190/176	162/159
5	339/303	333/299	327/307	331/304	325/303
20	399/344	390/343	389/347	389/345	387/346
50	423/374	420/373	420/376	420/374	420/375
Mass Loss (%)	**Temperature (°C) for Materials BPA.DA-MMA@Tb_2L_3 with Different Content of Metal Complex (B/P)**				
	0.1%	0.2%	0.5%	1%	2%
1	161/153	196/187	207/184	199/149	118/155
5	311/300	332/300	329/303	333/301	315/301
20	379/345	391/345	386/347	393/347	379/341
50	419/373	421/375	421/375	421/378	420/373

The thermal data are presented for the as-synthesized materials, i.e., in the form of blocks in an air atmosphere. Particular attention is drawn to the materials with 2% addition of dopants, for which the incorporated additives had the strongest impact on the material properties.

Taking into account the shape of the TG-DTG curves of the matrix and synthesized materials BPA.DA-MMA@Eu_2L_3 and BPA.DA-MMA@Tb_2L_3 in an air atmosphere, two main stages of decomposition were distinguished (Figures 4 and S4–S7, Tables 1 and 2). Mass loss of about 74% was recorded for the matrix and materials doped with 2% lanthanide complex in the temperature ranges 229–448, 163–454, and 119–454 °C, respectively.

As can be seen from the TG and DTG curves of the matrix and hybrid materials (Figures 4 and S4–S7), continuous mass losses (in the first stage) up to 311–340 °C connected with the release of humidity and/or water molecules from hydrated lanthanide complexes acting as dopants. The dehydration process in lanthanide(III) 1H-pyrazole-3,5-dicarboxylates occurred up to about 200 °C (Figure S8) [43]. For the pure matrix, mass losses of 1, 5, 20, and 50% on the TG curve were recorded at 228, 340, 399 and 420 °C, respectively. According to the thermal data corresponding to the first stage, mass losses in the materials of 1, 5, and 20% were found at lower temperatures compared to the free matrix (air). Mass losses of 50% in these materials were recorded at temperatures close to that observed for the matrix (Table 2).

At higher (air) temperatures, the decomposition process was reflected by significant mass changes recorded on the TG curves. The polymeric matrix displayed the highest thermal stability. The data presented in Table 2 indicate that the material doped with terbium(III) complex had lower thermal stability in comparison with corresponding materials containing the europium(III) dopant. The material with 2% addition of Tb_2L_3 complex exhibited the lowest stability. The observed slight differences in the thermal behaviors of the pristine matrix and hybrid materials are related to the thermal properties of the embedded lanthanide complexes [43] and their interaction with polymer framework. The discrepancies among corresponding hybrid materials containing the same percentage of metal complex may be related to the different thermal decomposition of metal complexes above 300 °C (Figure S8).

As follows from the DSC curves (in air), the first stage was dominated by endothermic effects driven by the overlapping melting and decomposition processes of polymeric matrix as well as the dopants' decomposition (Table 1, Figures S6 and S7). The top peaks of the first endothermic effects for the free matrix and hybrid materials with 2% added lanthanide complex were observed at very similar temperatures (401.2, 401.2, and 389.7 °C), while the observed values of endothermic effects differed. The lowest endothermic effect (11.3 J/g) was recorded for pure BPA.DA-MMA material. The presence of dopants caused an increase of endothermic effect values to 512.6 and 220.3 J/g for BPA.DA-MMA@2%Eu_2L_3 and BPA.DA-MMA@2%Tb_2L_3, respectively. At a slightly higher temperature, endothermic effects (129.8, 77.9, and 343.8 J/g) were observed in the temperature range 414–448 °C (peak) for the BPA.DA-MMA and hybrid materials. Regarding BPA.DA-MMA@2%Tb_2L_3, a third endothermic effect at 440.6 °C with ΔH of 130.5 J/g was observed (Table 1). The existence of these endothermic effects can be explained in terms of the pathway of the decomposition process, as well as the thermal properties of solid products' decomposition. The recorded difference in the energy of the endothermic effects may indicate diverse interactions between the matrix and the lanthanide complex incorporated in the materials. The higher value of the endothermic effects for BPA.DA-MMA@2%Eu_2L_3 in comparison with BPA.DA-MMA@2%Tb_2L_3 indicates stronger interactions between the europium complex and the polymeric matrix (Figure 1) due to the possible formation of hydrogen bonds and Van der Waals interactions [3].

Further heating in air resulted in the decomposition of unstable intermediate products, observed in the temperature ranges 449–592, 455–586, and 455–607 °C for the matrix and the BPA-MMA@2%Eu_2L_3 and BPA-MMA@2%Tb_2L_3 materials, respectively. The observed mass losses in the range of 25.2–26.7% (Table 1) were accompanied by very strong exothermic effects on the DSC curves due to the burning of organic solid residues (Figures S5–S7). Heating of the matrix in air resulted in a lack of any solid residues at 1000 °C, while for the investigated hybrid materials only traces (below 0.5%) of corresponding lanthanide oxides, i.e., Eu_2O_3 or Tb_4O_7, were noticed.

The impact of atmosphere temperature on the thermal stability of the block materials under investigation was determined by heating selected materials (BPA.DA-MMA, BPA.DA-MMA@2%Eu_2L_3, and BPA.DA-MMA@2%Tb_2L_3) in nitrogen. The profiles of the recorded TG curves were similar (Figure 4c). All materials exhibited higher thermal stability in nitrogen in comparison with the air atmosphere (Table S2) where additional oxidation reactions take place alongside the decomposition process [48]. Mass loss of 1% was recorded for the tested materials in the range 261–267 °C. The comparative thermal stabilities of the matrix and the investigated hybrid materials were very similar. At higher temperatures, significant mass loss of about 85% was observed due to the pyrolysis process. This stage took place in the temperature range 262–500 °C. Further heating resulted in slight changes of mass, recorded on the TG curves, and final solid residues were observed at 700 °C. The total mass loss for the free matrix was 91.8%, while for the BPA.DA-MMA@2%Eu_2L_3 and BPA.DA-MMA@2%Tb_2L_3 materials the corresponding final mass losses were 89.0% and 89.5%, respectively. In the case of solid residues of hybrid materials, in addition to unburnt carbon (free matrix), lanthanide oxides were also formed [18,35,36,38].

In addition to thermal analysis of the as-synthesized materials in the form of blocks, the powdered materials were also investigated. The profiles of the TG and DSC curves in air indicate clearly their different thermal stabilities and mechanisms of decomposition (Figures S4–S7). In general, the thermal stability of all the powdered materials was reduced compared with that of the blocks. Changes in the temperature associated with losses of 1, 5, 20, and 50% mass in comparison with block hybrid materials were in the ranges 3–50, 20–36, 34–55, and 43–48 °C, respectively. Temperatures resulting in 1% mass loss for the series of powdered BPA.DA-MMA@Eu_2L_3 materials were decreased by 34, 38, 28, 26, and 3 °C in comparison to corresponding block materials. For the series of powdered BPA.DA-MMA@Tb_2L_3 materials, temperatures for 1% mass loss were lowered by 8, 9, 23, 50, and 37 °C compared with block materials. The temperatures corresponding to 5, 20, and 50 % mass loss were similar for the respective powdered hybrid materials in the two series. The temperature associated with 1% of mass loss for powdered BPA-MMA decreased up to 121 °C (228 °C for the block), while for other analyzed points on the TG curves, the temperature alteration was not so drastic. It can be stated that the powdered hybrid materials containing lanthanide complexes did not demonstrate such a dramatic loss of stability relative to the powder matrix.

3.4. TG-FTIR Analysis of BPA-MMA Matrix and Hybrid Materials

Thermogravimetric measurements (TG) were taken and spectroscopic identification (FTIR) performed of the evolved gases in nitrogen and air for the matrix and materials (block form) with the highest concentration of dopant, i.e., 2 wt.%. The Gram–Schmidt plots representing the total intensities of all evolved gases as a function of time (temperature) during heating of the materials in nitrogen are provided in Figure S9. As a representative example, the 3D FTIR spectra of the volatile products of BPA.DA-MMA@2%Eu_2L_3 material decomposition are shown in Figure 5.

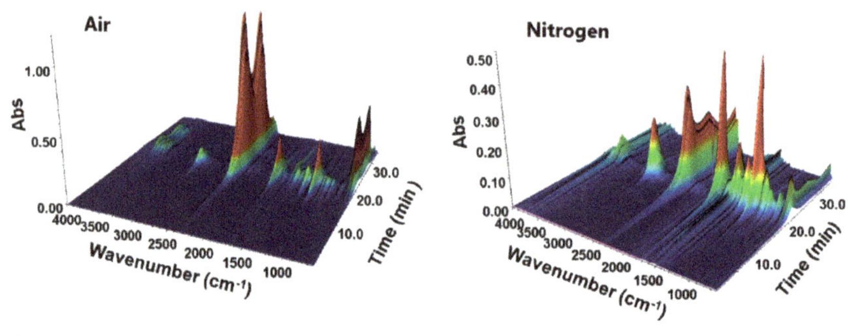

Figure 5. 3D projection of the FTIR spectra of volatile products of BPA.DA-MMA@2%Eu_2L_3 material decomposition in air and nitrogen atmosphere.

The FTIR spectra of volatile products of the investigated materials after heating in nitrogen indicated that water molecules evolved first. Diagnostic bands of weak intensity in the wavenumber ranges 4000–3500 and 1800–1300 cm^{-1} were due to the stretching and deformation vibrations of OH groups of evolved water molecules, and were observed up to about 270 °C. Next, further compounds such as carbon oxides and various organic compounds were released as products of the materials' degradation in nitrogen.

The infrared spectra show strong diagnostic bands with maxima at 2359 and 2343 cm^{-1} and others in the range 750–600 cm^{-1} ascribed to the stretching and deformation vibrations of carbon(IV) oxide molecules. The diagnostic double band with maxima at 2185 and 2107 cm^{-1} corresponds to the stretching vibrations of carbon oxide(II) [49]. Methacrylic acid and its derivatives such as methyl methacrylate and 2-hydroxyethylmethacrylate (Figure 6) are the main products of bonds breaking in the polymeric matrix [18,24,35,36].

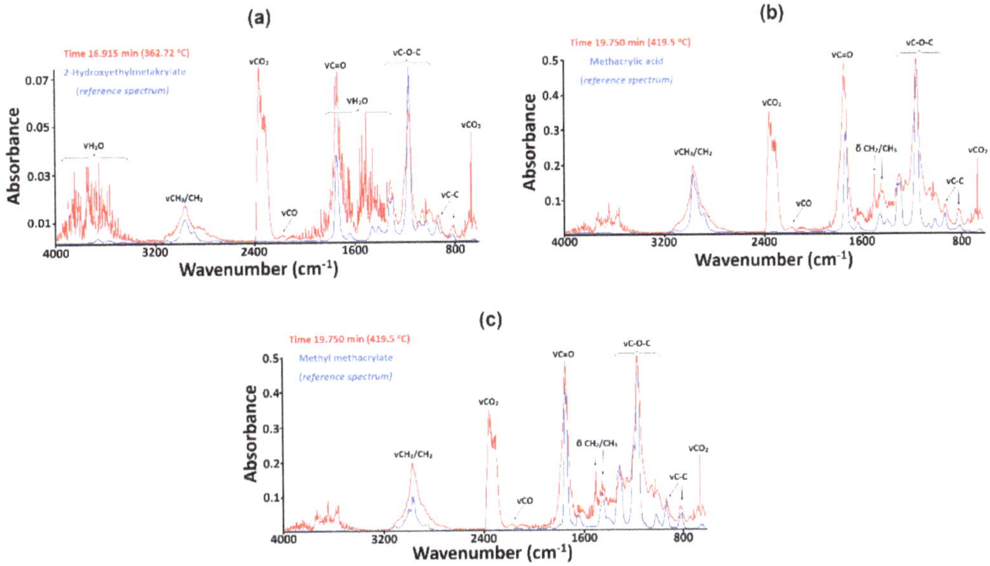

Figure 6. Experimental infrared spectra of gaseous products recorded in the nitrogen atmosphere at different temperatures are indicated in red, alongside FTIR spectra of references represented in blue: (**a**) 2-hydroxyethylmethacrylate, (**b**) methacrylic acid, and (**c**) methyl methacrylate.

The broad band in the range 3150–2800 cm^{-1} centered at 2972 cm^{-1} was attributed to the stretching vibrations of $=CH_2$ and $-CH_3$ moieties. The very intense band at 1748 cm^{-1} can be assigned to the stretching vibrations of carbonyl groups from the carboxylic and ester groups. Deformation vibrations of the CH_2/CH_3 groups were found at 1456 and 1362 cm^{-1}. The band at 1306 cm^{-1} and those at 1166 cm^{-1}, 1061, and 1031 cm^{-1} were found to correspond to the stretching vibrations of C-O-C ester groups. The absorption bands at 936 and 813 cm^{-1} are derived from the skeletal vibrations of C-C from the aliphatic moieties [44–47]. Figure 6 presents the experimental and reference FTIR spectra of identified compounds recorded at 362.72 and 419.50 °C. Further heating of the materials above 420 °C also resulted in the evolution of gaseous 4,4′-(propane-2,2-diyl) diphenol (bisphenol A, BPA) and its decomposition products such as methane and 4-propylphenol [50–53].

Alongside the bands from methacrylate compounds, water, and carbon oxides, the FTIR spectrum recorded at 447 °C displays characteristic diagnostic vibrations from bisphenol A and its decomposition compounds (Figure 7). The vibrations from the free hydroxyl groups of BPA produced a relatively strong band at 3649 cm^{-1}. The presence of two methyl groups in the released molecules of BPA is reflected in the stretching and deformation vibrations found at 2972, 1472, and 1362 cm^{-1}. The bands at 1602 and 1507 cm^{-1} and those at 826 and 746 cm^{-1} are derived from the stretching and out-of-plane deformation vibrations of $C_{Ar}C_{Ar}$ and $C_{Ar}H$ bonds from the aromatic rings. Additionally, the FTIR spectra were found to exhibit very intense double bands with maxima at 1258 and 1174 cm^{-1} assigned to the C_{Ar}-O-H groups from the phenol moieties. Methane molecules as the product of BPA molecules' degradation showed significant diagnostic bands in the range 3200–2950 cm^{-1} with a characteristic maximum at 3015 cm^{-1} due to the stretching vibrations of CH bonds. Evidence of 4-propylphenol formation as the bisphenol A decomposition product is reflected in the presence of the band at 2937 cm^{-1} from the stretching vibrations of CH_2 groups within the propyl substituent [51–54]. The reference spectrum of the compound fits well the experimental one.

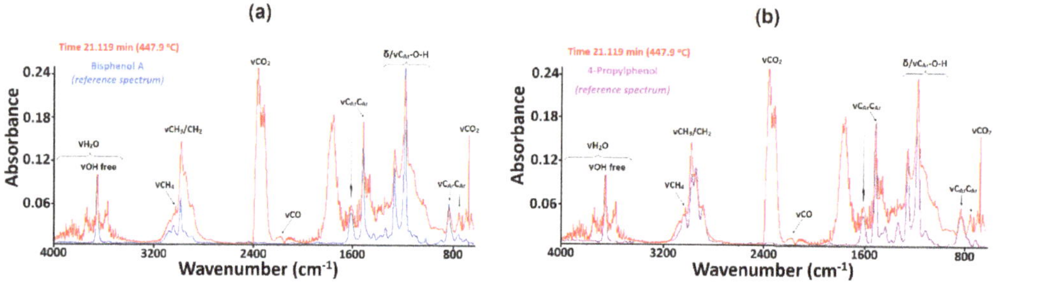

Figure 7. Experimental infrared spectra of gaseous products (shown in red) recorded at different atmosphere, along with FTIR spectra of references (shown in blue): (**a**) bisphenol A (air); (**b**) 4-propylphenol (nitrogen).

The FTIR spectra of the gaseous decomposition products of the pure matrix and the BPA.DA-MMA@2%Eu_2L_3 and BPA.DA-MMA@2%Tb_2L_3 materials after heating in air were also recorded. The profiles of the TG curve were very similar to those in the air atmosphere described above. The differences observed were associated with the diverse measurement conditions such as heating rate and rate of air flow. The TG curves' profiles confirm the two-stage decomposition process (Figure S4). The first significant mass change took place up to about 448 °C, with mass losses of 80.7 and 78.1% for the matrix and the hybrid materials, respectively. In this step, similarly to the observations under the nitrogen atmosphere, the FTIR spectra of evolved gases were dominated by characteristic bands from methacrylate derivatives, methane, and carbon oxides (Figure S10). It is worth mentioning that bisphenol A molecules were liberated in the inert atmosphere. A second mass loss of about 20% indicated by the TG occurred in the temperature range 450–600 °C. The recorded FTIR spectra show very strong bands from the carbon oxides while the bands from hydrocarbons are almost invisible. These gaseous products confirm that oxidizing processes of unstable solid residues occur mainly above 450 °C. For the investigated matrix, almost 100% mass loss was observed at 700 °C. Heating the hybrid materials resulted in 0.5% mass of residue due to the formation of suitable oxides (Eu_2O_3, Tb_4O_7).

3.5. Luminescent Properties of Lanthanide Complexes and Powdered Hybrid Materials

For all investigated samples, the luminescent properties were also analyzed in the short-wavelength spectral range and are discussed here. Due to the specific (powdered) type of original active media and the plates' lack of transparency, the typical measurements of the absorption characteristics were replaced by excitation spectra measurements. For all concentrations of Tb(III) and Eu(III) complexes in the hybrid materials the emission and excitation properties remained the same, differing only in the signal intensities, therefore only the results for the highest concentrations are presented.

Figure 8a shows a comparison of the excitation spectra of Tb_2L_3 complex and BPA.DA-MMA@2%Tb_2L_3 hybrid material recorded for the most intense green emission (544 nm). The strong broadband absorption bands in the UV region correspond to the most efficient excitation of terbium(III) ions via ligands (energy transfer from the triplet state of the ligand, preceded by absorption to the singlet state and intersystem crossing ISC to the triplet state, as shown in detail in Figure 9) [3,25,55,56]. Narrow lines related to the direct excitation of high-position energy levels of terbium(III) are also clearly visible. The recorded spectra for the complex and hybrid materials differed mainly in the position of maximal intensity of absorption through the ligand–ion energy transfer (which shifted towards shorter wavelengths for the composites).

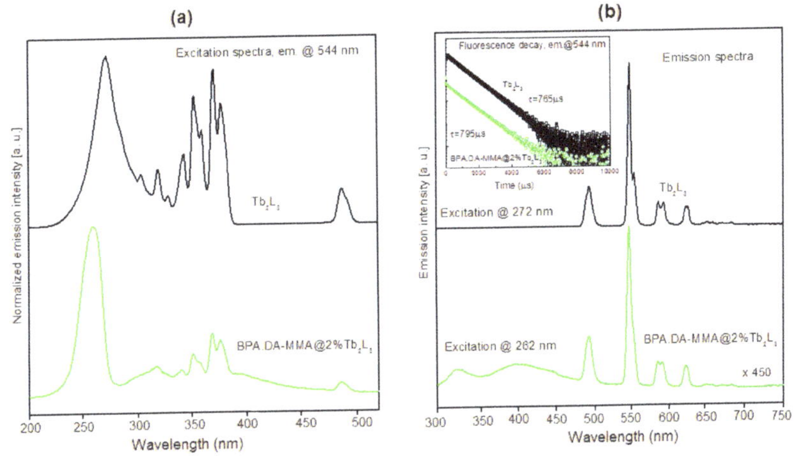

Figure 8. (**a**) Comparison of the excitation characteristics recorded for the Tb_2L_3 complex and the BPA.DA-MMA@2%Tb_2L_3 hybrid material (544 nm emission). (**b**) Comparison of the emission characteristics recorded for the Tb_2L_3 complex and the BPA.DA-MMA@2%Tb_2L_3 hybrid material (UV excitation).

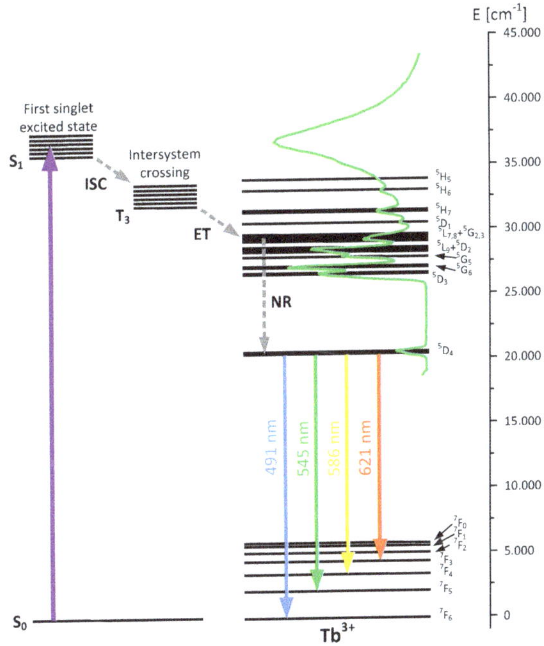

Figure 9. Energy transfer mechanism for Tb(III) sensitized fluorescence as most efficient excitation path together with excitation spectrum and emission transitions from $^5D_4$ level (ISC—intersystem crossing, ET—energy transfer, NR—non-radiative transitions).

On the luminescence spectra (Figure 8b), several emission lines typical of terbium(III) activated materials are clearly observable for both the complex and the composite material. The most intensive (centered at 544 nm) is evidently related to $^5D_4 \rightarrow {}^7F_5$ transition while the remaining four lines correspond to $^5D_4 \rightarrow {}^7F_{6,4,3,2}$ transitions [30,55,56] (shown schematically in Figure 9). The luminescence spectrum recorded for hybrid material differed only slightly from that observed for the terbium(III) complex. The spectral positions of all lines were the same, but the lines were slightly wider and the parasitic broadband luminescence extended from 290 nm to 470 nm, which was clearly observable and could be attributed to emission from the polymeric host. The luminescence decay profiles (shown in the insert) had the same character for all observable lines and were purely exponential, with a considerably lengthy time constant of 750–800 μs for both the complex and the hybrid materials.

In the case of Eu_2L_3 doped materials, in the excitation spectra recorded for all samples (Figure 10a) the ligand–ion energy transfer band was suppressed and the optimal mechanism of exciting europium ions was the direct excitation of Eu levels (the "antenna effect" being insufficient). This means that this particular ligand is not a good sensitizer for Eu(III) luminescence. For hybrid materials, the additional excitation band visible in the UV range (ca. 260 nm) is probably related to the polymer host [57].

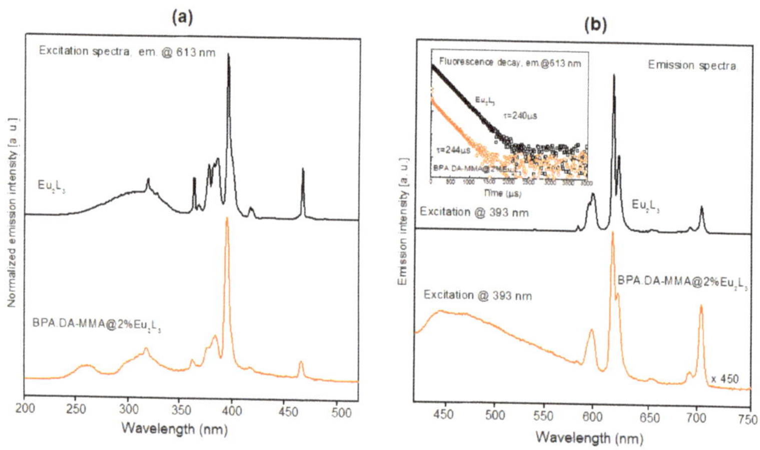

Figure 10. (**a**) Comparison of the excitation characteristics recorded for the Eu_2L_3 complex and BPA.DA-MMA@2%Eu_2L_3 composite material (recorded for 613 nm emission). (**b**) Comparison of emission characteristics recorded for the Eu_2L_3 complex and BPA.DA-MMA@2%Eu_2L_3 hybrid material (393 nm excitation).

The luminescence spectra of Eu_2L_3 complex and BPA.DA-MMA@2%Eu_2L_3 composites (Figure 10b) exhibited behaviors typical of the europium trivalent ion and included several emission lines in the visible range, related to optical transitions from the metastable $5D_0$ singlet to ^7F_J multiplets (shown schematically in Figure 11). The most intense line was centered at 613 nm, corresponding to the $^5D_0 \rightarrow {}^7F_2$ transition, responsible for the typical orange-red luminescence observed in europium-doped phosphors. It should be noted that quite strong broadband luminescence from the polymer host was observed in the hybrid material, significantly affecting the character of the emission spectrum. As in the previous case, the recorded fluorescence decays had the same character for all emission lines. The decay characteristics (presented in the inset of Figure 10b) are nearly exponential with the time constant of the order of 240 μs. For the hybrid material, rapid activity was clearly apparent in the initial part of the decay, related to the polymer host luminescence.

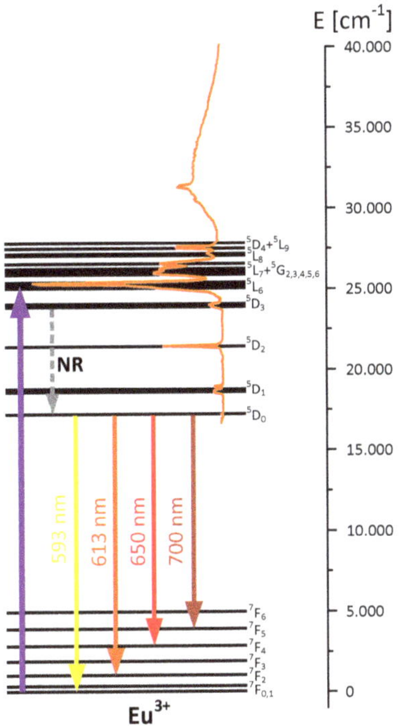

Figure 11. Scheme of the energy levels of Eu(III) together with most efficient excitation path, excitation spectrum, and emission transitions from $^5D_0$ level (NR—non-radiative transitions).

The excitation spectra of the tested lanthanide complexes indicate that the organic ligands used, i.e., 1H-pyrazole-3,5-dicarboxylates, do not show the optical features that would be supportive for the efficient transfer of the energy to lanthanide ions. Therefore, the desired "antenna effect" significantly enhancing the efficiency of luminescence was not confirmed in the investigated materials.

In the case of Tb_2L_3 compounds, the band related to excitation by ligand was clearly visible on the excitation spectrum (see figure below), corresponding with energy of ca. 37,000 cm^{-1}. However, the efficiency of excitation by ligand was nearly equal to the direct excitation efficiency, meaning that the position of the triplet level does not overlap well with the energy-accepting levels of terbium, and therefore does not support the efficient transfer of energy to the rare-earth ions.

In the case of the Eu_2L_3 complex, the excitation spectrum was dominated by f-f transitions of Eu(III) ions, and excitation via ligand hardly possible, which indicates the significant mismatch between the position of the triplet state and the rare-earth ion energy levels. Therefore, the energy transfer from ligand to Ln(III) ions was totally inefficient.

In general, when energy-absorbing states of lanthanide(III) ions are only slightly lower in energy than the triplet state of the ligand, one can expect strong metal-ion fluorescence. Although we were unable to measure the position of the triplet state, it is obvious that in the investigated case this condition was not met. We would guess that in both cases the triplet levels were located too high (30,000 cm^{-1} or above) to provide the efficient energy coupling between ligand and rare earth.

Methods of predicting and tailoring the processes of energy transfer between ligands and rare-earth ions will be the subject of further studies.

It should also be noted that with respect to emission intensity and fluorescence decay profiles, there were no significant differences between powders and bulk hybrid materials. Unfortunately, the emission intensity for all materials was much weaker than for the original lanthanide complexes, so the hybrid materials require further improvement in terms of optical quality and transparency. The fluorescence decay rate remained the same, independently from the concentration of the metal complexes and the form of the samples, indicating that active ions were effectively shielded from the highly energetic phonons of the polymer matrix and the ligand remained unchanged during the processing of the materials.

4. Conclusions

A series of novel luminescent hybrid materials have been presented, designed based on the top-down procedure. Coordination polymers of Eu(III) and Tb(III) ions with 1H-pyrazole-3,5-dicarboxylate ligand were incorporated in solid form into the polymeric BPA.DA-MMA matrix as luminescence donates. Material degradation pathways were found to be dominated by components of the polymer matrix and lanthanide(III) complex, as were their forms and heating conditions. The resulting series of hybrid materials demonstrated comparatively good thermal stability in the air atmosphere, but worse in comparison with the matrix. Materials with the addition of the europium complex were more stable than those with the terbium complex. In the inert atmosphere, an increase of thermal stability was observed in all the investigated materials as expected due to the limited oxidizing processes. The pyrolysis of the materials in nitrogen led to the evolution of products including water, carbon oxides, methacrylic acid and its derivatives, as well as bisphenol A and its decomposition compounds. In an air atmosphere, products of BPA decomposition were not observed, while a high intensity of carbon dioxide was apparent as an effect of the oxidation reactions. There were variations in the thermal behaviors of the block and powdered materials. Heating of the block materials in the air also allowed endothermic effects to be distinguished on the DSC curves, while for powdered materials mainly exothermic effects were observed.

All series of hybrid materials doped with different concentrations of europium(III) exhibited characteristic orange-red luminescence, while materials doped with terbium(III) emitted green light. The emission intensity and fluorescence decay profiles were independent of the form of the investigated materials. After improvement of optical properties, the resulting materials may be suitable for curing in thin layers and can be used as special protective coatings on a variety of materials.

Supplementary Materials: The following supporting information can be downloaded at: https://www.mdpi.com/article/10.3390/ma15248826/s1, Figure S1: ATR-FTIR spectra of BPA.DA-MMA (polymeric matrix) and BPA.DA-MMA@0.1–2%Eu_2L_3; Figure S2: ATR-FTIR spectra of BPA.DA-MMA (polymeric matrix) and BPA.DA-MMA@0.1–2%Tb_2L_3; Figure S3: The transmittance FTIR spectra; Figure S4: The TG curves of powder and block hybrid materials doped with europium complex (air); Figure S5: TG/DTG/DSC curves of polymeric matrix; Figure S6: TG/DTG/DSC curves of hybrid materials doped with europium(III) complex; Figure S7: TG/DTG/DSC curves of hybrid materials doped with terbium(III) complex; Figure S8: TG/DTG/DSC curves of doped complexes: (**a**) hexahydrate europium complex and (**b**) hexahydrate terbium complex (air); Figure S9: Gram–Schmidt plots; Figure S10: Comparison of gas (air/nitrogen) products at time of 19.750 min; Table S1: Parameters of materials synthesis; Table S2: Thermogravimetric results of free matrix and hybrid materials with 2 wt.% dopant (blocks) in air and nitrogen.

Author Contributions: Conceptualization, R.Ł., D.V. and B.P.; methodology, D.V., R.Ł., H.G., B.P., A.J. and R.P.; software, D.V., A.J. and R.P.; formal analysis, D.V., R.Ł., A.J. and R.P.; investigation, D.V., H.G., A.J. and R.P.; resources, D.V. and B.P.; data curation, D.V., R.Ł., A.J. and R.P.; writing—original draft preparation, R.Ł., D.V., A.J. and R.P.; writing—review and editing, R.Ł., D.V., A.J. and R.P.; visualization, D.V., A.J. and R.P.; supervision, R.Ł. and D.V.; project administration, D.V. and R.Ł. All authors have read and agreed to the published version of the manuscript.

Funding: This research received no external funding.

Institutional Review Board Statement: Not applicable.

Informed Consent Statement: Not applicable.

Data Availability Statement: All data are available from the corresponding author upon reasonable request.

Conflicts of Interest: The authors declare no conflict of interest.

References

1. Escudero, A.; Becerro, A.I.; Carrillo-Carrión, C.; Núñez, N.O.; Zyuzin, M.V.; Laguna, M.; González-Mancebo, D.; Ocaña, M.; Parak, W.J. Rare Earth Based Nanostructured Materials: Synthesis, Functionalization, Properties and Bioimaging and Biosensing Applications. *Nanophotonics* **2017**, *6*, 881–921. [CrossRef]
2. Capper, P.; Kasap, S. *Springer Handbook of Electronic and Photonic Materials*, 2nd ed.; Springer Handbooks; Kasap, S., Capper, P., Eds.; Springer International Publishing: Cham, Switzerland, 2017. [CrossRef]
3. Binnemans, K. Lanthanide-Based Luminescent Hybrid Materials. *Chem. Rev.* **2009**, *109*, 4283–4374. [CrossRef] [PubMed]
4. Bünzli, J.-C.G.; Piguet, C. Taking Advantage of Luminescent Lanthanide Ions. *Chem. Soc. Rev.* **2005**, *34*, 1048. [CrossRef] [PubMed]
5. Aghajamali, M.; Iqbal, M.; Purkait, T.K.; Hadidi, L.; Sinelnikov, R.; Veinot, J.G.C. Synthesis and Properties of Luminescent Silicon Nanocrystal/Silica Aerogel Hybrid Materials. *Chem. Mater.* **2016**, *28*, 3877–3886. [CrossRef]
6. Lezhnina, M.; Benavente, E.; Bentlage, M.; Echevarría, Y.; Klumpp, E.; Kynast, U. Luminescent Hybrid Material Based on a Clay Mineral. *Chem. Mater.* **2007**, *19*, 1098–1102. [CrossRef]
7. Kim, S.W.; Jyoko, K.; Masui, T.; Imanaka, N. Synthesis of Green-Emitting (La,Gd)OBr:Tb3+ Phosphors. *Materials* **2010**, *3*, 2506–2515. [CrossRef]
8. Wu, X.-Y.; Zhao, Q.; Zhang, D.-X.; Liang, Y.-C.; Zhang, K.-K.; Liu, Q.; Dong, L.; Shan, C.-X. A Self-Calibrated Luminescent Thermometer Based on Nanodiamond-Eu/Tb Hybrid Materials. *Dalton Trans.* **2019**, *48*, 7910–7917. [CrossRef]
9. Zhang, M.; Zhai, X.; Sun, M.; Ma, T.; Huang, Y.; Huang, B.; Du, Y.; Yan, C. When Rare Earth Meets Carbon Nanodots: Mechanisms, Applications and Outlook. *Chem. Soc. Rev.* **2020**, *49*, 9220–9248. [CrossRef]
10. De Faria, E.H.; Nassar, E.J.; Ciuffi, K.J.; Vicente, M.A.; Trujillano, R.; Rives, V.; Calefi, P.S. New Highly Luminescent Hybrid Materials: Terbium Pyridine−Picolinate Covalently Grafted on Kaolinite. *ACS Appl. Mater. Interfaces* **2011**, *3*, 1311–1318. [CrossRef]
11. Li, P.; Li, H. Recent Progress in the Lanthanide-Complexes Based Luminescent Hybrid Materials. *Coord. Chem. Rev.* **2021**, *441*, 213988. [CrossRef]
12. Nanko, M. Definitions and categories of hybrid materials. *Adv. Tech. Mat. Mat. Proc. J.* **2009**, *11*, 1–9.
13. Song, X.-Q.; Meng, H.-H.; Lin, Z.-G.; Wang, L. 2D Lanthanide Coordination Polymers: Synthesis, Structure, Luminescent Properties, and Ratiometric Sensing Application in the Hydrostable PMMA-Doped Hybrid Films. *ACS Appl. Polym. Mater.* **2020**, *2*, 1644–1655. [CrossRef]
14. Sosa, J.; Bennett, T.; Nelms, K.; Liu, B.; Tovar, R.; Liu, Y. Metal–Organic Framework Hybrid Materials and Their Applications. *Crystals* **2018**, *8*, 325. [CrossRef]
15. Feng, J.; Zhang, H. Hybrid Materials Based on Lanthanide Organic Complexes: A Review. *Chem. Soc. Rev.* **2013**, *42*, 387–410. [CrossRef]
16. Kalaj, M.; Bentz, K.C.; Ayala, S.; Palomba, J.M.; Barcus, K.S.; Katayama, Y.; Cohen, S.M. MOF-Polymer Hybrid Materials: From Simple Composites to Tailored Architectures. *Chem. Rev.* **2020**, *120*, 8267–8302. [CrossRef]
17. Pastore, V.J.; Cook, T.R. Coordination-Driven Self-Assembly in Polymer–Inorganic Hybrid Materials. *Chem. Mater.* **2020**, *32*, 3680–3700. [CrossRef]
18. Łyszczek, R.; Podkościelna, B.; Lipke, A.; Ostasz, A.; Puszka, A. Synthesis and Thermal Characterization of Luminescent Hybrid Composites Based on Bisphenol A Diacrylate and NVP. *J. Therm. Anal. Calorim.* **2019**, *138*, 4463–4473. [CrossRef]
19. Goliszek, M.; Podkościelna, B.; Klepka, T.; Sevastyanova, O. Preparation, Thermal, and Mechanical Characterization of UV-Cured Polymer Biocomposites with Lignin. *Polymers* **2020**, *12*, 1159. [CrossRef]
20. Sukanto, H.; Raharjo, W.W.; Ariawan, D.; Triyono, J.; Kaavesina, M. Epoxy Resins Thermosetting for Mechanical Engineering. *Open Eng.* **2021**, *11*, 797–814. [CrossRef]
21. Chen, W.; Fan, R.; Zhang, H.; Dong, Y.; Wang, P.; Yang, Y. Tunable White-Light Emission PMMA-Supported Film Materials Containing Lanthanide Coordination Polymers: Preparation, Characterization, and Properties. *Dalton Trans.* **2017**, *46*, 4265–4277. [CrossRef]
22. Zhang, H.; Fan, R.; Wang, P.; Wang, X.; Gao, S.; Dong, Y.; Wang, Y.; Yang, Y. Structure Variations of a Series of Lanthanide Complexes Constructed from Quinoline Carboxylate Ligands: Photoluminescent Properties and PMMA Matrix Doping. *RSC Adv.* **2015**, *5*, 38254–38263. [CrossRef]
23. Francisco, L.H.C.; Felinto, M.C.F.C.; Brito, H.F.; Teotonio, E.E.S.; Malta, O.L. Development of Highly Luminescent PMMA Films Doped with Eu3+β-Diketonate Coordinated on Ancillary Ligand. *J. Mater. Sci. Mater. Electron.* **2019**, *30*, 16922–16931. [CrossRef]
24. Wnuczek, K.; Puszka, A.; Klapiszewski, Ł.; Podkościelna, B. Preparation, Thermal, and Thermo-Mechanical Characterization of Polymeric Blends Based on Di(Meth)Acrylate Monomers. *Polymers* **2021**, *13*, 878. [CrossRef] [PubMed]
25. Eliseeva, S.V.; Bünzli, J.-C.G. Rare Earths: Jewels for Functional Materials of the Future. *N. J. Chem.* **2011**, *35*, 1165. [CrossRef]
26. Werts, M.H.V. Making Sense of Lanthanide Luminescence. *Sci. Prog.* **2005**, *88*, 101–131. [CrossRef]

27. Eliseeva, S.V.; Bünzli, J.-C.G. Lanthanide Luminescence for Functional Materials and Bio-Sciences. *Chem. Soc. Rev.* **2010**, *39*, 189–227. [CrossRef]
28. Stanley, J.M.; Holliday, B.J. Luminescent Lanthanide-Containing Metallopolymers. *Coord. Chem. Rev.* **2012**, *256*, 1520–1530. [CrossRef]
29. Zinna, F.; Arrico, L.; Funaioli, T.; Di Bari, L.; Pasini, M.; Botta, C.; Giovanella, U. Modular Chiral Eu(III) Complexes for Efficient Circularly Polarized OLEDs. *J. Mater. Chem. C* **2022**, *10*, 463–468. [CrossRef]
30. Allendorf, M.D.; Bauer, C.A.; Bhakta, R.K.; Houk, R.J.T. Luminescent Metal–Organic Frameworks. *Chem. Soc. Rev.* **2009**, *38*, 1330. [CrossRef]
31. Hasegawa, M.; Ohmagari, H.; Tanaka, H.; Machida, K. Luminescence of Lanthanide Complexes: From Fundamental to Prospective Approaches Related to Water- and Molecular-Stimuli. *J. Photochem. Photobiol. C Photochem. Rev.* **2022**, *50*, 100484. [CrossRef]
32. Hasegawa, Y.; Kitagawa, Y.; Nakanishi, T. Effective Photosensitized, Electrosensitized, and Mechanosensitized Luminescence of Lanthanide Complexes. *NPG Asia Mater.* **2018**, *10*, 52–70. [CrossRef]
33. Ha, J.M.; Hur, S.H.; Pathak, A.; Jeong, J.-E.; Woo, H.Y. Recent Advances in Organic Luminescent Materials with Narrowband Emission. *NPG Asia Mater.* **2021**, *13*, 53. [CrossRef]
34. Botelho, M.B.S.; de Queiroz, T.B.; Eckert, H.; de Camargo, A.S.S. Efficient Luminescent Materials Based on the Incorporation of a Eu(III)Tris-(Bipyridine-Carboxylate) Complex in Mesoporous Hybrid Silicate Hosts. *J. Lumin.* **2016**, *170*, 619–626. [CrossRef]
35. Łyszczek, R.; Gil, M.; Głuchowska, H.; Podkościelna, B.; Lipke, A.; Mergo, P. Hybrid Materials Based on PEGDMA Matrix and Europium(III) Carboxylates-Thermal and Luminescent Investigations. *Eur. Polym. J.* **2018**, *106*, 318–328. [CrossRef]
36. Gil-Kowalczyk, M.; Łyszczek, R.; Jusza, A.; Piramidowicz, R. Thermal, Spectroscopy and Luminescent Characterization of Hybrid PMMA/Lanthanide Complex Materials. *Materials* **2021**, *14*, 3156. [CrossRef]
37. Owusu-Ware, S.K.; Chowdhry, B.Z.; Leharne, S.A.; Antonijević, M.D. Quantitative Analysis of Overlapping Processes in the Non-Isothermal Decomposition of Chlorogenic Acid by Peak Fitting. *Thermochim. Acta* **2013**, *565*, 27–33. [CrossRef]
38. Papadopoulos, C.; Kantiranis, N.; Vecchio, S.; Lalia-Kantouri, M. Lanthanide Complexes of 3-Methoxy-Salicylaldehyde: Thermal and Kinetic Investigation by Simultaneous TG/DTG–DTA Coupled with MS. *J. Therm. Anal. Calorim.* **2010**, *99*, 931–938. [CrossRef]
39. Catauro, M.; Bollino, F.; Dell'Era, A.; Ciprioti, S.V. Pure Al2O3·2SiO2 Synthesized via a Sol-Gel Technique as a Raw Material to Replace Metakaolin: Chemical and Structural Characterization and Thermal Behavior. *Ceram. Int.* **2016**, *42*, 16303–16309. [CrossRef]
40. Catauro, M.; Dell'Era, A.; Vecchio Ciprioti, S. Synthesis, Structural, Spectroscopic and Thermoanalytical Study of Sol–Gel Derived SiO2–CaO–P2O5 Gel and Ceramic Materials. *Thermochim. Acta* **2016**, *625*, 20–27. [CrossRef]
41. Catauro, M.; Šiler, P.; Másilko, J.; Risoluti, R.; Vecchio Ciprioti, S. Synthesis, Structural, Morphological and Thermal Characterization of Five Different Silica-Polyethylene Glycol-Chlorogenic Acid Hybrid Materials. *Polymers* **2021**, *13*, 1586. [CrossRef]
42. Xing, Y.; Liu, Y.; Xue, X.; Wang, X.; Li, W. Zn-Er Heterometallic Carboxylate Framework Based on 3,5-Pyrazoledicarboxylic Acid with Nested Sandwich Structure and Its Luminescent Property. *Inorg. Chem. Commun.* **2017**, *84*, 153–158. [CrossRef]
43. Vlasyuk, D.; Łyszczek, R. Effect of Different Synthesis Approaches on Structural and Thermal Properties of Lanthanide(III) Metal–Organic Frameworks Based on the 1H-Pyrazole-3,5-Dicarboxylate Linker. *J. Inorg. Organomet. Polym.* **2021**, *31*, 3534–3548. [CrossRef]
44. Ullah, R.; Ahmad, I.; Zheng, Y. Fourier Transform Infrared Spectroscopy of "Bisphenol A". *J. Spectrosc.* **2016**, *2016*, 1–5. [CrossRef]
45. Silverstein, R.M.; Webster, F.X.; Kiemle, D.J.; Bryce, D.L. *Spectrometric Identification of Organic Compounds*, 8th ed.; John Wiley & Sons: Hoboken, NJ, USA, 2015.
46. Lange, R.S. Holly und P. Sohár: Absorption Spectra in the Infrared Region, Theoretical and Technical Introduction. 183 Seiten, zahlreiche Tab. Akadémiai Kiadó, Budapest 1975. *Nahrung* **1976**, *20*, 861–862. [CrossRef]
47. Nikolic, G.; Zlatkovic, S.; Cakic, M.; Cakic, S.; Lacnjevac, C.; Rajic, Z. Fast Fourier Transform IR Characterization of Epoxy GY Systems Crosslinked with Aliphatic and Cycloaliphatic EH Polyamine Adducts. *Sensors* **2010**, *10*, 684–696. [CrossRef]
48. Vohlídal, J. Polymer Degradation: A Short Review. *Chem. Teach. Int.* **2021**, *3*, 213–220. [CrossRef]
49. Głuchowska, H.; Łyszczek, R.; Mazur, L.; Kirillov, A.M. Structural and thermal investigations of Co(II) and Ni(II) coordination polymers based on biphenyl-4,4' -dioxydiacetate linker. *Materials* **2021**, *14*, 3545. [CrossRef]
50. Feng, J.; Hao, J.; Du, J.; Yang, R. Using TGA/FTIR TGA/MS and Cone Calorimetry to Understand Thermal Degradation and Flame Retardancy Mechanism of Polycarbonate Filled with Solid Bisphenol A Bis(Diphenyl Phosphate) and Montmorillonite. *Polym. Degrad. Stab.* **2012**, *97*, 605–614. [CrossRef]
51. Stellato, M.J.; Innocenti, G.; Bommarius, A.S.; Sievers, C. Pore Blocking by Phenolates as Deactivation Path during the Cracking of 4-Propylphenol over ZSM-5. *Catalysts* **2021**, *11*, 721. [CrossRef]
52. Jang, B.N.; Wilkie, C.A. The Thermal Degradation of Bisphenol A Polycarbonate in Air. *Thermochim. Acta* **2005**, *426*, 73–84. [CrossRef]
53. Kumagai, S.; Ono, S.; Yokoyama, S.; Kameda, T.; Yoshioka, T. Fate of Bisphenol A Pyrolysates at Low Pyrolytic Temperatures. *J. Anal. Appl. Pyrolysis* **2017**, *125*, 193–200. [CrossRef]
54. Łyszczek, R.; Rusinek, I.; Ostasz, A.; Sienkiewicz-Gromiuk, J.; Vlasyuk, D.; Groszek, M.; Lipke, A.; Pavlyuk, O. New Coordination Polymers of Selected Lanthanides with 1,2-Phenylenediacetate Linker: Structures, Thermal and Luminescence Properties. *Materials* **2021**, *14*, 4871. [CrossRef] [PubMed]

55. Gao, R.; Kodaimati, M.S.; Yan, D. Recent Advances in Persistent Luminescence Based on Molecular Hybrid Materials. *Chem. Soc. Rev.* **2021**, *50*, 5564–5589. [CrossRef] [PubMed]
56. Cotton, S. *Lanthanide and Actinide Chemistry: Cotton/Lanthanide and Actinide Chemistry*; John Wiley & Sons, Ltd.: Chichester, UK, 2006. [CrossRef]
57. Podkościelna, B. New Photoluminescent Copolymers of Naphthalene-2,7-Diol Dimethacrylate and N-Vinyl-2-Pyrrolidone: Synthesis, Characterisation and Properties. *J. Therm. Anal. Calorim.* **2014**, *116*, 785–793. [CrossRef]

materials

MDPI

Article

Thermal and Stress Properties of Briquettes from Virginia Mallow Energetic Crops

Marek Kurtyka [1], Magdalena Szwaja [2], Andrzej Piotrowski [2], Barbara Tora [3] and Stanislaw Szwaja [2,*]

1 Termo-Klima MK, Tartaczna 12, 40-749 Katowice, Poland
2 Faculty of Mechanical Engineering and Computer Science, Czestochowa University of Technology, Armii Krajowej 21, 42-200 Czestochowa, Poland
3 Faculty of Civil Engineering and Resource Management, AGH University of Science and Technology in Krakow, Mickiewicza 30, 30-059 Krakow, Poland
* Correspondence: stanislaw.szwaja@pcz.pl

Abstract: The article discusses the influence of briquetting/compaction parameters. This includes the effects of pressure and temperature on material density and the thermal conductivity of biomass compacted into briquette samples. Plant biomass mainly consists of lignin and cellulose which breaks down into simple polymers at the elevated temperature of 200 °C. Hence, the compaction pressure, compaction temperature, density, and thermal conductivity of the tested material play crucial roles in the briquetting and the torrefaction process to transform it into charcoal with a high carbon content. The tests were realized for samples of raw biomass compacted under pressure in the range from 100 to 1000 bar and at two temperatures of 20 and 200 °C. The pressure of 200 bar was concluded as the most economically viable in briquetting technology in the tests conducted. The conducted research shows a relatively good log relationship between the density of the compacted briquette and the compaction pressure. Additionally, higher compaction pressure resulted in higher destructive force of the compacted material, which may affect the lower abrasion of the material. Regarding heat transfer throughout the sample, the average thermal conductivity for the compacted biomass was determined at a value of 0.048 ± 0.001 W/(K·m). Finally, the described methodology for thermal conductivity determination has been found to be a reliable tool, therefore it can be proposed for other applications.

Keywords: thermal conductivity; compaction pressure; density; biomass; briquettes

Citation: Kurtyka, M.; Szwaja, M.; Piotrowski, A.; Tora, B.; Szwaja, S. Thermal and Stress Properties of Briquettes from Virginia Mallow Energetic Crops. *Materials* **2022**, *15*, 8458. https://doi.org/10.3390/ma15238458

Academic Editors: Agnieszka Kijo-Kleczkowska and Adam Gnatowski

Received: 27 October 2022
Accepted: 23 November 2022
Published: 28 November 2022

1. Introduction

This research considers the durability and thermal conductivity of biomass densified into briquettes, which is particularly important when using them in individual heating systems, e.g., heating boilers in domestic central heating installations. Therefore, the search for the most economically advantageous technologies and optimal parameters for compacting biomass into briquettes/pellets is a well-known research topic. The thermal and stress parameters of the compacted biomass should be of interest in both modeling and experimental analysis of biomass densified into briquettes and further into charcoal during the torrefaction process. However, the first studies regarding physical stress characterization of densified biomass were undertaken by Liu et al. [1]. Their research pointed to several important thermal and stress parameters of compacted biomass for its processing and applications in building construction. Literature in this field has been focused on the subject of searching for information on thermal conductivity and strength of the briquette/pellet materials. As indicated above, strength tests were carried out for briquettes/pellets, even though it is not a material whose strength properties will later be used as it is not used, for example, in building or construction material. In this regard, work by Gorzelany et al. [2] deserves attention, in which the compressive strength properties and durability assessment of pellets with a diameter of six and eight millimeters made of sawdust were discussed. They achieved compressive stresses from 5.4 to 7.9 MPa, and from 10.9 to 11.4 MPa for pellets with a diameter of six and eight millimeters, respectively. The authors also determined Young's modulus for the tested samples, which ranged between 82.27 MPa and 257.53 MPa. As noticed, the spread of Young's modulus was relatively large. The upper value was approximately three times higher than the lower value of the above-mentioned compartment. With regards to the investigation of the thermal conductivity of biomass-based briquettes, Sova et al. [3] conducted a theoretical analysis of the thermal conductivity of wood cells. They developed equations describing the effective transverse thermal conductivity. They also confirmed their modeling with experimental measurements. Rbihi et al. [4] determined thermal conductivity for cellulose. It was 0.0339 W/(K·m) at the density of 0.438 kg/m^3. Wang et al. [5] also conducted research on the thermal conductivity of various biomass-based materials. They obtained thermal conductivity between 0.3 and 0.32 W/(K·m).

With regard to the durability and stress analysis of the densified biomass, Markowski [6] proposed testing the durability of pellets in accordance with the PN-EN ISO 17831-1 standard. Details of the research methodology proposed in this standard are discussed in another chapter of this article. It should be noted that this standard has become an interpretation of the briquette/pellet durability analysis. Durability tests based on this standard were also carried out by Dyjakon et al. [7] and Lemos [8], who investigated the use of pellets in metallurgical furnaces. They both concluded that the mechanical durability of briquettes decreases with the increase in particle size. Another method of briquette durability measurement was proposed by Bembenek and Hryniewicz [9]. Their proposed methodology was based on the loss of mass caused by dropping a sample from a height of two meters onto a hard surface. The second method proposed by them, necessary for the analysis of the strength properties of the briquette, was compression on a Zwick Roell testing machine. Compressive strength tests of ash briquette samples were also conducted by Borowski [10]. He tested cylindrical samples compressed axially and the so-called Brazilian method, i.e., perpendicular to the cylinder axis, depending on the amount of binder used. Tests of briquette durability and briquette density depending on the compaction temperature in the range of 200–250 °C were carried out by Niedziółka and Szpryngiel [11]. They confirmed that the increase in temperature increased both tested values; however, they did not refer to the direct relationship between durability and density. Chin [12] and O'Dogerty [13] presented a different approach to the analysis of compressed biomass. They analyzed the dependence of the density of the compacted biomass on the compression pressure. According to their results, this dependence is in log correlation. Mechanical strength analysis of briquettes depending on the compaction pressure was also carried out by Nikiforov et al. [14]. They concluded that the compaction pressure of 20 MPa was sufficient for the briquette to have a sufficiently high density and burn stably. The relationship between density and compression pressure was also investigated by Plistil and his team [15]. In this work, they found a positive dependence between the density and the compressive breaking force on the compaction pressure. Research on the compressive strength of the briquette was carried out, among others, by Rahman et al. [16]. However, this work concerned fine coal briquettes and presented only the results of the compression test, without further analysis, including the determination of Young's modulus and Poisson's ratio. Kaliyan [17] investigated the efficiency of the compaction process in order to produce a sustainable pellet. He came to the conclusion that parameters, such as strength and durability of compacted products, can be determined by testing the material depending on the binder composition and compaction pressure. According to him, mechanical resistance was defined as compressive and impact resistance, as well as water resistance. In contrast, durability according to Kaliyan was defined as abrasion resistance [17]. In conclusion, the author raised an important problem of developing standards for acceptance criteria for the strength and durability levels of briquettes/pellets. For this purpose, they provided the guidelines needed to develop such standards. Chłopek [18] conducted experimental studies of the process of densification and consolidation of composite fuels, which were mixtures of coal and biomass. He examined a number of models in the field of granulation and compaction theory. Based on his research, it was shown that blends containing traditional solid fuels and biomass can be successfully subjected to pressure agglomeration in a granulator with a flat matrix, assuming the correct selection of process parameters. Roman [19] presented an analysis of the course of biomass compaction by measuring the resistance of the tested briquetted material. He related the resistance to the stresses and deformation of the material pressed on a press with a pressure of up to 100 kN. The work of Rejdak and others is also interesting in this regard [20]. They tested the strength and density of briquetted material from energy willow (Salix Viminalis) depending on the material temperature and compaction pressure during the production process. They used the Brazilian tensile strength method for compressive strength tests. Świętochowski et al. [21] presented the results of their research on the determination of Young's modulus depending on the deformation method of the compressed material and its diameter. They found that the mean values of the modulus of elasticity and the maximum stresses were in most cases lower for the larger sample sizes (briquettes) than for the pellet size samples. The procedure and methodology similar to that described in ISO 17831 was also applied by Temmerman [22]. He tested the durability defined as briquette abrasion in a drum machine.

In terms of the standards for the durability of briquettes, several standards have been developed that introduce certain methods and formulas for fuel presented as densified biomass in the form of both briquettes and pellets. The ISO 17831 standard has been introduced for biomass which is produced for conversion to heat and/or electricity. The literature review has been carried out with particular emphasis on the research on the durability of compacted biomass. It can be observed that most studies focused on linking durability with biomass density and compaction parameters (pressure and temperature). The concept of the durability of densified biomass has a number of definitions, most of which are reduced to the phenomenon of material abrasiveness and the methods of its measurement. In this case, the vast majority of tests are based on the research methodology described in the ISO 17831 standard. It was considered that the problem of testing the durability and strength properties of compacted biomass was not fully developed. Hence, it was assumed that the research process was properly prepared, and it requires thorough analysis and development of the research methodology, which will contribute to the practical usefulness of the research. Firstly, it was concluded that testing the durability (in many publications called strength) of the briquette using the methodology presented in the ISO 17831 standard will not be adequate to meet expectations regarding the results and conclusions. It does not give results that can be used to find the relationship between density and the compaction parameters (pressure and temperature).

Summing up, the main objective of the research in this article is to determine the relationship between compaction pressure and both density and destructive force at elevated temperatures, and to determine the thermal conductivity of the compacted material. These data are necessary to analyze the compaction and torrefaction processes and for the optimization of these processes. Moreover, it was found that the results of the so-called durability/strength tests with the methodology outlined in ISO 17831 cannot be used to analyze the actual material properties of biomass compacted into briquettes. The tests presented in the article were carried out on Virginia mallow (Sida Hermaphrodita) feed material. It is an energetic crop with the potential for combustion. It is feedstock for biogas fermentation plants as well as fodder for cattle and other domestic animals [23,24].

Finally, let us emphasize that our goal in this article was not only to show the properties of Virginia mallow towards compaction and in the further the torrefaction process, but also to show the methodology and some trends appearing between properties for briquetted material. It can be concluded that based on this proposed research methodology, other energy crops and/or waste biomass can be analyzed in terms of their effective compaction.

2. Materials and Methods

2.1. Rod Sample Preparation

The samples used for tests were characterized by the following dimensions:

- diameter: 25 mm
- length: 100 mm.

The input material for the production of samples were plants of the Virginia mallow (Sida hermaphrodita) crop presented in the form of chips with the size of 0.5 to 5 mm compacted at various pressures of 100, 200, 400, 600, 800, and 1000 bar and at a temperature of 20 and 200 °C. For the measurements of thermal conductivity, the briquetting parameters were as follows: compaction pressure 200 bar and temperature 200 °C.

2.2. Methodology for Thermal Conductivity Determination

The thermal conductivity was determined on the basis of measurements and modeling of heat conduction at a transient flow inside the sample, assuming that the heated sample was an infinitely long cylinder located horizontally in an oven, taking this into account in modeling the heat flow through it. This results in a forced airflow, which ensures the Biot number is high enough to consider mainly heat conduction, rather than convection, as the leading type of heat transfer during the experiment. The Biot number expresses the ratio of heat conduction resistance through the solid mass to heat convection resistance between the body and the environment. The Biot number has been estimated to be approximately 9.

Simplifications:

- The heat flux q'' (W/m^2) was calculated from the conductivity Equation (1) assuming the temperature gradient inside the rod close to the rod surface

$$q'' = k_{mean} \cdot \frac{dT}{dr}, \tag{1}$$

- The heat flux q'' was constant across the pellet
- k_{mean} is the material conductivity determined by Equation (2), calculated for the temperature gradient inside the rod close to the rod surface

$$k_{mean} = \frac{k(T_{s-1}) + k(T_s)}{2}, \tag{2}$$

- T_s represents the temperature taken from the endpoint near the surface of the rod at quasi-steady-state conditions. This means conditions under which the temperature has been stabilized, whereas T_{s-1} is the temperature inside the rod at a distance of a single step in the discrete mesh
- The contact resistance between the thermocouple and the material surface has been ignored

It was assumed that the cylinder was infinitely long and the axis was symmetrical, which ensures 1D heat conduction. Temperature T is a function of 2 variables: time t and radial coordinate r, $T = T(t, r)$.

The temperature determination can be obtained by solving Equation (3) for heat conduction in a cylindrical coordinate system (r, θ, z).

$$\frac{1}{r}\frac{d}{dr}\left(r \cdot k\frac{dT}{dr}\right) + \frac{1}{r^2}\frac{d}{d\theta}\left(k\frac{dT}{d\theta}\right) + \frac{d}{dz}\left(k\frac{dT}{dz}\right) + \dot{e}_{gen} = \rho c\frac{dT}{dt}, \tag{3}$$

where: k represents the thermal conductivity coefficient; ρ represents density; and c represents specific heat.

From the general form of heat conduction (Equation (3)), a simplified form can be obtained that applies to the experiment case. Assuming there is no heat generation inside the rod body, homogeneous properties (k, ρ, and c), and the fact that the temperature distribution depends only on the radial coordinate (r), Equation (3) can be written as follows (Equation (4)):

$$\frac{d^2T}{dr^2} + \frac{1}{r}\frac{dT}{dr} = \frac{1}{\alpha}\frac{dT}{dt}, \tag{4}$$

where $\alpha = \frac{k}{\rho c}$ represents thermal diffusivity ((k represents the thermal conductivity coefficient; ρ represents density; and c represents specific heat).

They have been adopted as follows:

$$\rho = 750\ \mathrm{kg/m^3}$$

$$c = 1650\ \mathrm{J/(kg{\cdot}K)}.$$

The parameter k was taken as a parameter varying from 0.03 to 0.08 in this study. The problem is considered to be axially symmetric, hence the temperature distribution neither depends on z nor θ coordinates.

As depicted in Figure 1, the Dirichlet boundary conditions related to the symmetry of the cylinder and external rod surface temperature are as follows (Equations (5)–(7)):

- Surface temperature changing with time (measured by the thermocouple surface):

$$T_1 = T_{surface} = T_{measured}\ at\ r = r_{max}, \tag{5}$$

$$T_2 = T_{center} = T_{measured}\ at\ r = 0, \tag{6}$$

- The initial conditions are as follows:

$$T(r,t) = T_{initial}\ at\ t = 0\ and\ r = r_{max}, \tag{7}$$

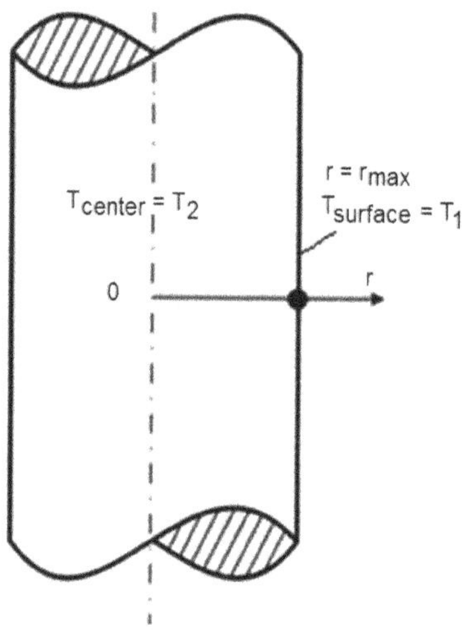

Figure 1. Temperature distribution inside an infinite rod.

Additionally, the calculations took into account the Neumann boundary condition concerning the temperature change inside the bar center (Equation (8)).

$$\frac{dT}{dr} = 0\ at\ r = 0 \tag{8}$$

Inside the sample (rod), a hole with a diameter of 2 + 0.2 mm and a depth of 75 mm was drilled. Then, a conductive paste was injected into the drilled hole. This was undertaken to reduce the thermal contact resistance between the material and the thermocouple. The thermocouple with a diameter of 2 mm was inserted into the drilled hole. The thermocouple wiring system was insulated to minimize heat conduction through the thermocouple itself. In order to measure the rod surface temperature, the thermocouples were mounted on the external rod surface (Figure 2). The heating and cooling experiments were conducted for the top configuration (surface thermocouple on the top of the rod) and the bottom configuration (surface thermocouple on the bottom of the rod). The tests were repeated 4 times. Figure 2 shows the placement of the thermocouples on the sample (rod) in the top and bottom configurations.

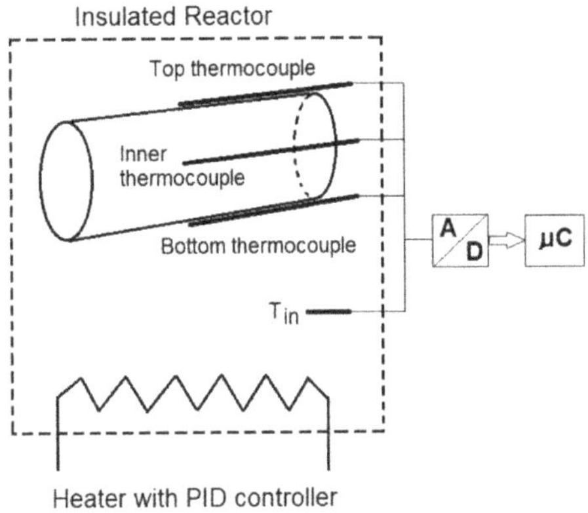

Figure 2. The location of thermocouples in the sample.

2.2.1. Test Procedure

The oven was set to a temperature in the range of 100 to 110 °C. After the oven temperature had stabilized, the rod specimen was placed inside the oven. The heating experiment was carried out for a time that allowed the temperature inside the rod to stabilize. The temperature data were collected continuously by an analog-to-digital converter coupled with a data logger and a computer. In Figure 2, the schematic diagram of the constructed system for the heating experiment is presented. The cooling experiment was conducted outside the oven. The heated rod was removed from the oven and hung on threads, maintaining the same configuration as in the heating experiment. The rod was subjected to free air convection and ambient temperature conditions to cool the rod sample.

2.2.2. Heating and Cooling Tests

The normalized temperature was introduced to make a comparison between tests which can vary with respect to the final temperature.

The normalized temperature for heating tests was calculated using Equation (9):

$$\theta_h = \frac{T(r,t) - T_{initial}}{T_\infty - T_{initial}}, \tag{9}$$

where:

$T_{initial}$ represents the temperature for $t = 0$,

$T\infty$ represents the final temperature after the rod has reached steady-state conditions.

The normalized temperature for the cooling tests was calculated as follows (Equation (10)):

$$\theta_c = \frac{T(r,t) - T_\infty}{T_{initial} - T_\infty}, \tag{10}$$

where:

$T_{initial}$ represents the temperature for time $t = 0$,

T_∞ represents the temperature after the rod has reached steady-state conditions.

Hence, Equation (4) remains the same form after temperature normalization (Equation (11)):

$$\frac{d^2\theta}{dr^2} + \frac{1}{r}\frac{d\theta}{dr} = \frac{1}{\alpha}\frac{d\theta}{dt}, \quad (11)$$

With regard to the steel coating of the thermocouple and its potential impact on measurement accuracy, it was found that the average thermal conductivity of stainless steel is in the range of 13–17 W/(K·m). This is several orders of magnitude higher than the material conductivity. It can therefore be concluded that the impact of thermocouple conductivity is almost negligible, and thus can be ignored.

2.3. Methodology for Stress Analysis

Experimental tests were carried out in the field of material compressive strength, where the basic measured quantities were:

- force acting on the sample
- sample deformation

Tensile tests were not carried out. This was due to the high brittleness of the tested material and difficulties in the method of effective fastening, which had no effect on the tensile test.

For the experimental tests, samples in the form of cylindrical briquettes pressed at various pressures and temperatures were used. A Zwick Roell Z100 testing machine was used to determine the correlations between compaction pressure and material density, as well as the maximum destructive force. The pressure of 200 bar was adopted as the most economically viable in the briquetting technology.

In terms of the compaction temperature, 2 temperatures were adopted:

- 200 ± 1 °C
- ambient temperature (approximately 20 ± 0.5 °C)

The justification for the selection of the compaction temperature of 200°C results directly from the analysis of the target application of the torrefaction briquetting technology on an industrial scale.

2.4. Compressive Strength Test

After briquetting, the samples were tested for compressive strength in a plane perpendicular to the symmetry axis (Figure 3a) of the sample cylinder and in a plane perpendicular to it (Figure 3b).

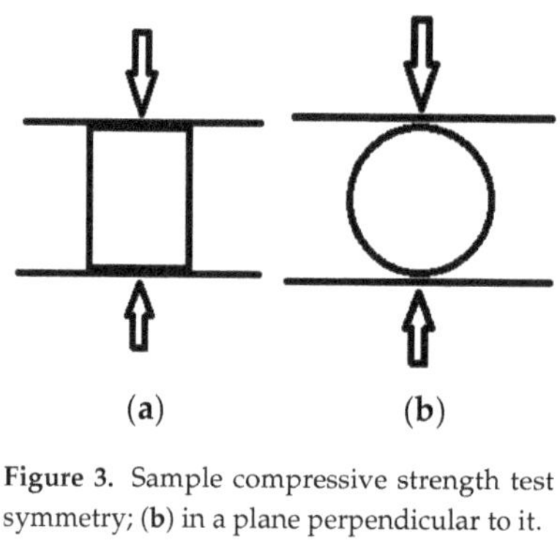

Figure 3. Sample compressive strength test scheme: (**a**) in a plane perpendicular to the axis of symmetry; (**b**) in a plane perpendicular to it.

Investigation on the correlation between the density of the compacted biomass and the compaction pressure was carried out. For this purpose, the displacement of the press punch (piston) was measured as a function of the compaction force exerted by the punch on the material to be compacted.

The deformation ε as a result of compaction was calculated using Equation (12):

$$\varepsilon = \frac{\Delta h}{h_p} \cdot 100\% \quad (12)$$

where:

Δh represents the change in height of the sample during compaction under a given pressure, h_p represents the initial sample height before compaction.

The initial sample height h_p was assumed to be the specimen height after initial punch pressure at 5 bar. The deformation calculated in this way was used to investigate the relationship with the compaction pressure. Due to the expected logarithmic relationship, the compression pressure and strain graphs are plotted on a logarithmic scale as a function of the compression pressure.

3. Results and Discussion

3.1. Results from Conductivity Tests

Figure 4 presents the temperature data from exemplary tests for heating and cooling cases.

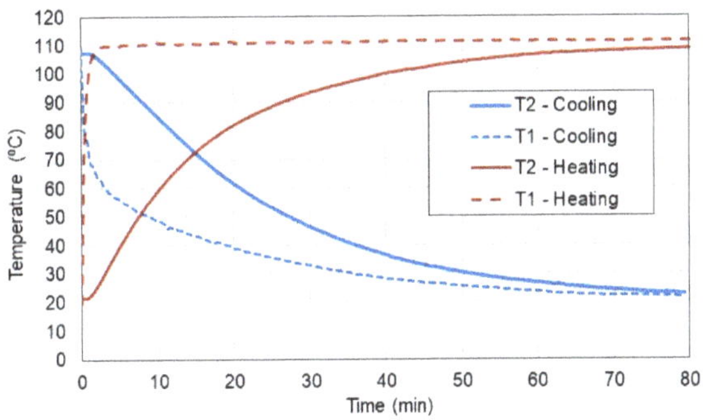

Figure 4. Inner T2 and outer T1 temperatures for heating and cooling tests.

As can be seen in Figure 4, the temperature T2 inside the sample varies typically to the theoretically described heat transfer to or from the sample and asymptotically reaches the final temperature. However, to recognize the effect of cooling and heating it is recommended to plot the normalized temperature versus time to compare all data sets as shown in Figure 5. From the normalized temperature diagrams, it can be concluded that the heat transfer by conduction in the heating tests was higher in comparison to the cooling tests. Of course, it can be riposted that the heat transfer by convection at the surface of the sample was higher in the heating test compared to the cooling test, although, both these cooling and heating tests were performed in an environment without forced air (no turbulence) near the surface of the sample. Hence, this is another observation of this study.

Figure 6 shows an example of the temperature distribution T2 inside the sample as a function of time during the cooling test. In the figure, the inner temperature T2 from the tests was compared with the inner temperature calculated from Equation (4) (marked as T2 calc) for the three various conductivity coefficients: 0.36, 0.48, and 0.60 W/(K·m).

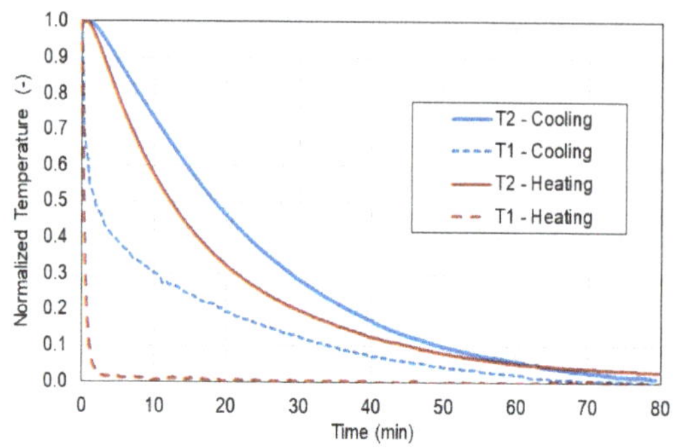

Figure 5. Normalized temperature for heating and cooling tests.

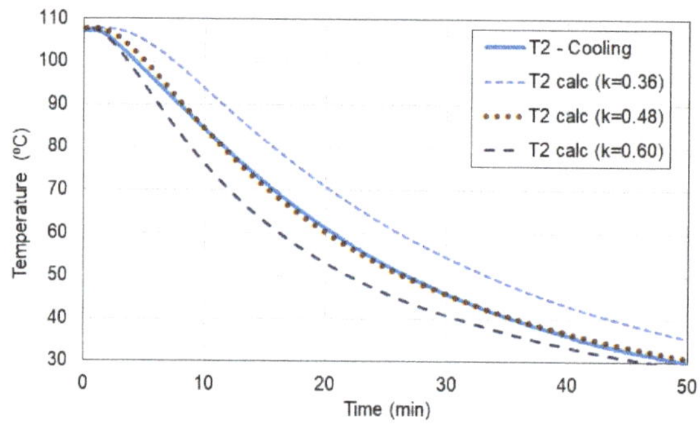

Figure 6. Comparison of inner temperature from experiment and calculations.

Based on this procedure, the conductivity coefficient k was determined for the four cooling and the four heating tests. The average conductivity k was determined as follows:

- for cooling tests: k = 0.048 ± 0.001 W/(K·m)
- for heating tests: k = 0.049 ± 0.003 W/(K·m)

As observed, the results from the cooling tests seemed to be more reliable due to the lower uncertainty of 0.001 W/(K·m), hence, for further analysis, the cooling tests were proposed as more reliable in this case.

3.2. *Stress Analysis*

Figure 7a shows exemplary compression test results of samples made of briquettes compacted under various pressures from 100 to 1000 bar at a temperature of 200 °C in the axis of symmetry of a cylindrical sample (as shown in Figure 3a). The results of deformation from the performed strength test as a function of load (Figure 3b) were not taken into account, because the obtained results were burdened with a high measurement uncertainty resulting from a large discrepancy in the anisotropy of the compacted samples. Moreover, in this case, the alignment of the samples on the machine had a significant influence on the results. In the graphs (Figure 7a), the maximum compressive/destructive force is marked with a circle. After reaching this value, the compressed material was destroyed as a result of its scattering. Based on the observation of the destruction of the sample, it could be noticed that the samples did not show any plasticity properties and the first visual signs of sample crushing occurred when the destructive force reached its maximum. Hence, it can be concluded that the destruction of the internal structure of the sample probably occurred a little earlier, just before reaching the maximum destructive force. This effect was clearly visible for the sample compressed at a pressure of 100 bar. Figure 7b shows the maximum destructive force as a function of compaction pressure for the briquettes presented in Figure 7a. As can be seen, there is a linear correlation between the maximum destructive force and the compaction pressure. This observation can be used in further analysis of abrasion and compacted briquettes. These tests were repeated three times. Based on these tests, mean values were determined. However, the plots in Figure 7, on the other hand, are plots from individual compression tests shown as an example to confirm the trend between the compaction pressure and the maximum destructive force. On the other hand, averaging these runs vs. the press punch displacement led to a large discrepancy regarding clearance removal at the starting point of each compression test.

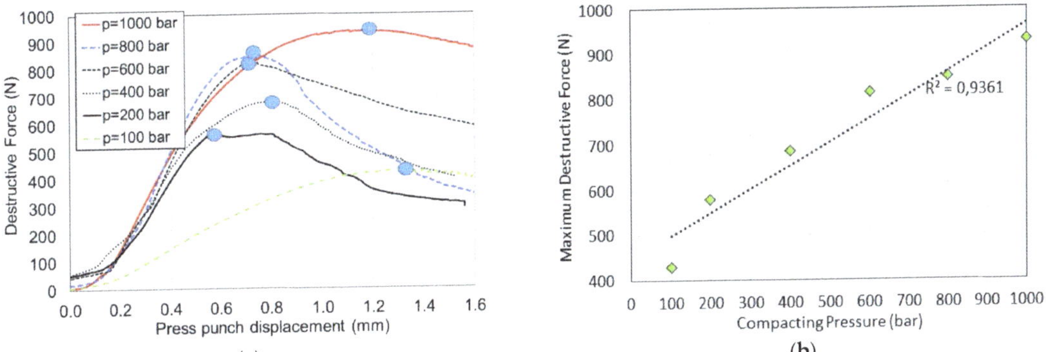

(**a**) (**b**)

Figure 7. (**a**) The destructive force vs. press punch displacement for tests with various compaction pressures for briquettes at the temperature of 200 °C; (**b**) maximum destructive force vs. compaction pressure for briquettes.

According to Equation (12), the deformation in relation to the compaction pressure is presented in Figure 8. As can be observed, the logarithmic correlation appears when the compaction pressure does not exceed 350 bar, which causes the deformation of the briquetted samples of 55 and 60% depending on compaction temperature of 20 and 200 °C, respectively.

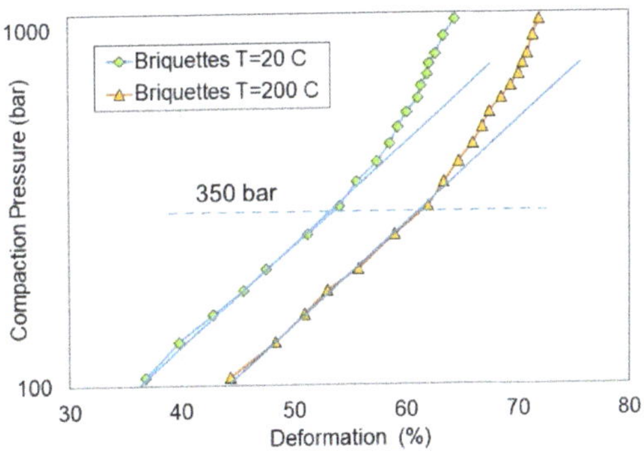

Figure 8. Compaction pressure course vs. deformation of briquettes.

In terms of the relationship between the density of the compacted briquette sample and the compaction pressure, the thesis of Chin and O'Dogerty confirmed that the logarithmic dependence (Equation (13)) best fits the experimental results, linking the density with the compaction pressure [12,13]. However, the values of the a and b coefficients, which significantly exceeded the ranges proposed by these scientists, have not been confirmed. The dependence of the density as a function of compaction pressure with trend lines and their equations is shown in Figure 9 and in Table 1. Chin and O'Dogerty proposed to present this relationship in an analytical form using Equation (13) [12,13]:

$$\rho = a \cdot \ln(p) + b \left(kg/m^3\right) \quad (13)$$

where *a* and *b* are empirically determined coefficients.

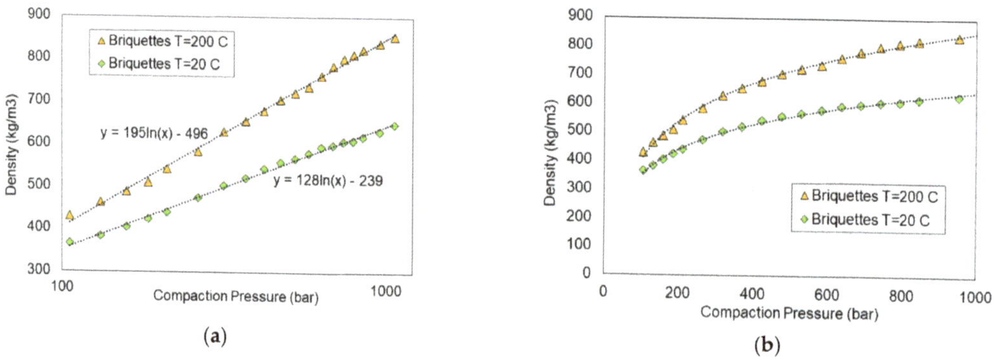

Figure 9. Density of compacted biomass versus pressure and temperature of compaction (**a**) in log scale; (**b**) in linear scale.

Table 1. Coefficients of Equation (13) and the regression coefficient R^2.

	Coefficient a	**Coefficient b**	R^2
Briquettes at T = 200 °C	195	−496	0.997
Briquettes at T = 20 °C	128	−239	0.997

For the graphs presented in Figure 9a, the trend lines were determined in accordance with the logarithmic relationship and the coefficients a and b for Equation (13) were calculated. The regression coefficient (R2) for all cases is very high, almost close to one, which may indicate a very good correlation of the experimental results with the logarithmic function presented in Table 1.

Moreover, it can be seen in Figure 9 that the density of the compacted biomass also depends on the temperature. The change of the compaction temperature from 20 to 200 °C significantly increased the density by approximately 300 kg/m^3 at a compaction pressure of 200 bar. Moreover, it can be seen that the compaction process at a pressure higher than 400 bar is not economically feasible since the density no longer increases as much as it does with an increase in pressure in the range of up to 400 bar. One can conclude from Figure 9b that the increase in density is by 15% (from 530 to 610 kg/m^3 at 20 °C), whereas the compaction pressure had to increase twice from 400 to 800 bar. Furthermore, we can suggest the optimal compaction pressure due to energy consumption is 200 bar. By comparing the results of the briquette's density at 450 kg/m^3 obtained under 200 bar with a density of 610 kg/m^3 at 800 bar compaction pressure, a density increase of 35% can be observed. However, the energy required for briquetting at 800 bar is much higher and the briquetting process becomes less economical. Moreover, the destructive force is also high enough for the compaction pressure of 200 bar that a crushing or abrasion will not remarkably deteriorate the compacted material. Therefore, it was considered inadvisable to increase the compaction pressure, which further increases the cost of compaction due to the need for more energy.

4. Conclusions

The following conclusions were drawn from the research work:

- The average thermal conductivity was calculated from the equation of heat conduction in the unsteady state for cylindrical coordinates. The obtained thermal conductivity was 0.048 ± 0.001 W/(K·m). As mentioned in Section 1, thermal conductivity determined by others was between 0.030 and 0.039 W/(K·m) for various biomasses densified at various pressures.
- The cooling experiment for thermal conductivity determination allowed for a more precise adjustment of the theoretical temperature to the experimental data. This can be caused by the lower impact of the thermocouple wiring system on heat conduction and additionally, the impact of the heat convection may be marginal.
- The proposed methodology for determining the thermal conductivity for biomass-based briquettes can be considered a useful tool for other low-conductivity materials.
- The destructive force is strictly in line with the compaction pressure.
- The temperature of the compaction process of 200 °C can be considered optimal regarding direct briquetting after the torrefaction process has taken place at a typical temperature range between 200 and 260 °C.
- The compaction pressure of 200 bar has been proposed as an optimal compaction pressure due to economic considerations related to the energy required for the compaction process.

Author Contributions: Conceptualization, M.K. and B.T.; methodology, M.K., M.S., and B.T.; software, M.S. and S.S.; investigation, M.K., A.P., and M.S.; validation, M.K., A.P., and B.T.; formal analysis, M.K., A.P., and S.S.; writing—original draft preparation, M.K. and A.P.; visualization, M.K.; supervision, B.T. and S.S. All authors have read and agreed to the published version of the manuscript.

Funding: This research received no external funding.

Institutional Review Board Statement: Not applicable.

Informed Consent Statement: Not applicable.

Data Availability Statement: Not applicable.

Conflicts of Interest: The authors declare no conflict of interest.

References

1. Liu, H.; Li, Q.; Quan, H.; Xu, X.; Wang, Q.; Ni, S. Assessment on the Properties of Biomass-Aggregate Geopolymer Concrete. *Appl. Sci.* **2022**, *12*, 3561. [CrossRef]
2. Gorzelany, J.; Zardzewiały, M.; Murawski, P.; Matłok, N. Analysis of Selected Quality Features of Wood Pellets. *Agric. Eng.* **2020**, *24*, 25–34. [CrossRef]
3. Sova, D.; Porojan, M.; Bedelean, B.; Huminic, G. Effective Thermal Conductivity Models Applied to Wood Briquettes. *Int. J. Therm. Sci.* **2018**, *124*, 1–12. [CrossRef]
4. Rbihi, S.; Laallam, L.; Sajieddine, M.; Jouaiti, A. Characterization and Thermal Conductivity of Cellulose Based Composite Xerogels. *Heliyon* **2019**, *5*, e01704. [CrossRef] [PubMed]
5. Wang, J.; Hou, S.; Shen, Z.; Wen, J.; Qi, C. Thermal Characteristics and Simulation of Enzymatic Lignin Isolated from Chinese Fir and Birch. *Forests* **2022**, *13*, 914. [CrossRef]
6. Markowski, J. Assessment of the Susceptibility of Biomass to Attempts at Improving the Durability of Pellets. *Nafta-Gaz* **2020**, *76*, 965–968. [CrossRef]
7. Dyjakon, A.; Sobol, Ł.; Krotowski, M.; Mudryk, K.; Kawa, K. The Impact of Particles Comminution on Mechanical Durability of Wheat Straw Briquettes. *Energies* **2020**, *13*, 6186. [CrossRef]
8. Lemos, L.R.; da Rocha, S.H.F.S.; de Castro, L.F.A.; Assunção, G.B.M.; Silva, G.L.R. da Mechanical Strength of Briquettes for Use in Blast Furnaces. *REM-Int. Eng. J.* **2019**, *72*, 63–69. [CrossRef]
9. Bembenek, M. *Badania Wpływu Kształtu Powierzchni Roboczej Walców Na Efekty Pracy Prasy Walcowej*; Praca dokt.; Akademia Górniczo-Hutnicza im. St.Staszica w Krakowie: Kraków, Poland, 2010.
10. Borowski, G. Porównanie Dwóch Sposobów Określenia Wytrzymałości Brykietów z MateriałóW Drobnoziarnistych. *Postępy Nauk. i Tech.* **2012**, *12*, 115–121.

11. Niedziółka, I.; Szpryngiel, M. Wpływ Temperatury Na Gęstość i Trwałość Brykietów Wytworzonych w Brykieciarce Ślimakowej. *Inżynieria Rol.* **2013**, *3*, 301–310.
12. Chin, O.C.; Siddiqui, K.M. Characteristics of Some Biomass Briquettes Prepared under Modest Die Pressures. *Biomass Bioenergy* **2000**, *18*, 223–228. [CrossRef]
13. O'Dogerty, M.J. A Review of the Mechanical Behaviour of Straw When Compressed to High Densities. *J. Agric. Engng. Res.* **1989**, *44*, 241–265. [CrossRef]
14. Nikiforov, A.S.; Prikhodko, E.V.; Kinzhibekova, A.K.; Nurkina, S.M. Study of Strength Characteristics of Fuel Briquettes from Organic Waste. *AIP Conf. Proc.* **2020**. [CrossRef]
15. Plíštil, D.; Brožek, M.; Malaťák, J.; Roy, A.; Hutla, P. Mechanical Characteristics of Standard Fuel Briquettes on Biomass Basis. *Res. Agric. Eng.* **2012**, *51*, 66–72. [CrossRef]
16. Rahman, A.N.E.; Aziz Masood, M.; Prasad, C.S.N.; Venkatesham, M. Influence of Size and Shape on the Strength of Briquettes. *Fuel Process. Technol.* **1989**, *23*, 185–195. [CrossRef]
17. Kaliyan, N.; Vance Morey, R. Factors Affecting Strength and Durability of Densified Biomass Products. *Biomass Bioenergy* **2009**, *33*, 337–359. [CrossRef]
18. Chłopek, M. *Ciśnieniowa Aglomeracja Kompozytowych Paliw Stałych*; Rozprawa d.; Akademia Górniczo-Hutnicza im. Stanisława Staszica: Kraków, Poland, 2015.
19. Roman, K. The Characteristics of Briquetting Process with Resistance Sensors Usage. *Przegląd Elektrotechniczny* **2019**, *1*, 100–103. [CrossRef]
20. Rejdak, M.; Czardybon, A.; Ignasiak, K.; Sobolewski, A.; Robak, J. Compaction Studies of Torrefied Willow. *J. Ecol. Eng.* **2017**, *18*, 183–192. [CrossRef]
21. Swietochowski, A.; Lisowski, A.; Dabrowska-Salwin, M. Strength of Briquettes and Pellets from Energy Crops. In Proceedings of the 15th Internal Scientific Conference "Engineering for Rural Development", Latvia University of Agriculture, Jelgava, Latvia, 25–27 May 2016; pp. 547–551.
22. Temmerman, M.; Rabier, F.; Jensen, P. Comparative Study of Durability Test Methods for Pellets and Briquettes. *Biomass Bioenergy* **2006**, *30*, 964–972. [CrossRef]
23. Jablonowski, N.D.; Kollmann, T.; Meiller, M.; Dohrn, M.; Müller, M.; Nabel, M.; Zapp, P.; Schonhoff, A.; Schrey, S.D. Full Assessment of Sida (Sida Hermaphrodita) Biomass as a Solid Fuel. *GCB Bioenergy* **2020**, *12*, 618–635. [CrossRef]
24. Kitczak, T.; Jarnuszewski, G.; Łazar, E.; Malinowski, R. Sida Hermaphrodita Cultivation on Light Soil—A Closer Look at Fertilization and Sowing Density. *Agronomy* **2022**, *12*, 2715. [CrossRef]

materials

Article

Numerical Analysis of Transient Pressure Damping in Viscoelastic Pipes at Different Water Temperatures

Qiang Sun [1], Zhilin Zhang [1], Yuebin Wu [2,3,*], Ying Xu [4] and Huan Liang [5]

1 School of Civil Engineering, Institute of Artificial Environment Control and Energy Application, Northeast Forestry University, Harbin 150040, China; sunqiang@nefu.edu.cn (Q.S.); zhangzhilin@nefu.edu.cn (Z.Z.)
2 School of Architecture, Harbin Institute of Technology, Harbin 150090, China
3 Laboratory of Cold Region Urban and Rural Human Settlement Environment Science and Technology, Harbin Institute of Technology, Harbin 150090, China
4 School of Energy and Architecture Engineering, Harbin University of Commerce, Harbin 150028, China; joexying@126.com
5 Industrial Control Energy Saving Business Division, Beijing Huada Zhibao Electronic System Co., Ltd., Beijing 100020, China; lianghuan0720@163.com
* Correspondence: ybwu@hit.edu.cn; Tel.: +86-189-446-528-62

Citation: Sun, Q.; Zhang, Z.; Wu, Y.; Xu, Y.; Liang, H. Numerical Analysis of Transient Pressure Damping in Viscoelastic Pipes at Different Water Temperatures. *Materials* **2022**, *15*, 4904. https://doi.org/10.3390/ma15144904

Academic Editors: Agnieszka Kijo-Kleczkowska and Adam Gnatowski

Received: 25 May 2022
Accepted: 11 July 2022
Published: 14 July 2022

Abstract: Water temperature affects the peak pressure damping of transient flows in viscoelastic pipes. Owing to the viscoelastic properties of pipes, the accuracy of peak pressure damping simulations hinges on both viscoelastic and frictional factors. In simulations, the influence of both factors on peak pressure damping at different water temperatures is unclear. In this study, the Kelvin–Voigt model with both a quasi-steady friction model and modified Brunone model was employed. Based on experimental data, the accuracy of simulated peak pressure damping was verified at four different water temperatures (13.8, 25, 31, and 38.5 °C). From the perspective of energy transfer and dissipation, the influence of viscoelastic and frictional factors on peak pressure damping were clarified, and the applicability of different friction models was determined based on the contributions of viscoelastic and frictional factors to peak pressure damping. The numerical results indicate that the viscoelastic properties of pipes have a greater impact on peak pressure damping than their frictional properties at 25, 31, and 38.5 °C. Higher temperatures result in a delay in the rate of work and a decrease in the frequency of work performed by viscoelastic pipes. Viscoelastic properties play a more important role than frictional ones in calculating peak pressure damping as the water temperature increases. In addition, the one-dimensional quasi-steady friction model can accurately simulate peak pressure damping within a specified water temperature range.

Keywords: energy analysis; one-dimensional friction model; peak pressure damping; transient flow; viscoelastic pipe

1. Introduction

Viscoelastic pipes such as those made of polyvinyl chloride (PVC), low-density polyethylene (LDPE), and high-density polyethylene (HDPE) are widely used in urban supply systems. When valves suddenly close or pumps stop abruptly, the resultant water hammer phenomenon can threaten the safe operation and maintenance of a hydraulic system. Accurately simulating this phenomenon in viscoelastic pipes is thus very important [1]. Both friction and viscoelasticity have significant impacts on the calculation of the peak pressure damping of transient flows in viscoelastic pipes, especially at different water temperatures [2]. This is because the water temperature affects not only the fluid characteristics [3], such as the density and viscosity of the water, but also the range of linear viscoelasticity with respect to the viscoelastic pipes [4–6]. Therefore, it is necessary to investigate the influence of viscoelastic and frictional factors in the numerical simulation of the peak pressure damping of transient flows in viscoelastic pipes at different water temperatures.

Several numerical models have been used to investigate friction and viscoelasticity. With respect to viscoelasticity, the Kelvin–Voigt (K–V) model has been used to describe the viscoelastic behavior of the pipe wall [7]. As for friction, one-dimensional (1D) quasi-steady and unsteady friction models [8–11] and two-dimensional friction models [12–15] have been used to describe the wall shear stress in transient flows. Generally, the classic transient flow model with a quasi-steady friction model is capable of accurately simulating the maximum value in terms of the pressure fluctuation in elastic pipes; however, it cannot accurately describe the peak pressure damping in most instances [1]. Cao et al. [8] modified the 1D instantaneous acceleration-based (IAB) model to better predict the wave peak and valley. Urbanowicz [9] improved the 1D model with a weighting function to calculate the unsteady-state component. Pezzinga et al. [10] compared the numerical results of 1D elastic and viscoelastic models used to simulate transient flows in HDPE pipe systems. The results showed that the viscoelastic model was more accurate when simulating the peak pressure damping than the elastic model. Abdeldayem et al. [11] compared the accuracy of different unsteady fiction models in engineering practice. They concluded that the modified IAB model is relatively suitable for simulating pressure fluctuation in transient flows. Covas et al. [16] investigated the effects of the viscoelastic parameters of the K–V model and verified the accuracy of the transient flow. Duan et al. [17,18] discussed the characteristics of viscoelastic pipes under transient flow with respect to energy transfer and dissipation based on energy analysis.

Another important factor that affects the peak pressure damping of transient flows is friction. Duan et al. [19] investigated the influence of the scale of the pipe system, particularly the pipe length and diameter, on the unsteady friction of transient flows in elastic pipes. The numerical results showed that a higher scale ratio between the pipe length and pipe diameter resulted in an increase in the role of unsteady friction on the peak pressure damping of transient flows in elastic pipes. Based on the experimental data reported by Covas et al. [14], Seck et al. [20] modified the unsteady friction correlation coefficient to improve the accuracy of the unsteady friction model. The results showed that the inclusion of viscoelasticity and unsteady friction generated dramatic peak pressure damping.

The water temperature affects not only the viscoelasticity of the pipe but also the speed and attenuation of the pressure wave of the transient flow in viscoelastic pipes. Neuhaus et al. [21] conducted transient flow experiments at different temperatures and compared the accuracy of simulation results under different tests. Saidani et al. [22] investigated the effect of temperature on a transient cavitation flow. Mousavifard [23] studied the influence of temperature on turbulence parameters during transient cavitation flows in viscoelastic pipes. The results showed that at higher temperatures, the flexibility of the pipe and the delay of the velocity distribution increased, whereas the velocity gradient near the wall decreased.

As mentioned above, some researchers have explored the effects of viscoelasticity and friction on the peak pressure damping of transient flows in viscoelastic pipes; however, the influence of viscous-elasticity and friction on peak pressure damping at different water temperatures has not yet been thoroughly studied. Therefore, this study analyzed simulation results for the peak pressure damping of transient flows in viscoelastic pipes at different water temperatures; the simulations employed the modified K–V model with the quasi-steady friction model and modified Brunone model. The mean absolute percentage error (MAPE) and difference evaluation methods were used to estimate the accuracy of the numerical simulations of the peak pressure damping. Then, from the perspective of energy transfer and dissipation, the influence of friction and viscoelasticity on peak pressure damping was investigated. The contributions of thviscoelastic, quasi-steady friction, and unsteady friction factors to the peak pressure damping in viscoelastic pipes are discussed separately.

2. Establishment of Mathematical Models

2.1. Governing Equations

The 1D transient flow equation consists of continuity and momentum equations, which can be expressed in the following form [14]:

$$\frac{\partial H}{\partial t} + \frac{a^2}{gA}\frac{\partial Q}{\partial x} + \frac{2a^2}{g}\frac{\partial \varepsilon_r}{\partial t} = 0 \tag{1}$$

$$\frac{\partial Q}{\partial t} + gA\frac{\partial H}{\partial x} + \frac{\pi D \tau_w}{\rho} = 0 \tag{2}$$

where H is the piezometric head, Q is the flow rate, a is the pressure wave speed, A is the pipe cross-sectional area, x is the coordinate along the pipeline axis, t is the time, τ_w is the wall shear stress, and ε_r is the retarded strain. The dependent variables are H, Q, τ_w, and ε_r, which are functions of x and t, respectively.

In Equations (1) and (2), the time derivative of the retarded strain, $\frac{\partial \varepsilon_r(t)}{\partial t}$, and the wall shear stress, τ_w, cannot be calculated directly; thus, further numerical discretization is required.

2.1.1. Numerical Discretization of the Wall Shear Stress

The wall shear stress, τ_w, is defined as having two parts in transient flows: τ_q is the steady-state component, and τ_u is the unsteady-state component [1].

$$\tau_w = \tau_q + \tau_u \tag{3}$$

For the quasi-steady friction (QF) model, the steady-state component, τ_q, is calculated as follows [4]:

$$\tau_q(t) = \frac{f}{8}\rho V^2 \tag{4}$$

where V is the average velocity in the section, f is the Darcy–Weisbach friction factor, and ρ is the density of the liquid.

The unsteady-state component, τ_u, is typically neglected in the QF model. However, rapid transient events and high pulsating frequencies require the accurate representation of unsteady friction [24–27].

For the unsteady friction (UF) model, the modified Brunone model was employed in this study. It is assumed that instantaneous local acceleration and convective acceleration affect unsteady friction [28–30]. Hence, the unsteady-state component is calculated as follows:

$$\tau_u(t) = \frac{k^* \rho D}{4}\left(\frac{\partial V}{\partial t} + sign(V)a\left|\frac{\partial V}{\partial x}\right|\right) \tag{5}$$

where $\partial V/\partial x$ is calculated as follows:

$$\frac{\partial V}{\partial x} = \frac{\Delta V}{\Delta x} \tag{6}$$

Through theoretical analysis of simulation results and experimental data, Vardy and Brown [27] provided a method to calculate the value of the empirical coefficient of the Reynolds number as follows:

$$k^* = \sqrt{C^*}/2 \tag{7}$$

$$C^* = \begin{cases} 0.00476 & \mathrm{Re} \le 2300 \\ 7.41/\mathrm{Re}^{\lg(14.3/\mathrm{Re}^{0.05})} & \mathrm{Re} \ge 2300 \end{cases} \tag{8}$$

2.1.2. Numerical Discretization of the Retarded Strain

The Kelvin–Voigt (K–V) model (Figure 1) is the most commonly used model to describe the creep compliance and relaxation modulus of materials. The creep function can be expressed as follows [31]:

$$J(t) = J_0 + \sum_{k=1}^{N} J_k(1 - e^{-t/\tau_k}) \tag{9}$$

where $J_0 = 1/E_0$ is the instantaneous elastic creep, $J_k = 1/E_k$ is the creep compliance of the spring of the kth element, $\tau_k = F_k/E_k$ is the retarded component of the kth element, F_k is the viscosity of the kth element, E_k is the modulus of elasticity of the spring of the kth element, and N is the number of K–V elements.

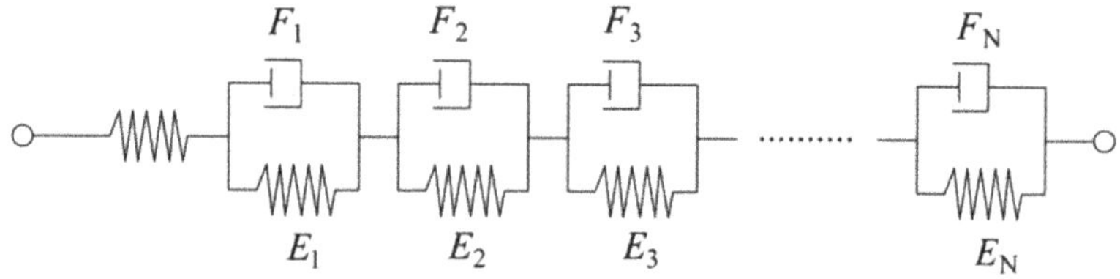

Figure 1. Kelvin–Voigt model.

The total strain, ε, is given by the sum of the instantaneous strain, ε_e, and the retarded component, ε_r [31].

$$\varepsilon = \varepsilon_e + \varepsilon_r \tag{10}$$

Using the K–V model with N elements, the retarded strain is the sum of the single-element deformation [16].

$$\partial\varepsilon_r(t) = \sum_{k=1\cdots N} \partial\varepsilon_{rk}(t) \tag{11}$$

According to the Boltzmann superposition principle [32], the time derivative of the retarded strain in Equation (1) is expressed as follows:

$$\frac{\partial\varepsilon_r(t)}{\partial t} = \sum_{k=1\cdots N} \frac{\partial\varepsilon_{rk}(t)}{\partial t} \tag{12}$$

Considering the creep function (Equation (9)), the retarded strain can be written as follows [31]:

$$\varepsilon_r(x,t) = \sum_{k=1\cdots N} \varepsilon_{rk}(x,t) = \sum_{k=1\cdots N} \frac{\gamma\alpha D}{2e} \int_0^t \left[H(x,t-t') - H_0(x)\right] \frac{J_k}{\tau_k} e^{\frac{-t'}{\tau_k}} dt' \tag{13}$$

where γ is the bulk weight.

2.2. *Numerical Scheme*

The governing equations are solved using the method of characteristics (MOC). This method transforms the partial differential equations into ordinary differential equations by calculating along the characteristic lines $dx/dt = \pm a$, as shown in Figure 2. The ordinary differential equations are given as follows:

$$\frac{dH}{dt} \pm \frac{a}{gA}\frac{dQ}{dt} \pm \frac{a}{g}\frac{4\tau_w}{\rho D} + \frac{2a^2}{g}\frac{\partial\varepsilon_r}{\partial t} = 0 \tag{14}$$

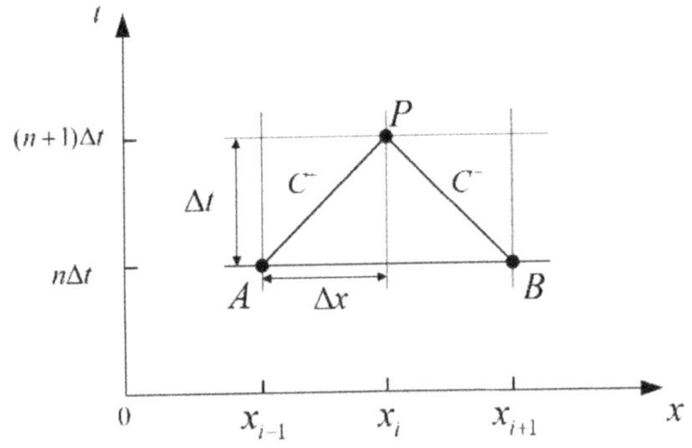

Figure 2. Rectangular grid system.

By integrating Equation (14) on the characteristic lines between times $n\Delta t$ and $(n + 1)\Delta t$, the discretized forms can be obtained as follows:

$$H(x,t) - H(x \pm \Delta x, t - \Delta t) \mp \frac{a}{gA}[Q(x,t) - Q(x \pm \Delta x, t - \Delta t)] \mp \frac{a}{g}\frac{4\tau_w(x,t)}{\rho D} + \frac{2a^2}{g}\frac{\partial \varepsilon_r(x,t)}{\partial t} = 0 \tag{15}$$

By substituting Equation (13) into Equation (15), the characteristic equations can be obtained as follows:

$$(1 + \frac{2a^2\Delta t}{g}\sum_{k=1}^{N}\frac{\alpha D}{2e}\gamma\frac{J_k}{\Delta t}e^{-\Delta t/\tau_k})H_i^{n+1} + \frac{a}{gA}Q_i^{n+1} + \frac{4a\Delta t}{\rho g D}\tau_{w,i-1}^{n} = C_p \tag{16}$$

$$(1 + \frac{2a^2\Delta t}{g}\sum_{k=1}^{N}\frac{\alpha D}{2e}\gamma\frac{J_k}{\Delta t}e^{-\Delta t/\tau_k})H_i^{n+1} - \frac{a}{gA}Q_i^{n+1} - \frac{4a\Delta t}{\rho g D}\tau_{w,i-1}^{n} = C_m \tag{17}$$

where the coefficients C_p and C_m are written as

$$C_p = H_{i-1}^n + \frac{a}{gA}Q_{i-1}^n - \frac{4a\Delta t}{\rho g D}\tau_{w,i-1}^n + \frac{2a^2\Delta t}{g}\sum_{k=1}^{N} V_E \tag{18}$$

$$C_m = H_{i+1}^n - \frac{a}{gA}Q_{i+1}^n + \frac{4a\Delta t}{\rho g D}\tau_{w,i+1}^n + \frac{2a^2\Delta t}{g}\sum_{k=1}^{N} V_E \tag{19}$$

$$V_E = \frac{\alpha D}{2e}\gamma\frac{J_k}{\Delta t}e^{-\Delta t/\tau_k}H_0 + \frac{\alpha D}{2e}\gamma\frac{J_k}{\Delta t}(1 - 2e^{-\Delta t/\tau_k})H(x, t - \Delta t) + \frac{e^{-\Delta t/\tau_k}}{\tau_k}\varepsilon_{rk}(x, t - \Delta t) \tag{20}$$

By combining Equations (16) and (17), the pressure head, discharge, and retarded strain at $(n + 1)\triangle t$ can be obtained as follows:

$$H_i^{n+1} = \frac{C_p + C_m}{2(1 + \frac{2a^2\Delta t}{g}\sum_{k=1}^{N}\frac{\alpha D}{2e}\gamma\frac{J_k}{\Delta t}e^{-\Delta t/\tau_k})} \tag{21}$$

$$Q_i^{n+1} = \frac{gA}{2a}(C_p - C_m) \tag{22}$$

$$\varepsilon_{rki}^{n+1} = \gamma\tau_k\left\{H_i^{n+1}[\frac{\alpha D}{2e}\frac{J_k}{\tau_k}e^{-\frac{\Delta t}{\tau_k}} - \frac{\alpha D}{2e}\frac{J_k}{\Delta t}(1 - e^{\frac{\Delta t}{\tau_k}})] + H_i^n[\frac{\alpha D}{2e}\frac{J_k}{\Delta t}(1 - e^{\frac{\Delta t}{\tau_k}}) - \frac{\alpha D}{2e}\frac{J_k}{\tau_k}e^{-\frac{\Delta t}{\tau_k}}] + H_0\frac{\alpha D}{2e}\frac{J_k}{\tau_k}e^{-\frac{\Delta t}{\tau_k}}(1 - e^{\frac{\Delta t}{\tau_k}}) + \frac{e^{-\frac{\Delta t}{\tau_k}}}{\tau_k}\varepsilon_{rki}^n\right\} \tag{23}$$

3. Simulation Results and Accuracy Evaluation

An experimental facility comprising a typical reservoir-pipe-valve system was considered to verify the effects of viscoelasticity friction on the pressure head under transient flow at different water temperatures. The experimental facility is composed of three parts: a constant pressure tank, an LDPE pipe, and an end quick-closing valve, as shown in Figure 3. The total length of the pipeline is 43.1 m, the inner diameter is 41.6 mm, and the wall thickness is 4.2 mm. Both ends of the pipe section are anchored using fixed brackets. The volume of the pressure vessel is 9 m^3. The vessel is equipped with heating and water temperature control. The transient events are caused by the closure of the downstream quick valve, and the valve closing time is 12 ms. The experimental data were reported by Gally et al. [29,33] at four different water temperatures, as summarized in Table 1. The temperature refers to the water temperature in the pipes; the initial velocity is the average velocity of the section under steady flow conditions; the reservoir pressure is the piezometric head of a constant pressure tank in the steady-state condition; the Reynolds number is a dimensionless value that determines the flow state of the fluid in the pipes; and the Darcy–Weisbach friction factor is a coefficient that is used to calculate the on-way resistance along the pipeline. In the numerical simulations of transient flows, the pipeline is divided into 64 elements of equal length and the piezometric head is calculated by MATLAB\2014b. The pressure wave speeds obtained by calibration in Cases 1, 2, and 3 are 370–390 m/s, and the corresponding values in Case 4 are 450–470 m/s.

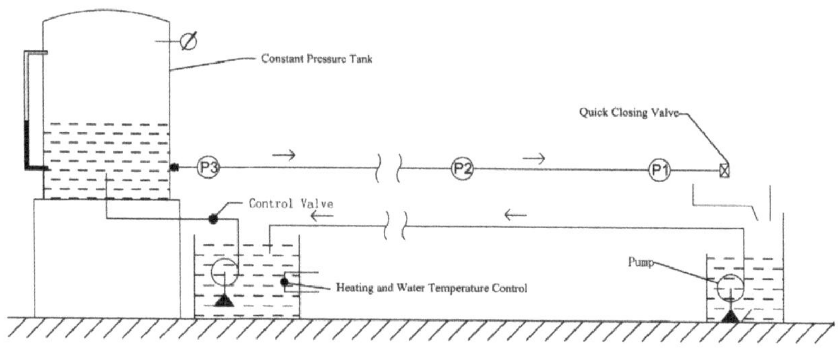

Figure 3. Schematic diagram of the experimental setup [29].

Table 1. Experimental data for different cases [29].

Test	Temperature (°C)	Reservoir Pressure ($\times 10^5$ Pa)	Initial Velocity (m/s)	Pressure Wave Speeds (m/s)	Water Kinematic Viscosity ($\times 10^{-6}$ m^2/s)	Reynolds Number	Water Bulk Modulus ($\times 10^9$ Pa)	Water Density (kg/m^3)	Darcy–Weisbach Friction Factor
Case 1	13.8	1.0593	0.49	379.4	1.17	17,422	2.14	999.3	0.0268
Case 2	25	1.0661	0.55	369.4	0.892	25,650	2.24	997.1	0.0244
Case 3	31	1.0670	0.57	381.2	0.784	30,245	2.27	995.3	0.0234
Case 4	38.5	1.0649	0.56	470.9	0.675	34,513	2.295	992.6	0.0227

The creep compliance, $J(t)$, is calculated using dynamic tests performed on a Rheovibron apparatus [29,33]. The results for this creep parameter in LDPE at different water temperatures are listed in Table 2. J_k and τ_k are the coefficients of the constitutive equation of the viscoelastic pipe (i.e., Equation (9)).

Table 2. Creep function coefficients in different cases [29].

Test	T(°C)	J_0 10^{-9} Pa^{-1}	J_1 10^{-9} Pa^{-1}	J_2 10^{-9} Pa^{-1}	J_3 10^{-9} Pa^{-1}	τ_{k1} 10^{-4} s	τ_{k2}(s)	τ_{k3}(s)
Case 1	13.8	1.414	0.516	0.637	0.871	0.56	0.0166	1.747
Case 2	25	1.542	0.754	1.046	1.237	0.89	0.0222	1.864
Case 3	31	1.791	1.009	1.397	1.628	1.15	0.0221	1.822
Case 4	38.5	2.239	1.479	2.097	3.57	1.24	0.0347	3.077

3.1. Simulation Results for the Pressure Head

Four experimental cases [33] were analyzed, and the main difference between these cases was the water temperature in the transient flow. The water temperature affects not only the density, but also the kinematic viscosity. To analyze the accuracy of the different friction models, numerical results were calculated using the QF, no-friction (NF), and UF models.

Figure 4 presents a comparison of the experimental and simulation results at four water temperatures. The QF, NF, and UF models have strong similarities in terms of the period and phase of pressure fluctuation. However, with respect to the peak pressure values, there are some distinctions between the different friction models at the four water temperatures. At 13.8 °C, the peak pressures calculated by the models are higher than the experimental data. At 25, 31, and 38.5 °C, all of the friction models can predict the peak pressure damping of the transient flow. However, for the maximum value in terms of the peak pressure, i.e., 38.5 °C (Figure 4), the difference between the simulation and experimental results is greater than that in the other three cases. The periods of the pressure curves have clear distinctions under different conditions, and these differences are greatest at 38.5 °C.

For the peak pressures at different periods, the degree of matching between the simulation and experimental results is not perfect. As we can see from Figure 4a, the simulated trough of pressure fluctuation is lower than the experimental pressure because of the inaccuracy of the coefficients of the creep function. We will discuss this in the following section. However, the pressure attenuation of the different friction models is similar above 13.8 °C.

3.2. Simulation Results for the Strain

To further investigate the performance of different friction models at different water temperatures, the retarded strain based on Equations (13) and (23) was simulated. Owing to the large differences in pressure between the simulated and experimental results at 13.8 °C and 25 °C, the variations in the total strain and retarded strain simulated by different friction models are shown in Figure 5. It can be seen that at 13.8 °C and 25 °C, the attenuation of the total strain is greater than that of the retarded strain. That is because the total strain is the sum of the instantaneous strain and retarded strain. When the peak pressure rapidly decays, it results in a dramatic decrease in instantaneous strain. The total and retarded strains obtained by the QF model are higher than those obtained by the UF model. This change is most pronounced at the beginning of the transient flow. However, in the latter stages of the transient flow, the strains of the QF and NF models gradually approach each other. This is also consistent with the conclusions reported by Duan [34]. In addition, it can be observed that the maximum strains of the QF model are larger than those of the UF model. In particular, for each case in Figure 5, the total strain and retarded strain exhibit a greater delay at higher water temperatures.

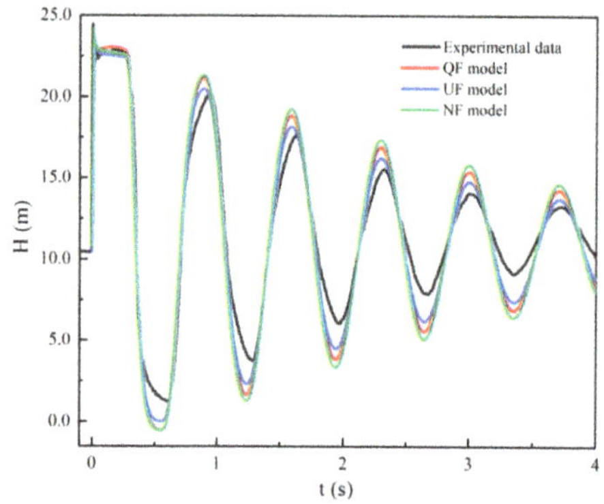

(**a**) In the case of 13.8 °C

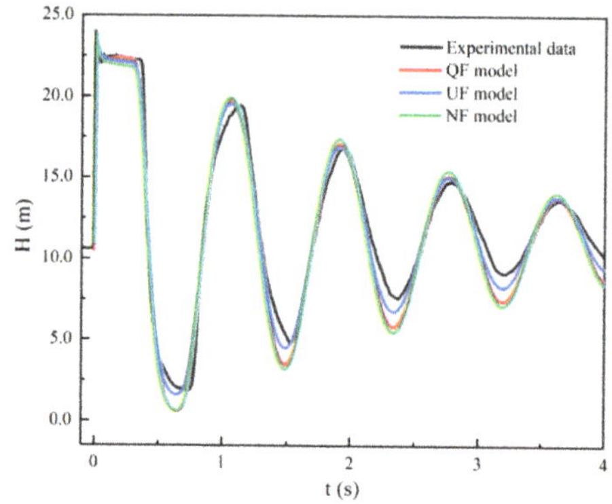

(**b**) In the case of 25 °C

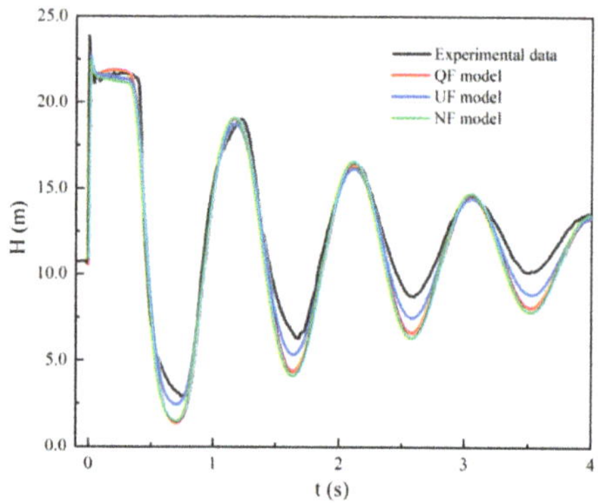

(**c**) In the case of 31 °C

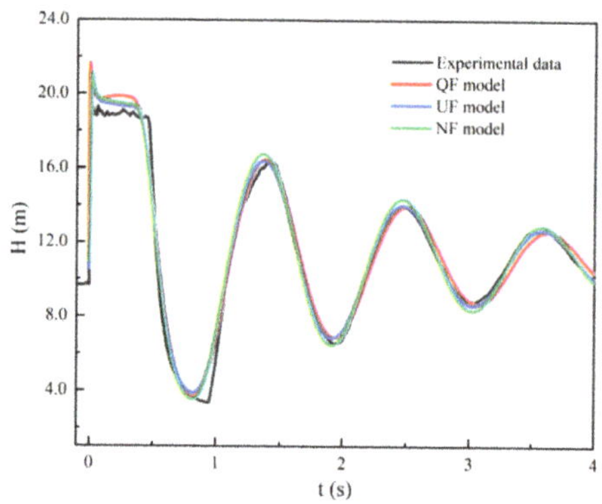

(**d**) In the case of 38.5 °C

Figure 4. Pressure heads calculated by different friction models at four water temperatures: (**a**) 13.8 °C; (**b**) 25 °C; (**c**) 31 °C; (**d**) 38.5 °C.

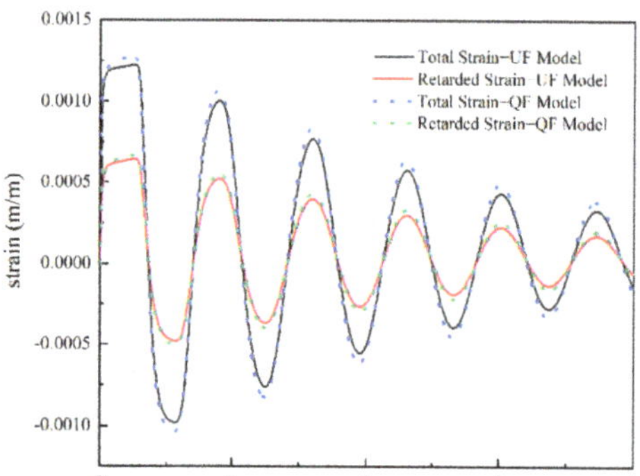

(**a**) In the case of 13.8 °C

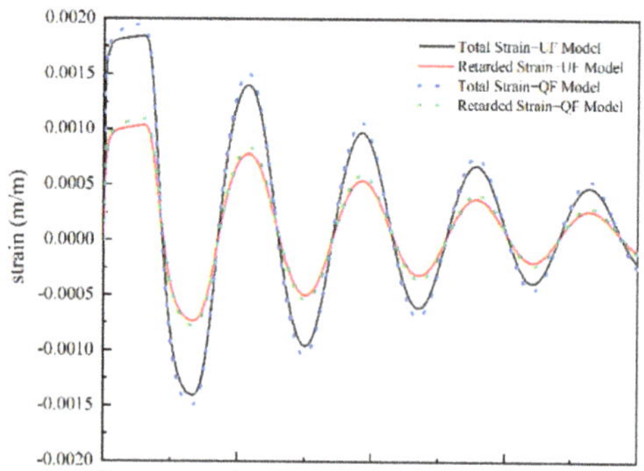

(**b**) In the case of 25 °C

Figure 5. Comparison of total and retarded strains at different water temperatures: (**a**) 13.8 °C; (**b**) 25 °C.

3.3. *Accuracy of the Numerical Results*

3.3.1. Mean Absolute Percentage Error

In Figure 4, the QF and NF models are very similar above 25 °C, particularly in the description of the pressure peaks and phases. The MAPE reflects the accuracy of the peak pressure damping simulations. To assess the effects of the friction of transient flows at different water temperatures, the MAPE evaluation method was carried out as follows:

$$\mathrm{MAPE} = \frac{\sum_{i=1}^{k} \frac{\left|H_{i,num} - H_{i,\exp}\right|}{H_{i,\exp}}}{k} \times 100\% \quad (24)$$

where $H_{i,num}$ is the ith simulated peak and valley pressure, $H_{i,\exp}$ is the ith experimental peak and valley pressure, and k is the number of pressure extremums.

Figure 6 shows the differences in the peak pressure damping calculated by the QF and UF models for the four cases. The percentage of deviation between the peak pressure calculated by the UF model and the corresponding experimental value at 13.8 °C is approximately 22.94%; the percentage of corresponding difference in the QF model is 31.97%. Similarly, the difference in MAPE values calculated using the 1D friction models at 25 and 31 °C is approximately 10% (i.e., at 25 °C, QF model: 18.54%, UF model: 7.59%; at 31 °C, QF model: 17.12%, UF model: 9.42%). However, at 38.5 °C, there was a slight deviation (i.e., QF model: 3.14% and UF model: 3.6%).

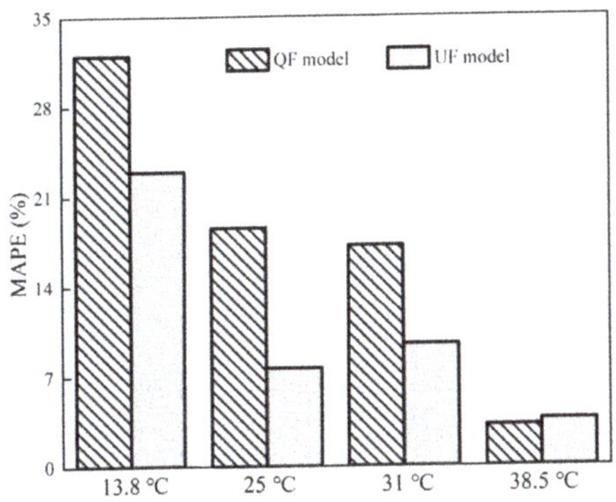

(a)

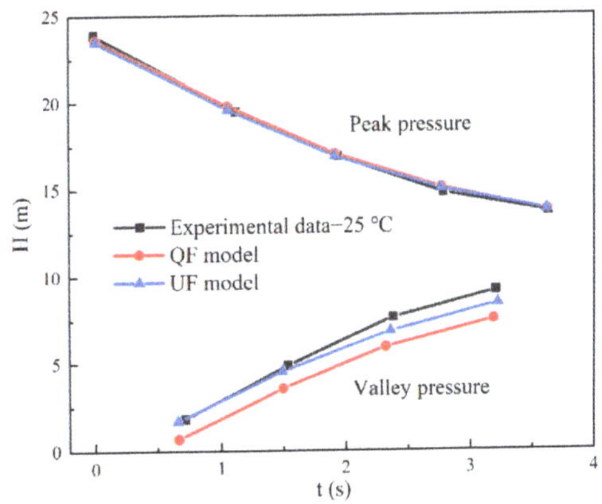

(b)

Figure 6. Comparison of the peak pressure damping results computed using the QF and UF models: (**a**) MAPE values; (**b**) peak pressure damping at 25 °C.

Combined with the pressure curve in Figure 4, the MAPE values were analyzed. At 13.8 °C, the peak and valley values of pressure fluctuation obtained by the QF and UF models display large differences relative to the experimental ones. Thus, these differences result in a large MAPE value. At 25 °C, there is no significant difference between the pressure peak values of the two models and the experimental pressure peak values. However, with respect to the pressure valley value, the simulation results of the UF model are closer to the experimental results than those of the QF model. This is the reason that the MAPE value of the QF model is greater than that of the UF model, as shown in Figure 6b. The variability of pressure peak and valley values at 31 °C is similar to that at 25 °C. Thus, from the simulation accuracy in terms of pressure peak attenuation (Figure 4b,c), the MAPE value is acceptable, with it being in the range of 20%. This illustrates that the unsteady-state component contributes to a lesser degree to the difference in the numerical calculation of the pressure at 38.5 °C.

Figure 6b also shows a comparison of the peak pressure damping results simulated by the QF and UF models at 25 °C. The calculation results of the QF and UF models are coincident; however, they are higher than the experimental results. This means that the unsteady-state component has little effect on the peak pressure damping.

On the other hand, the creep compliance coefficients and retarded times of "creep function" have also been shown to affect the peak and valley pressure of transient flow in viscoelastic pipes [35]. We studied the modification of the creep curve and pressure curve when the coefficient varies at 25 °C, referring to the method proposed by Urbanowicz [35]. The detailed values of the quantitative analysis are summarized in Table 3.

Table 3. Tested creep function details.

Case	J_0 10^{-9} Pa^{-1}	J_1 10^{-9} Pa^{-1}	J_2 10^{-9} Pa^{-1}	J_3 10^{-9} Pa^{-1}	τ_{k1} 10^{-4} s	τ_{k2}(s)	τ_{k3}(s)
00	1.542	0.754	1.046	1.237	0.89	0.0222	1.864
01	1.542	**0.9802**	**1.3598**	**1.6081**	0.89	0.0222	1.864
02	1.542	**0.5278**	**0.7322**	**0.8659**	0.89	0.0222	1.864
03	1.542	0.754	1.046	1.237	**0.089**	**0.00222**	**0.1864**
04	1.542	0.754	1.046	1.237	**8.9**	**0.222**	**18.64**

The Two Tests were analyzed in detail as follows:

Test A: There were variations in term of J_k values only and constancy in terms of the τ_k values as well as initial J_0 values. The modified function coefficients J_k in case 01 were obtained by multiplying the corresponding coefficients in case 00 by 1.3, and the function coefficients J_k in case 02 were multiplied by the corresponding coefficients in case 00 by 0.7.

Test B: There was an increase and decrease in terms of τ_k values and constancy in terms of the J_k values. To obtain function coefficients in case 03, the retarded time, τ_k, of function coefficients in case 00 was multiplied by the value 0.1. On the contrary, the function coefficients in case 04 were obtained by multiplying the initial τ_k of function coefficients in case 00 by 10.

Figure 7 shows the simulation results for Tests A and B, which were obtained by modifying the function coefficients in cases 01, 02, 03, and 04 at 25 °C. The simulation pressure in case 02 has lower valley values and higher peak values compared to that in case 01. In contrast with cases 01 and 02, the pressure decays faster in cases 03 and 04. From the obtained results (Figure 7), it can be seen that the variations in creep compliance coefficients and retarded time result in a mismatch between the numerical simulation and experimental results.

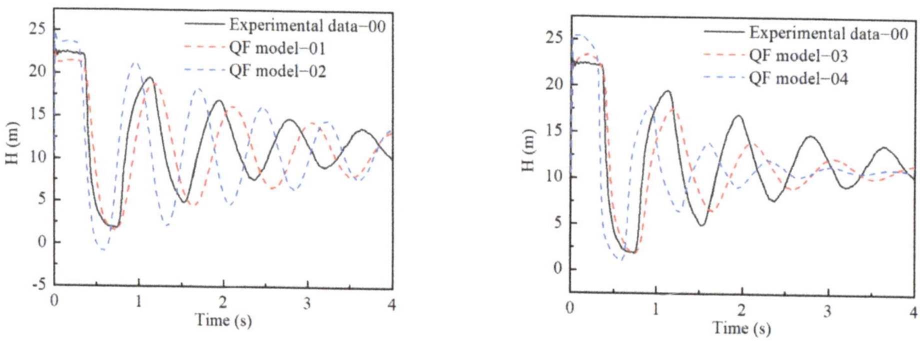

(a) (b)

Figure 7. *Cont.*

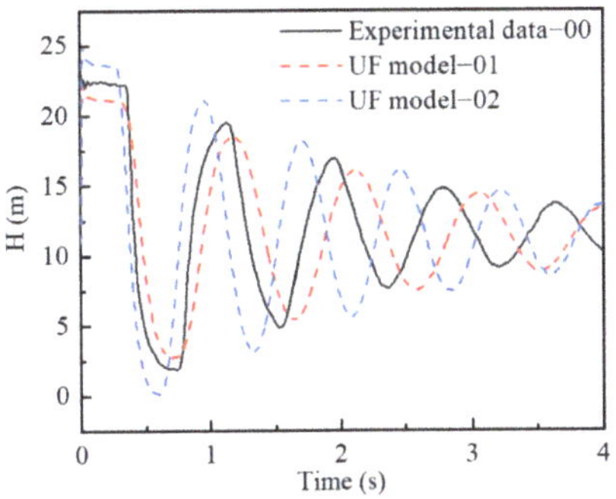

(c)

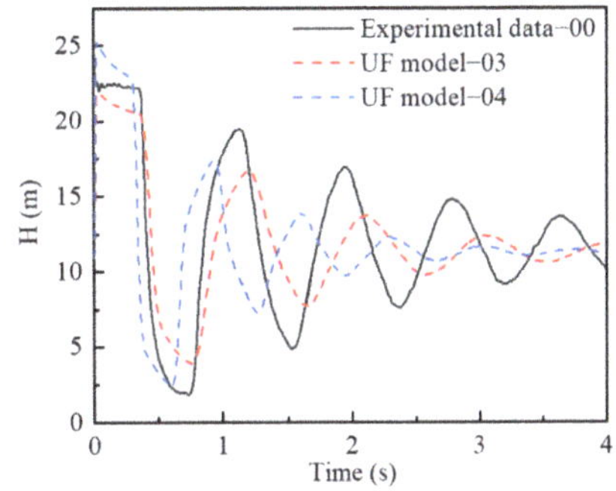

(d)

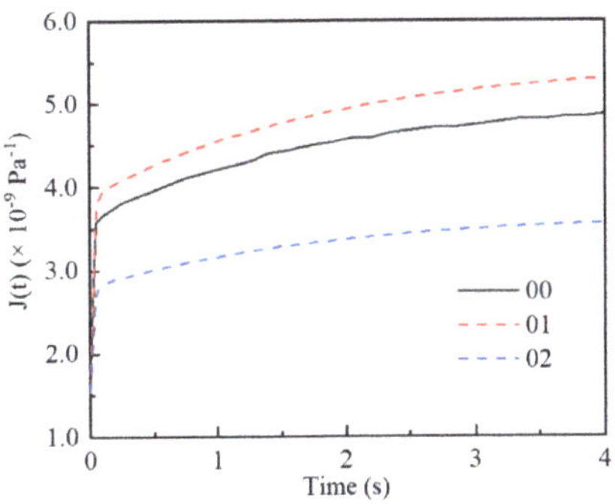

(e)

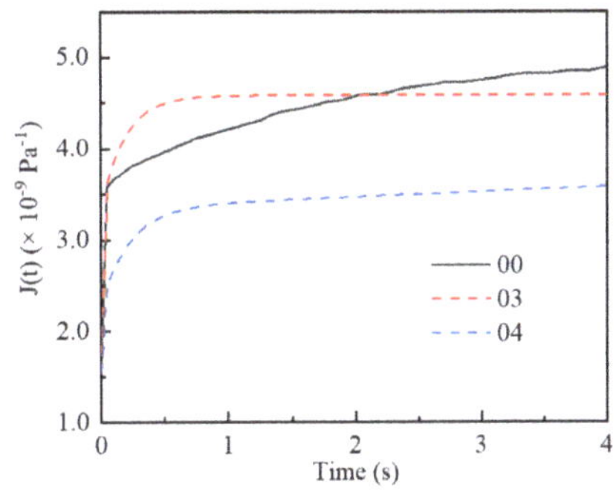

(f)

Figure 7. Pressure head and creep function: (**a**) Test A of QF model; (**b**) Test B of QF model; (**c**) Test A of UF model; (**d**) Test B of UF model; (**e**) Creep function of Test A; (**f**) Creep function of Test B.

3.3.2. Dimensionless Difference Value

To further analyze the variation in the peak pressure damping under different water temperatures, the dimensionless forms of the pressure and time was used. The dimensionless pressure head was defined as $H^* = (H - H_0)g/aV_0$, and the dimensionless time was expressed as $t^* = at/L$.

The difference evaluation method facilitates further analysis of the peak pressure and averages over a half-cycle period. The dimensionless pressure curve is divided by the centerline of the dimensionless pressure $(H^* = 0)$ into a few half-cycles, and the interval of dimensionless time within each period is $|t^*|$ $(t_i^* \le t^* < t_{i+1}^*, i = 1, \cdots, m)$. The percentage dimensionless difference value, E, is calculated as follows [36]:

$$E = \frac{1}{t_{m+1}^* - t_1^*} \sum_{i=1}^{m} W_i A_i \times 100\% \quad (25)$$

$$W_i = \max\left(\left|H_{i,num}^* - H_{i,\exp}^*\right|\right)_{t_i^*}^{t_{i+1}^*}; A_i = \int_{t_i^*}^{t_{i+1}^*} \left|H_{i,num}^* - H_{i,\exp}^*\right| dt^* \quad (26)$$

where $H_{i,num}^*$ is the ith simulated dimensionless peak and valley pressure, $H_{i,\exp}^*$ is the ith experimental dimensionless peak and valley pressure, m is the number of halfcycles, and $\max(\,)_{t_i^*}^{t_{i+1}^*}$ is the maximum value in the period $(t_i^* \le t^* < t_{i+1}^*, i = 1, \cdots, m)$.

Figure 8 presents a comparison of the dimensionless difference value, E, at different water temperatures. The different values in terms of E calculated by the QF and UF models are similar (approximately 0.027%) at 38.5 °C. At 13.8 °C, the maximum E values of the QF and UF models are due to the difference between the simulated pressure peak and valley values and the experimental results. At 31 °C, the larger E values of the QF model are caused by the difference between the simulated pressure valley value and the experimental results. The deviation may be caused by improper experimental operation, the external ambient temperature, or a specific historical accumulation of stress in the material itself. Therefore, the E values are acceptable, with them being in the range of 0.4%.

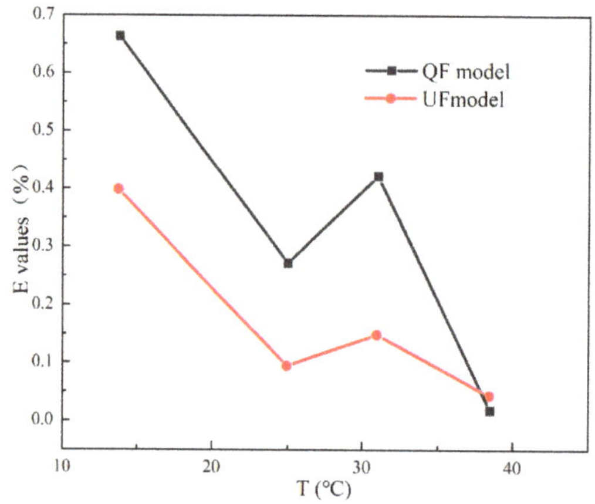

Figure 8. Comparison of dimensionless difference values at different water temperatures.

4. Analysis and Discussion

4.1. Energy Analysis of the Numerical Results

To quantitatively describe the contribution of viscoelasticity and friction to the peak pressure damping from the perspective of energy dissipation and transfer, energy analysis was used. According to the literature [36], the energy relation for the governing equations of transient flows can be expressed as follows:

$$\frac{dM}{dt} + \frac{dG}{dt} + D_f + W_R + W_L = 0 \tag{27}$$

where $M(t)$ is the total internal energy, $G(t)$ is the total kinetic energy, D_f is the total rate of frictional dissipation, W_R is the power on the pipe wall, and W_L is the power from the ends of the pipeline.

The total rate of frictional dissipation, D_f, power on the pipe wall, W_R, and the rest term of Equation (27) are calculated as follows:

$$D_f(t) = \frac{4}{D}\int_0^L \tau_w(x,t)Q(x,t)dx \tag{28}$$

$$W_R(t) = \int_0^L \rho g H(x,t) q_R(x,t)dx \tag{29}$$

$$M(t) = \frac{\rho g^2 A}{2a^2}\int_0^L H^2(x,t)dx \tag{30}$$

$$G(t) = \frac{\rho}{2A}\int_0^L Q^2(x,t)dx \tag{31}$$

$$W_L(t) = \rho g[H(L,t)Q(L,t) - H(0,t)Q(0,t)] \tag{32}$$

where $q_R = 2A\partial\varepsilon/\partial t$ is the radial flow per unit pipe length.

4.1.1. Analysis of Viscoelasticity and Friction

Figure 9 shows the numerical results for the work performed by the viscoelastic term and the friction dissipation rates of the UF model in the four cases. As shown in Figure 9, the rate of work on the pipe wall due to the viscoelastic effect shows a cyclical fluctuating trend, and it has both positive and negative values. When the transient pressure head increases, the fluid performs positive work on the pipe wall. At this time, part of the energy in the viscoelastic pipe is dissipated as heat energy, whereas the other part is stored in the pipe owing to the viscoelastic characteristics of the pipe. When the pressure head decreases, the pipe wall performs negative work on the fluid. These results are consistent with those of previous studies [37,38]. In particular, for each case shown in Figure 9, the rate of work on the pipe wall owing to the viscoelastic effect has a greater delay at higher water temperatures.

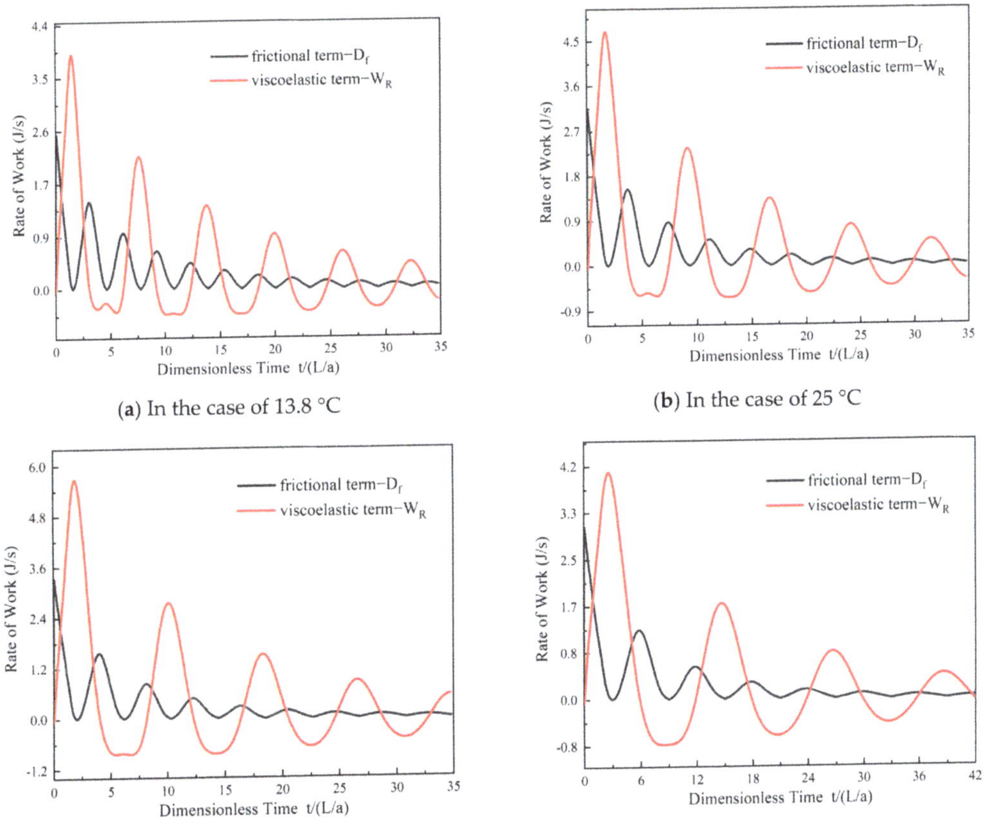

(**a**) In the case of 13.8 °C

(**b**) In the case of 25 °C

(**c**) In the case of 31 °C

(**d**) In the case of 38.5 °C

Figure 9. Energy results for the total rate of frictional dissipation and power on the pipe wall at different temperatures: (**a**) 13.8 °C; (**b**) 25 °C; (**c**) 31 °C; (**d**) 38.5 °C.

The work performed by frictional dissipation is always positive, and its value is reflected in the corresponding peak pressure damping. The work performed by the friction is larger than that performed by the viscoelasticity in the first stage of pressure fluctuation (dimensionless time = 0 to L/a), which implies that the maximum peak pressure of the transient flow (Figure 4) is mainly affected by friction. In the subsequent pressure fluctuation, the work performed by the viscoelasticity is larger than that performed by the friction.

4.1.2. Analysis of the QF and UF Models

Figure 10 shows the dissipation rates of the transient flow in the QF and UF models at 13.8, 25, 31, and 38.5 °C. The results show that in the first stage of the transient flow, there is no significant difference between the friction effects of the QF and UF models in the four cases. In the subsequent stages of the transient flow, there is a trend of rapid decay in the work performed in the results of the UF and QF models. The decay trend of the work performed in the results of the QF model is steeper than that of the UF model. This indicates that the UF component has a negligible influence on the peak pressure damping of the transient flow in viscoelastic pipes.

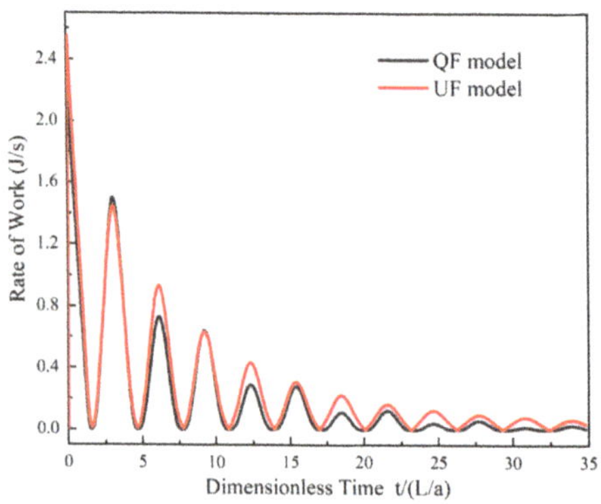

(**a**) In the case of 13.8 °C

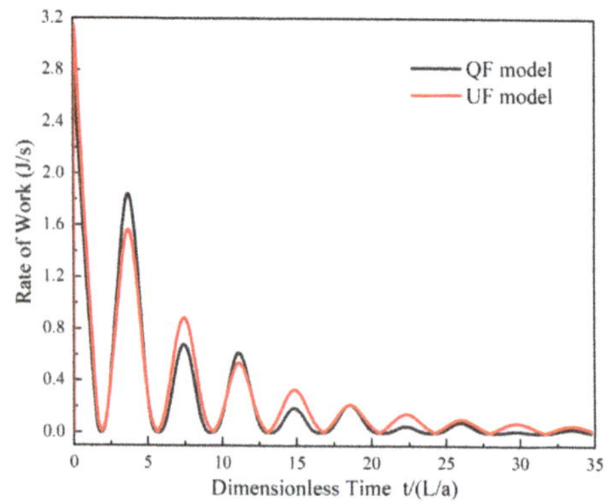

(**b**) In the case of 25 °C

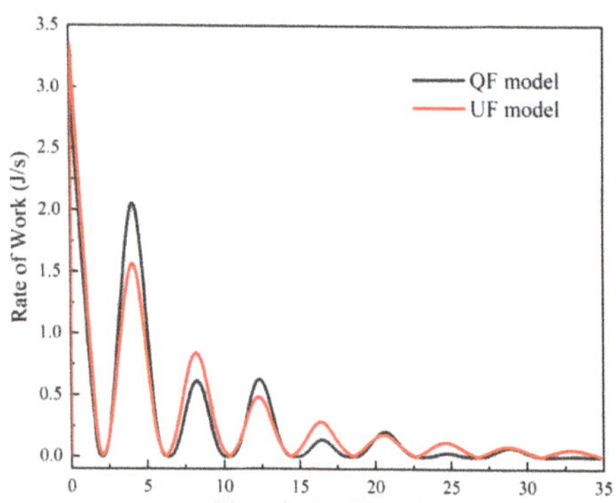

(**c**) In the case of 31 °C

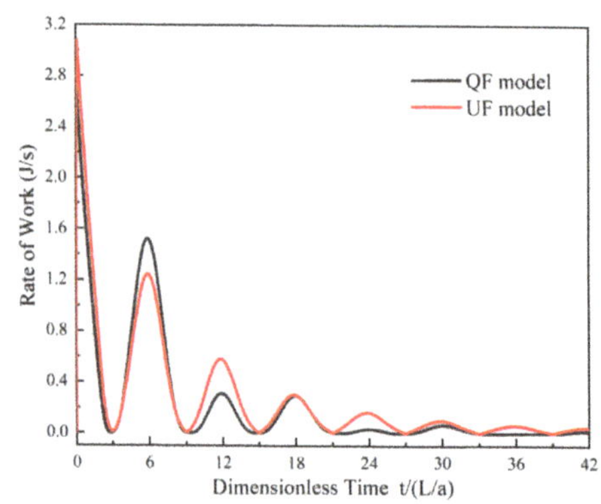

(**d**) In the case of 38.5 °C

Figure 10. Comparison of the dissipation rates of the QF and UF models: (**a**) 13.8 °C; (**b**) 25 °C; (**c**) 31 °C; (**d**) 38.5 °C.

4.2. Discussion of the Contribution of the Viscoelastic Component at Different Water Temperatures

In the following quantitative analysis, to distinguish the effect of viscoelasticity from the QF and UF components, the contributions of the QF, UF, and viscoelastic (VE) components to the pressure damping are defined as follows [19]:

$$\gamma_p = \frac{\Delta H_p^T - \Delta H_p^S}{\Delta H_p^T} \times 100\% \quad (33)$$

where γ is the contribution of the pressure damping, ΔH is the total pressure head attenuation relative to the initial transient pressure head, the superscript T indicates experimental results, superscript S indicates the simulated pressure, and subscript p indicates the point at the pressure peak.

Figure 11 shows the contributions of the QF + VE, UF + VE, and VE components of the transient flow model to the peak pressure damping in the four cases. As shown in Figure 10, the contribution to the peak pressure damping is the greatest in the case of the transient flow model with the UF and VE components. The minimal contribution to the peak pressure damping under transient flow occurs when only the VE component is considered. At 13.8 °C, the contribution of the UF component is greater than that of the QF component (with a difference of approximately 0.3%). However, in the other three cases, the contribution of the QF component gradually approaches the corresponding value of the UF component (with a difference of approximately 0.05%). This indicates that a higher water temperature weakens the effect of the UF component on the peak pressure damping. This is because the variation in water temperature affects the density and viscosity of the liquid, which influences the calculation of the unsteady friction near the pipe wall. On the other hand, as the water temperature increases, the creep function coefficients enlarge, which leads to extreme pressure damping [29,35].

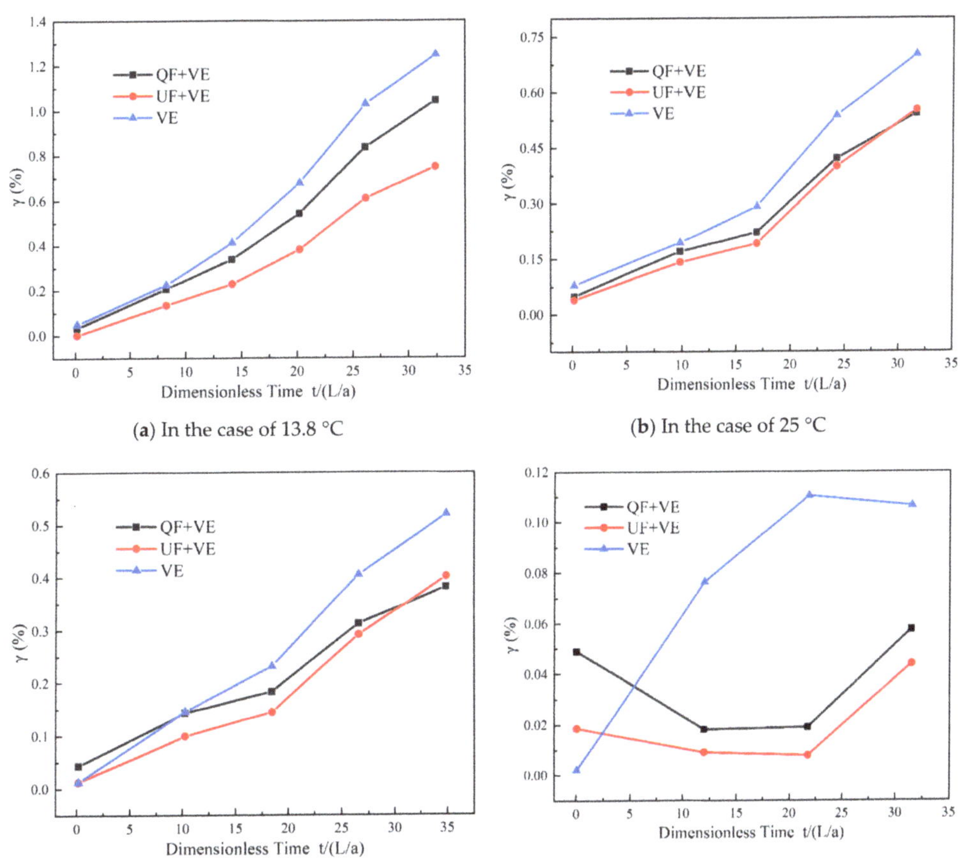

(**a**) In the case of 13.8 °C

(**b**) In the case of 25 °C

(**c**) In the case of 31 °C

(**d**) In the case of 38.5 °C

Figure 11. Comparison of the contributions of QF, UF, and VE components: (**a**) 13.8 °C; (**b**) 25 °C; (**c**) 31 °C; (**d**) 38.5 °C.

37. Cao, H.; Nistor, I.; Mohareb, M.E. Effect of boundary on water hammer wave attenuation and shape. *J. Hydraul. Eng.* **2020**, *146*, 04020001. [CrossRef]
38. Urbanowicz, K. Analytical expressions for effective weighting functions used during simulations of water hammer. *J. Theor. Appl. Mech.* **2017**, *55*, 1029–1040. [CrossRef]

materials

MDPI

Article

Material and Structural Functionalization of Knitted Fabrics for Sportswear

Ivana Salopek Čubrić [1], Vesna Marija Potočić Matković [1,*], Željka Pavlović [1] and Alenka Pavko Čuden [2]

[1] Department of Textile Design and Management, Faculty of Textile Technology, University of Zagreb, Prilaz baruna Filipovića 28a, HR-10000 Zagreb, Croatia; ivana.salopek@ttf.unizg.hr (I.S.Č.); zeljka.pavlovic@ttf.unizg.hr (Ž.P.)

[2] Department of Textiles, Graphic Arts and Design, Faculty of Natural Sciences and Engineering, University of Ljubljana, Snežniška 5, SI-1000 Ljubljana, Slovenia; alenka.cuden@ntf.uni-lj.si

* Correspondence: marija.potocic@ttf.unizg.hr; Tel.:+385-01-3712-573

Abstract: Comfort is an important quality criterion, especially for sportswear. It influences well-being, performance and efficiency. The necessary dissipation of heat and air flow, at high metabolic rates, must be designed and planned in advance. The influence of structure, density, mass and thickness of fabric were considered as well as yarn material composition, yarn linear density, yarn evenness and yarn hairiness. The influence of the mentioned parameters on thermal properties and air permeability was calculated. From the correlation analysis, it can be concluded that yarn's linear density, yarn short fibers hairiness, and mass per unit area of knitted fabric has the greatest impact on heat resistance. The yarn linear density, the yarn hairiness of the longer protruding fibers, and the thickness of the knitted fabric have the greatest impact on air permeability. A statistically significant model of multiple linear regression equations was offered to predict the thermal comfort of knitted fabric.

Keywords: sportswear; knitted fabric; thermal properties; air permeability; comfort

Citation: Salopek Čubrić, I.; Potočić Matković, V.M.; Pavlović, Ž.; Pavko Čuden, A. Material and Structural Functionalization of Knitted Fabrics for Sportswear. *Materials* **2022**, *15*, 3306. https://doi.org/10.3390/ma15093306

Academic Editors: Agnieszka Kijo-Kleczkowska and Adam Gnatowski

Received: 25 March 2022
Accepted: 29 April 2022
Published: 5 May 2022

1. Introduction

Knitwear has always been associated with relaxed clothing and fashion. Now synonymous with the American concept of simple and relaxed casual clothing, sportswear developed as a strong element of the twentieth-century knitwear trade. The simple sweater became emblematic of certain sports like the cricket, golf or sweater [1]. In the Cambridge dictionary, sportswear is defined as "informal clothing designed for comfort" [2], while in Merriam-Webster it is broadly defined as "clothing designed for casual or informal wear" [3]. Recently, a term "athleisure" has also been used, merging the terms "athlete" and "leisure", meaning "casual clothing designed to be worn both for exercising and for general use" [4] and "a style of clothing that is comfortable and suitable for doing sports, but also fashionable and attractive enough to wear for other activities" [5]. Therefore, the crossover between sport and leisure has been blurred, as sport and leisure are essentially about fun and well-being [6], which is reflected in the terminology as well as in the design and marketing of multifunctional and multipurpose sportswear or athleisure wear.

Today, the blurring boundaries between sportswear and urban wear, and a large number of luxury and high-end ready-to-wear brands developing sports-inspired ranges, are a growing source of novel design orientations for performance clothing. Sports-inspired garments are now standard everyday wear, especially among younger generations, defining what consumers will wear in the future. The interactions between these two areas of influence, sports and fashion, are an important source of new trends in both industries. As for sports companies, a large part of their business is selling after-sports apparel to their core clientele. They are also trying to attract a broader clientele that is looking for a sporty silhouette but not necessarily engaged in a specific sporting activity [6].

The reason for the merging of sportswear and casual or urban fashion could be the dramatic increase in participation in sports activities and interest in health and fitness in the last three decades of the twentieth century, which has expanded the market for sports-specific apparel. The global sportswear market is expected to grow at a CAGR of 5.8% over the next five years. At the same time, the share of sports apparel in amateur sports is around 85%. Performance sportswear has become increasingly sophisticated in terms of styling and detailing, benefiting from rapid developments in fibre and fabric technology and modern garment construction methods. These influences have been adopted in products ranging from extreme sports to related market sectors such as adventure travel, corporate wear and health. It can be said that sportswear is becoming a very popular style statement and fashion trend [6,7].

Sportswear has, in a sense, come full circle from the influence of sportswear on high fashion and streetwear, to the influence of streetwear on the design and marketing of the collections of sportswear companies. Today, for example, many major ski-wear brands have launched their own urban collections, adding a Nordic high-fashion twist to technical outdoor designs [8]. People like to wear sportswear, which means they want to feel comfortable in it.

Comfort is an important quality criterion. It influences not only the wearer's well-being but also their performance and efficiency. Comfort is also an important sales aspect [6]. However, it is very complex and difficult to define. According to Fourt and Hollies [9], comfort includes thermal and non-thermal components and is related to wearing situations such as work, non-critical and critical conditions. The physiological responses of the human body to a given combination of clothing and environmental conditions are predictable when the system reaches a steady state. According to Slater [10], comfort is a pleasant state of physiological, psychological, neurophysiological and physical harmony between the human and the environment.

Modern knitwear encompasses a wide range of products, from functional and protective workwear and activewear for sports or other physical activities that border on technical textiles in their properties, to sidewalk and high-fashion knitwear with distinct esthetic value and many intermediate variations. In all of them, thermal comfort is a very important property that must be designed and planned in advance. This can be achieved through the material, mechanical and chemical functionalization of the knitted structure and style.

2. Thermal Comfort of Knitted Sportswear

Thermal comfort is the state of mind that expresses satisfaction with the thermal environment, which means that a person feels neither too cold nor too hot [11]. As an interface between the human body and the environment, textiles play an important role in the heat exchange between the body and the environment [12]. The contribution of clothing to thermal comfort is to assist the human body in maintaining comfortable thermophysiological conditions in a wide range of environments. The thermophysiological comfort of clothing is therefore the ability of clothing to assist the human body in maintaining thermal comfort [13]. From a functional point of view, comfort includes a suitable shape and fit that provides freedom of movement, thermal regulation with moisture-wicking and quick-drying properties, and lightweight protection and durability. A suitable mix of all these aspects contributes to an overall "feel good factor" for the wearer [14].

The fabrics for activewear and sportswear are specially designed to achieve the necessary dissipation of heat and moisture at high metabolic rates, both in terms of geometry, packing density and the structure of the constituent fibres in yarns as well as in terms of the structure of the fabric [6].

The thermal insulation of a fabric is determined as its barrier function or resistance to heat transport. Therefore, insulation is physically defined as the thermal resistance of the fabric. At a steady state, there is a temperature difference between the two sides of the fabric and a defined heat flow through the fabric. Thus, the thermal resistance, usually referred to as R_{ct} (in $m^2\ KW^{-1}$), can be determined using the following equation:

$$R_{ct} = \frac{(T_s - T_a)}{Q} \cdot A \tag{1}$$

where T_s is the temperature at the skin surface (in K); T_a is the temperature of the environment (in K); Q is the heat flux through the fabric (in W); A is the area of the fabric (in m^2).

The thermal properties of clothing, especially high-performance, athletic, or undergarments, are critical to the comfort of the wearer. Many researchers have studied the relationship between thermal resistance and the structural characteristics of fabrics, such as structure/texture, thickness, mass per unit area, thread or loop density in woven and knitted fabrics, loop length in knitted fabrics, etc., and have focused on the association of thermal properties with the properties of specific structures, yarns and fibres.

The influence of different knitted structures, namely single jersey, 1 × 1 rib and interlock, was investigated by Oglakcioglu and Marmarali. The results showed that double jersey structures may be preferred for winter garments to protect against cold due to their high thermal insulation values. 1 × 1 rib fabrics are still more commonly used due to their warmer feel on first contact. On the other hand, single jersey structures should be chosen for active sports or summer clothing due to their better moisture management properties [15]. Amber et al. investigated the thermal properties and moisture transport of socks knitted in single jersey, half terry and terry structures that differed in fibre type and yarn structure. The results indicate that, apart from the effects on regain, the fibre effects are generally small and probably masked by the effects of fabric structure [16]. Uçar and Yilmaz analysed the natural and forced convective heat transfer properties of 1 × 1, 2 × 2 and 3 × 3 rib knitted fabrics from acrylic yarns. It was found that a decrease in the number of ribs led to a decrease in heat loss due to an increase in the amount of air trapped between the front and reverse loops. The results also showed that as the tightness of the fabric increases, the heat loss decreases due to lower air permeability [17]. Erdumlu and Saricam investigated the thermal comfort properties of flat knitted fabrics with different structures, tightness, thickness and porosity made of acrylic fibres specifically designed for winter clothing. Thermoregulatory properties, breathability and thermophysiological properties and their relationship with the fabric's structural parameters were investigated. The results showed that 2 × 2 rib structures provide the optimal conditions in terms of thermoregulation, breathability and thermophysiological comfort. The thickness and porosity of the fabrics should be adjusted accordingly, as thickness improves thermal insulation and porosity improves breathability [18]. The effect of position and number of tuck loops on the thermal comfort of knitted fabrics was studied by Senthilkumar and Suganthi. It was found that the greater the distance between successive tuck points, the better the air, heat and moisture transfer properties. Bi-layer knitted fabrics with loose structure allowed less thickness and mass per unit area and exhibited better thermal comfort properties [19]. Yang et al. studied bi-layer knitted structures made from filament yarns on a circular knitting machine. It was found that bi-layer knitted fabrics with meshes on one side had better air permeability, moisture management, drying performance and thermo-physiological properties than bi-layer knitted fabrics with trim and symmetrical structure without meshes. Fabrics composed of yarns with finer fibers exhibited better thermal comfort properties [20].

In addition to the structure itself, researchers have also focused on investigating other properties of knitwear that may be related to thermal properties. Ozdil et al. evaluated the influence of yarn count, twist coefficient, combing process and tightness on the thermal properties of 1 × 1 cotton rib fabrics. It was found that with the increase in yarn twist, the thermal absorptivity and water vapour permeability of the fabrics also increased. Higher thermal absorptivity gives the fabric a cooler feel. Thermal resistance decreases as the yarn twist coefficient increases. The effect of yarn twist on fabric conductivity is insignificant. Thermal resistance is lower in fabrics made of combed yarns, while thermal conductivity, thermal absorbency and water vapour permeability are higher compared to fabrics made of carded yarns [21]. Prakash et al. studied the effect of loop length on thermal comfort properties of bamboo knitted fabrics and observed an increase in air and water vapour permeability with a decrease in yarn linear density and an increase in loop length [22]. Mishra et al. investigated the thermo-physiological comfort properties of single knit fabrics and their derivatives made from 100% cotton yarns at three different yarn linear densities and after different relaxation stages. It was found that the physical properties of the fabric were affected by the change in linear density of the yarn and by the dry or wet relaxation stages. The percentage or number of tuck stitches, the location of tuck stitches, and the ratio of tuck stitches to knit stitches have a strong influence on the physical and thermo physiological properties of single knitted fabrics, even when the other knitting parameters remained the same [23].

Other researchers investigated the influence of different fibre and yarn types on thermal comfort. Oglakcioglu et al. investigated the properties of single jersey knitted structures made with channeled and hollow polyester fibres. The results showed that channeled PES fabrics are suitable for hot climates and high physical activities because they have high air permeability, high moisture transfer and dry quickly. Air permeability and thermal properties are improved by combining lyocell yarn with polyester technical yarns. Combining Lyocell or cotton with technical yarns deteriorates the moisture management properties and increases the drying times [24]. Sampath et al. investigated the thermal comfort properties of microdenier polyester filament, filament polyester and spun polyester, blends of polyester and cotton, and 100% cotton. The test results showed that the finishing treatment for moisture management resulted in higher thermal conductivity and thermal absorptivity, lower wet thermal resistance and improved water vapour permeability for all the fabrics studied. It was observed that microdenier polyester fabrics offered faster heat transfer, faster evaporation of sweat from the skin through the fabric and also a cooler feel at first touch [25]. Gericke and Pol investigated the selected comfort properties of cotton, regenerated bamboo and viscose rayon. No empirical evidence was found that regenerated bamboo fibres are superior to cotton and viscose fibres in terms of thermal and moisture management properties [26]. Gun investigated the thermal comfort properties of plain knitted fabrics made from modal viscose yarns with microfibers and conventional fibres. It was found that fabrics made of microfibers have lower air permeability, higher values of thermal conductivity, thermal absorptivity and maximum heat flux, and lower values of thermal resistance and thermal diffusivity compared to conventional fibres. Fabrics made of microfibers also provide a cooler feel compared to fabrics made of conventional fibres [27]. Lizák et al. focused on the thermal transport properties of polypropylene knitted fabrics. It was found that the rib structure is the most suitable for heat insulation due to the number of pores between the fibres and the low areal density of the fabric. It was demonstrated that heat loss is higher in a cardigan (tuck) structure [28]. In one of their studies on the thermal properties of fabrics made of natural and regenerated bamboo cellulose fibres, Majumdar et al. found that air and water vapour permeability increased but thermal conductivity decreased as the percentage of bamboo fibres increased [29]. Another study by Biviainyte et al. on double layer fabrics knitted from cotton, bamboo and four types of synthetic yarns found that fabrics knitted from a combination of cotton and synthetic yarns had a higher thermal conductivity coefficient than those knitted from a combination of bamboo and synthetic yarns. The knitted structure has the greatest influence on the thermal properties. Combined structures have a higher thermal conductivity coefficient than plain plated structures [30]. Jhanji et al. found that the thermal resistance and thermal conductivity of plated fabrics knitted with polyester yarn of different linear density in the inner layer decreased as the polyester yarn became finer. Similar trends were observed for fabrics knitted with viscose yarn of different linear density in the inner layer and cotton yarn of fixed linear density in the outer layer. It was also found that as the loop length of the fabric increases, the thermal resistance and thermal conductivity decrease, regardless of the fiber type of the inner layer, i.e., polyester, viscose or nylon [31]. Kaplan and Yilmaz investigated the thermal conductivity and air permeability of double-sided knitted fabrics combining functional yarns such as Thermosoft®, Nilit Heat®, Vi-loft® and wool with standard polyester (PET) and polypropylene (PP) in a false rib structure. The results show that filament yarns have lower areal density and thickness, while staple yarns have higher air permeability. Despite the general trend of using hydrophobic fibers for the inner layers of knitted fabrics, wool and other functional man-made fibers such as Thermosoft®, Heat®, and Viloft® can also be used for the inner surfaces that come into contact with the skin to improve sensory performance [32]. In their study, Kumar et al. investigated nine different types of knitted fabrics made from machine-spun yarns of Eri-silk, worsted wool, and bamboo fibres in three different knitted structures such as single jersey, single pique, and honeycomb. The thermo-physiological properties such as thermal conductivity, thermal resistance, thermal absorptivity, air permeability and water vapor permeability were investigated. It was found that the Eri-silk based knitted honeycomb structure was rated as better than the other samples [33].

The literature review shows that mainly knitted fabrics made of cotton [15,21,24–26,29–31], polyester [15,19,24,25,30–32] and regenerated cellulose [19,22,24,26,27,29–31] have been studied. A few studies were conducted with polyacrylonitrile [16–18], wool [16,32] and polypropylene [26,32], while individual studies were conducted with silk [33]. In terms of yarn parameters, the yarn linear density was mostly varied. In some cases, the influence of yarn twist [16,21] and plying of different yarns [22] were also studied. Single, rib and interlock knits produced on circular knitting machines have been studied most frequently [15–29,31]. In some cases, different loop lengths were designed for each knitted structure [17,18,22,27]. Thermal properties were most commonly tested using Alambeta and Permatest devices. Individual tests have been carried out based on specially designed devices [17,30].

3. Experimental

In order to design a multipurpose knitted structure suitable for indoor and outdoor work and leisure as well as sports activities, the aim of the research was to investigate the thermal and air permeability properties of different single and double knitted structures made from different yarns on a 12-gauge knitting machine to achieve the optimal material and structural functionalization of knitted fabrics. For the present study, yarns/materials and plain, miss and tuck knitted structures were selected, which, according to the literature reviewed, have not been studied in detail for thermophysiological comfort by other researchers according to the literature reviewed. In addition, flat knitting technology was chosen for the fabrication of the samples, which is suitable for both indoor and outdoor athleisure fabrication and is more flexible in terms of patterning and adjustment of structural parameters of knitted fabrics compared to circular knitting technology, which is mostly used for finer body-hugging athleisure production.

3.1. *Sample Preparation*

3.1.1. Yarn Selection

Commercially available materials most commonly used in the manufacture of knitted sportswear and casual outerwear were selected for the study: cotton, wool, PAN and a wool/PAN blend. In addition, PA filament was used as a possible material for a more porous knit for active sportswear.

Yarns were designated as follows: Y1–100% cotton; Y2–100% wool; Y3–100% PAN; Y4–50% Wool/50% PAN; Y5–100% PA.

3.1.2. Knitted Samples Preparation

The knitted fabrics were produced on a flat knitting machine, a Shima Seiki SES 122 RT, gauge 12E in five structures: plain single jersey, single miss, single tuck/single piqué, plain double jersey (all needles in function) and Milano rib under the same cam settings.

Knitted structures were designated as follows: SP-single plain; SM-single miss; ST-single tuck; DP-double plain; DM-double Milano. The combinations of yarn and structure designations were used for the knitting samples, i.e., cotton single jersey-Y1 SP. Samples of 100% cotton (Y1) and 100% PA filament (Y5) were knitted with double threads while samples of 100% wool (Y2), 100% PAN (Y3) and wool/PAN blend yarn (Y4) were knitted with single threads (Tables 1 and 2).

Table 1. Single and double knitted structures selected for the knitted samples.

	Single Structures			**DOUBLE Structures**	
Appearance front					
Appearance back					
Yarn path	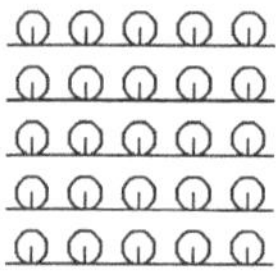	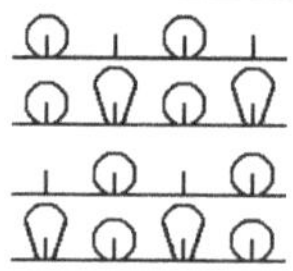	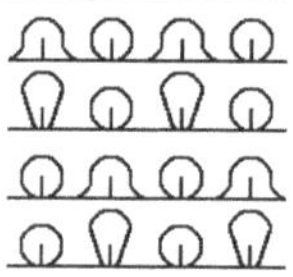	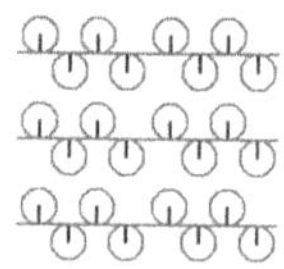	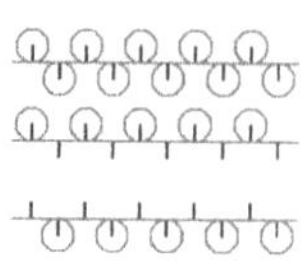
Structure	Single plain	Single miss	Single tuck	Double plain	Double Milano
Designation	SP	SM	ST	DP	DM

Table 2. Yarn designation and characteristics.

Yarn Designation	Yarn Material Composition	Yarn Linear Density (tex)	Yarn Twist (m^{-1})	Number of Threads for Knitting	Yarn Tenacity ($cNtex^{-1}$)	Yarn Elongation at Break (%)
Y1	100% Cotton	29.1	661	2	13.90	5.20
Y2	100% Wool	68.6	88	1	5.25	2.90
Y3	100% PAN (bouclé)	72.3	10	1	4.66	12.55
Y4	50% Wool 50% PAN	71.4	114	1	12.23	14.30
Y5	100% PA (filament)	15.4	-	2	40.76	301.27

After knitting, the samples were relaxed for 48 h under standard conditions before testing. The yarn paths and photographs of the tested knitted structures are shown in Table 1.

3.2. Testing Methods

3.2.1. Yarn Testing

Within the scope of the experiments presented in this manuscript, a number of yarn properties were tested: yarn linear density, twist, evenness, hairiness and tensile properties.

The yarn linear density was determined using the skein method, following the procedure described in ISO 2060 [34]. The number of twists was determined on a torsion meter-

Twist tester (Mesdan lab). The determination was conducted using the untwist/retwist method [35]. The coefficient of mass variation, a parameter that characterizes the yarn's unevenness, was measured using the Keisokki evenness tester 80, type B. During the measurement, the sensibility level was set to +50/+50/+200 (thin places/thick places/neps). The number of protruding hairs was measured using the Zweigl G 565 hairiness tester. For the measurement, the following lengths of protruding hairs were defined: 2, 4, 6 and 8 mm. The testing speed was set to 50 m/min. Finally, the yarn tensile properties (including breaking elongation and tenacity) were measured on the dynamometer Statimat M (Textechno). The measurements were performed following ISO 2062 and using method B: automatic. The yarn samples were taken directly from the conditioned packages [36].

The characteristics of the yarns are listed in Table 2: material composition, linear density, and twist, tenacity and elongation at break as well as the number of threads used to knit the samples on the 12-gauge knitting machine to achieve appropriate fabric density and feel.

3.2.2. Knitted Fabric Testing

The basic structural parameters of the knitted fabric affecting the thermal properties were determined as fabric thickness, horizontal density and vertical density, and mass per unit area. The fabric thickness was measured according to the standard ISO 5084 under 20 g/cm^2 pressure with 0.01 mm accuracy [37]. The horizontal and vertical densities (Dh and Dv) were determined by counting the loops over a length of 50 mm. The densities were then converted to the number of loops per cm. The mass per unit area was measured using the specimen sized at 100 $\times$ 100 mm and by weighting them on an analytic balance Kern ALJ 220-4.

A sweating guarded hotplate (MTNW, Seattle, DC, USA) was used to measure the fabric heat resistance and water vapour resistance of textiles. The device itself consists of a metal plate attached to a conductive block, a heating device, sensors, a temperature controller, and the appropriate software. Its structure allows the simulation of the processes occurring next to the human skin and is related to the transfer of heat and sweat [38–40]. For the measurement of sweat transfer, the semipermeable membrane is placed on the heated plate and the sample is placed on the membrane. In this way, the transfer of sweat from human skin through the placed material is simulated. In this experiment, the tests were performed in accordance with ISO 11092 [41]. For the measurements of material heat resistance, the air temperature was set at 20 °C, relative humidity at 65%, and air velocity at 1 ms^{-1}. For the measurement of water vapour resistance, the air temperature was set to 35 °C and the humidity to 40% of the relative humidity. In this case, the air velocity was also 1 ms^{-1}.

The air permeability of the samples was measured using the air permeability tester Air-Tronic B. The measurements were performed according to EN ISO SIST EN ISO 9237 [42] with a constant pressure drop of 50 Pa. For the samples knitted from yarns Y1 (100% cotton), Y2 (100% wool), Y3 (100% PAN) and Y4 (wool-PAN blend), the test area was 20 cm^2. For the sample knitted from Y5 (100% PA filament), the test area was 5 cm^2 due to its porosity.

4. Results

4.1. Results of Yarn Testing

The results of yarn linear density and yarn twist measurements are shown in Table 2. It can be seen that the PA filament yarn (Y5) differs significantly from the other (worsted) yarns, not only by its smooth appearance but also by a much lower linear density, resulting in much more porous knitted structures.

The results of the measurements of the yarn tensile properties, consisting of elongation at break and tenacity, are shown in Table 2. Among the worsted yarns, the wool/PAN blend yarn shows the higher elongation at break followed by PAN and cotton yarn. The wool yarn has the least elongation. The filament yarn PA is clearly different from all the worsted yarns as it shows extreme elongation at break of more than 300%. PA filament shows the highest tenacity, followed by a much lower tenacity of 100% cotton and wool/PAN blend yarn. 100% wool and 100% PAN yarn show the lowest tenacity.

The results of the measurement of yarn evenness and hairiness are shown in Table 3. Only the measurements for worsted yarns were carried out (samples Y1–Y4). The results show that the yarn made from wool/PAN blend has the highest hairiness, followed by yarn made from 100% wool and 100% PAN. The cotton yarn is significantly smoother. The yarn made of 100% wool is the most uneven, followed by the yarns made of PAN and cotton. The yarn from the wool/PAN blend has the highest evenness of all yarns.

Table 3. Yarn evenness and hairiness.

Yarn Designation	Yarn Material Composition	Yarn Evenness–CV_{mass} (%)	Yarn Hairiness			
			n1 (2 mm)	n2 (4 mm)	n3 (6 mm)	n4 (8 mm)
Y1	100% Cotton	11.75	952	147	23	5
Y2	100% Wool	19.04	1604	494	156	88
Y3	100% PAN (bouclé)	12.04	1350	718	346	283
Y4	50% Wool 50% PAN	9.47	1804	535	174	95
Y5	100% PA (filament)	-	-	-	-	-

The cotton yarn has a much higher twist (661 m^{-1}) than other worsted yarns (10–114 m^{-1}), which leads to a strong skewing of the single structures and could thus affect the usability of the cotton yarn for the production of knitwear.

4.2. Results of Knitted Fabric Testing: Structural Parameters

Table 4 shows the results of the measurements of the structural parameters of knitted samples, which are horizontal density, vertical density, mass per unit area and fabric thickness. As expected, the values of vertical density are higher than the values of horizontal density for all samples. Single tuck structure, i.e., the single piqué from each analysed yarn, has the lowest horizontal and vertical density of all the structures tested, which means that the single piqué structure is the most porous of all the structures, due to the tuck loop in every other wale and course.

As expected, the mass per unit area ranges from 92.68 for a single plain structure of PA yarn to 471.54 for double plain wool/PAN structure. Single structures have lower mass than double structures for all types of knitted fabrics, which is in line with expectations. The greatest thickness was measured for the plain double structure in all the materials analysed, followed by the Milano rib structure. Among the single structures, the thickest is the single tuck structure, i.e., single piqué, while the thinnest is the plain single structure for all materials (Table 4).

Fabric thickness has a great influence on thermal resistance: the thicker the fabric, the more thermally resistant. At this point it should be noted that the thickness of the knitted fabric was measured according to the standard ISO 5084 under a pressure of 20 g/cm^2, which means that the actual thickness of the unloaded knitted fabric is likely to be higher due to the voluminosity and compressibility of the examined knitted fabrics. On the other hand, the general trend in clothing is towards lighter end products [43]. When designing knitted sportswear or leisurewear, therefore, the choice of a single or double structure must be adapted to the intended use of the garment and the season or environmental conditions.

Table 4. Knitted samples designation and structural properties.

Material Composition	Fabric Designation	Knitted Structure	Horizontal Density (cm^{-1})	Vertical Density (cm^{-1})	Mass per Unit Area (gm^{-2})	Thickness (mm)
Y1 100% Cotton	Y1 SP	single plain	7.98	8.32	260.38	1.23
	Y1 SM	single miss	7.88	8.32	287.17	1.37
	Y1 ST	single tuck/single piqué	5.32	6.68	267,35	1.54
	Y1 DP	double plain	6.40	9.08	433.40	1.92
	Y1 DM	double Milano rib	7.26	8.82	430.49	1.80
Y2 100% Wool	Y2 SP	single plain	6.00	8.00	238.19	0.98
	Y2 SM	single miss	6.70	7.76	281.49	1.22
	Y2 ST	single tuck/single piqué	4.66	6.64	286.40	1.85
	Y2 DP	double plain	5.76	8.76	438.49	2.00
	Y2 DM	double Milano rib	6.54	8.42	453.85	1.77
Y3 100% PAN	Y3 SP	single plain	5.80	7.90	233.73	1.22
	Y3 SM	single miss	6.36	7.92	267.36	1.52
	Y3 ST	single tuck/single piqué	4.62	5.72	250.79	1.78
	Y3 DP	double plain	5.16	8.72	459.36	2.28
	Y3 DM	double Milano rib	6.08	8.64	453.15	2.01
Y4 Wool/PAN	Y4 SP	single plain	5.88	7.64	234.20	1.00
	Y4 SM	single miss	6.44	7.62	263.41	1.24
	Y4 ST	single tuck/single piqué	4.68	6.80	299.42	1.89
	Y4 DP	double plain	5.82	8.88	471.54	2.06
	Y4 DM	double Milano rib	6.44	8.08	455.06	1.76
Y5 100% PA	Y5 SP	single plain	5.52	8.20	92.68	0.58
	Y5 SM	single miss	5.88	6.36	132.51	0.77
	Y5 ST	single tuck/single piqué	3.88	3.86	132.51	1.40
	Y5 DP	double plain	5.36	5.88	231.54	1.70
	Y5 DM	double Milano rib	6.02	5.16	216.42	1.44

4.3. Results of Knitted Fabric Testing: Thermal Properties, Air Permeability, and Water Vapour Resistance

Table 5 and Figure 1 show the results of the heat resistance measurements of the knitted samples measured on the hot plate.

All knitted samples made of PA filament yarn show the lowest heat resistance, while knitted samples made of 100% wool, 100% PAN and wool/PAN blend show the highest heat resistance. There are also greater differences in heat resistance between different structures knitted from these yarns (Figure 1). The heat resistance values of the structures knitted from 100% cotton and 100% PA filament yarns, which at the same time have lower linear densities and were knitted with double threads, show smaller differences. For wool, PAN and wool/PAN blend yarns, the single tuck/single piqué, and double plain structures are the most heat resistant.

Table 5. Heat resistance of knitted samples.

Yarn/Structure		Heat Resistance (m^2 KW^{-1})				
		SP	SM	ST	DP	DM
		Single Plain	Single Miss	Single Tuck	Double Plain	Double Milano rib
Y1	100% Cotton	0.0444	0.0504	0.0424	0.0464	0.0419
Y2	100% Wool	0.0584	0.0694	0.0834	0.0824	0.0704
Y3	100% PAN	0.0564	0.0614	0.0844	0.0714	0.0614
Y4	Wool/PAN	0.0754	0.0634	0.0854	0.0812	0.0724
Y5	100% PA	0.0122	0.0160	0.0172	0.0123	0.0162

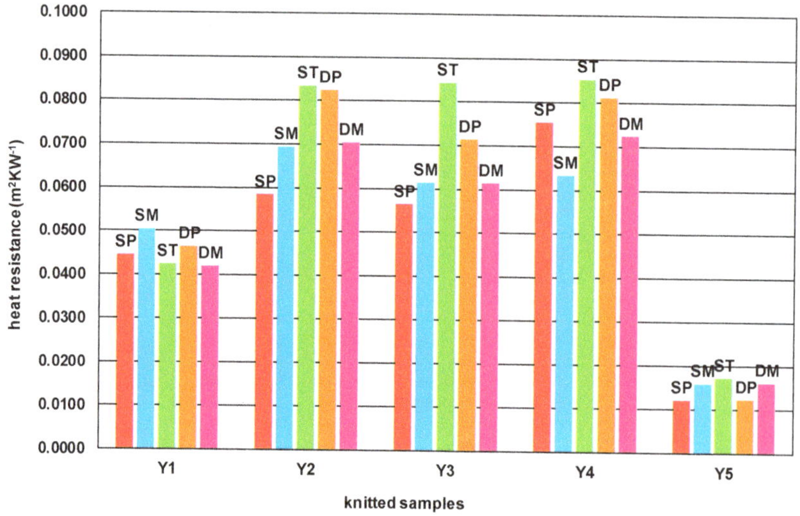

Figure 1. Heat resistance of knitted samples in different single and double structures.

Table 6 and Figure 2 show the results of the air permeability measurements of the knitted samples.

Table 6. Air permeability of knitted samples.

Yarn/Structure		Air Permeability ($l\ min^{-1}$)				
		SP	SM	ST	DP	DM
		Single Plain	Single Miss	Single Tuck	Double Plain	Double Milano rib
Y1	100% Cotton	140.12	119.72	206.14	82.44	77.14
Y2	100% Wool	100.40	88.20	193.30	70.92	66.20
Y3	100% PAN	78.76	64.68	110.66	56.78	51.84
Y4	Wool/PAN	126.18	108.74	194.86	64.90	74.60
Y5	100% PA	145.88	130.18	146.28	93.92	91.02

As can be seen in Figure 2, single-knit structures show much greater air permeability compared to double-knit structures, which is due to their smaller thickness. The tuck structure stands out as having the lowest horizontal and vertical density among the single structures, despite having the greatest thickness, and the greatest porosity due to the enclosed tuck loops.

The air permeability results of knitted fabrics made of worsted yarns are not directly comparable with the air permeability results of filament yarns (Figure 2) because of the different measurement conditions. However, all yarns show a similar relationship between the air permeability of different structures, with the exception that the single tuck made of PA yarn is less noticeable compared to other structures. Presumably, the reason for this is the high porosity of all structures knitted from PA filament.

Table 7 and Figure 3 show the results of the water vapour resistance measurements of the knitted samples. The measured values differ significantly and range from 0.5570 to 5.7542 m^2 Pa W^{-1}. As can be seen in Figure 3, knitted samples made of PA filament yarn show the lowest water vapour resistance. There are also differences in water vapour resistance between different structures. As expected, double structures show higher water vapour resistance than single structures, since their compact structure blocks a certain amount of water vapour that is supposed to pass through the material. Considering the obtained results, the single structures of 100% PA (especially when forming single plain, miss or tuck structures) are considered desirable for wearing during intense activities where the human body produces significantly more sweat. Such structures facilitate the perception of optimal thermophysiological comfort, as the sweat produced can be more easily removed from the body. This result is important for the behaviour of structures and is to be considered when developing material for a particular application.

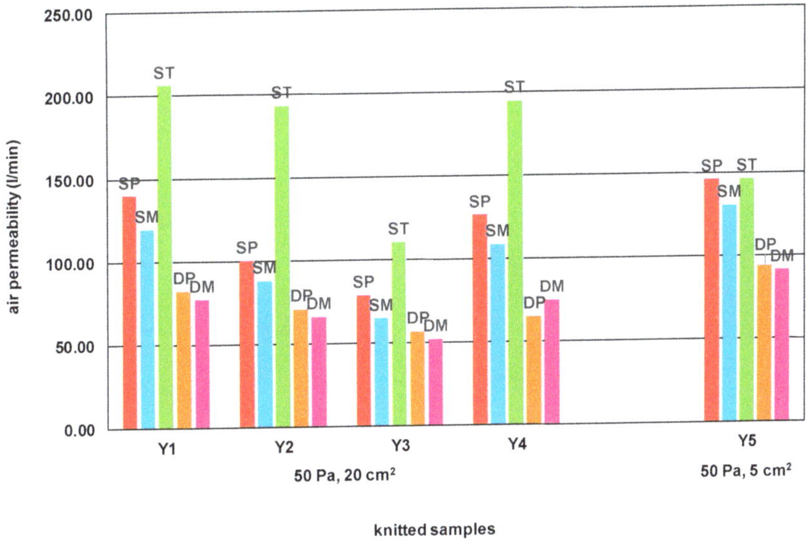

Figure 2. Air permeability of knitted samples in different single and double structures.

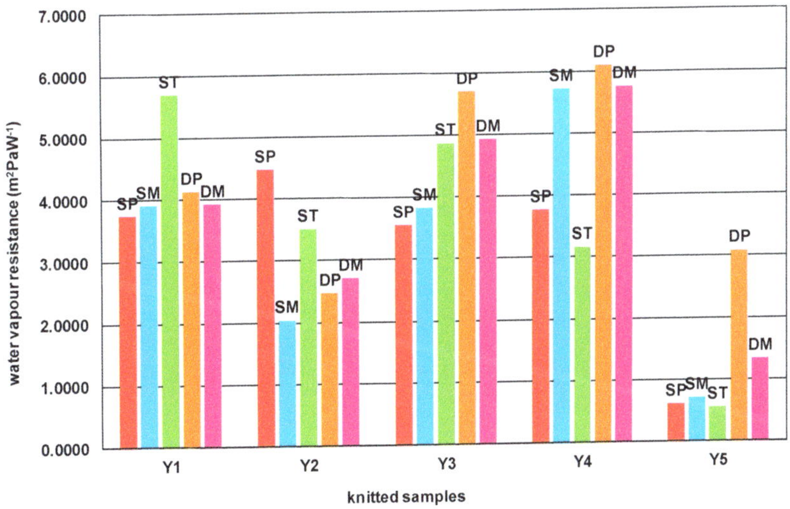

Figure 3. Water vapour resistance of knitted samples in different single and double structures.

Table 7. Water vapour resistance of knitted samples.

Yarn/Structure		Water Vapour Resistance (m^2 Pa W^{-1})				
		SP	SM	ST	DP	DM
		Single Plain	Single Miss	Single Tuck	Double Plain	Double Milano rib
Y1	100% Cotton	3.7448	3.9135	5.6966	4.1322	3.9222
Y2	100% Wool	4.4758	2.0459	3.5071	2.4678	2.7070
Y3	100% PAN	3.5522	3.8274	4.8722	5.6936	4.9208
Y4	Wool/PAN	3.7760	5.7246	3.1658	6.0944	5.7542
Y5	100% PA	0.6112	0.7075	0.5570	3.0855	1.3329

5. Discussion

For the study, three single and two double basic structures were made from different yarns used commercially for sports and casual wear. The selected structures are suitable for indoor and outdoor activities, active sports and après-sports and leisure. The single plain structure is the simplest and most economical weft structure to produce and has maximum covering power. It usually has a recovery of 40% in width after stretching [44]. Due to its low-cost production, single jersey and its variants account for a large proportion of industrial production. All of the single plain knitted fabrics (made from all the selected yarns) had the least thickness. Compared to the other structures, their heat resistance was lowest when knitted from wool, PAN and PA yarns, and moderate when knitted from cotton and wool/PAN blend. All single plain knits have high air permeability due to their low thickness and rather open structure compared to single miss and double structures, which tend to be compact due to the underlying floats in single structure and folded wales in double structures. Air permeability is greater for single structures made of cotton and wool/PAN blend than for 100% wool and 100% PAN.

Missed structures show texture. The held loops give the impression of holes, and the floats produce a range of texture effects depending on their arrangement [45]. Structures containing floats tend to show dull horizontal lines. Knitted fabrics in miss-structures are narrower than equivalent all-knit fabrics because the wales are pulled closer together by the floats, which reduces the elasticity in width and improves the stability of the knitted fabric [44]. The miss-structure used in this study, also called weft locknit [44,46], only has short floats on the reverse side, but as can be seen in Table 1, it gives the impression of an inlaid or woven structure. All the single miss knits (from all the selected yarns) had the second lowest thickness. Compared to the other structures, their heat resistance was moderate when knitted from wool, PAN, wool/PAN blend and PA yarns and greatest when knitted from cotton yarn. All plain miss knits have lower air permeability compared to plain and tuck single structures and greater air permeability than double structures. This may be due to their relatively close structure compared to the other two single structures studied and their small thickness compared to double structures.

Tuck structures increase fabric width because tuck loops pull the held loops downward, spreading them outward and providing the additional yarn for width stretch. Fabric distortion and three-dimensional relief are caused by the accumulation of tuck loops, the displacement of wales and by the different number of tuck and knit loops per wale [44]. The most commonly encountered textured tuck structures have a cellular appearance on the reverse side of the fabric and a small piqué effect on the front. Tuck structures are usually flexible and have a bubbly surface [45]. The tuck structure used in this study was single piqué, also called Lacoste or Fred Perry [44]. Pique knit fabric has raised yarns that form various diamond-like shapes. This typically gives the fabric more body, making it perfect for polo-shirts, shirt dresses and more structured knit fashion apparel [47]. The examined single tuck structure had the greatest thickness of all the single structures due to the convexity of the loop arcs. It also had the greatest thermal resistance of all structures except those made of cotton. This could be due to the air trapped between the loops in the knitted structure. The tuck structure studied also showed the highest air permeability due to the porosity associated with the tuck loops and the low fabric density.

The double plain structure, or as it is also called, 1 × 1 rib, has a vertical cord appearance because the face loop wales tend to move over and in front of the reverse loop wales. Relaxed 1 × 1 rib is theoretically twice as thick and half as wide as an equivalent plain fabric, but has twice as much recoverable stretch in width and a heavier structure. The structure is elastic, form-fitting, and retains heat better than plain structures [44]. In this study, the double plain structure or 1 × 1 rib had the greatest thickness of all the structures due to the overlapping of the front and reverse loop wales. The heat resistance is highest in the knits of wool, PA and wool/PAN blend yarn. It is moderate in the knits made of cotton yarn and lowest in the knits made of PA filament. The double plain structure is less permeable to air compared to the three single structures analysed, which is due to its thickness.

Milano rib is a non-jacquard double structure [44]. The fabric shows courses of small loops on both sides of the fabric, alternating with courses of slightly larger loops, with the upper arcs of adjacent loops perfectly aligned as in 1 × 1 rib fabric. The structure shows a higher horizontal density and a lower vertical density than the other double structures studied, e.g., double plain structure or 1 × 1 rib. The fabric also has a lower thickness than the double plain structure. The heat resistance is lower than with a double plain structure for all the materials studied. Milano rib structures made of wool, PAN and wool/PAN blend have the highest heat resistance, while the one made of PA has the lowest. At the same time, Milano rib structures have the lowest air permeability of all the structures studied, except for the one made of wool/PAN blend, which is slightly less permeable than the double plain structure.

As we can conclude from the discussion above, the structure of the knitted fabrics (single plain, single miss, single tuck, double plain, and double Milano rib) certainly has an impact on comfort, heat resistance and air permeability. However, the structure itself is impossible to quantify and process statistically.

Certain elements of knitted fabric (horizontal density, vertical density, mass per unit area and thickness) can be quantified (Table 4). When the results of horizontal density and vertical density are statistically analysed, the correlation of the variables of horizontal density and vertical density with heat resistance is not statistically significant. Similar results are obtained from the correlation analysis of horizontal density, vertical density and air permeability.

However, the effect of thickness on heat resistance is moderate to strongly correlated and proportional. As the thickness of the knitted fabric increases, the heat resistance increases. Mass per unit area is strongly correlated with heat resistance, more so than thickness (Table 8). It is enough to look at Figure 1 to see that the properties of the yarns have a noticeable effect on heat resistance. The correlation analysis confirms that the correlation coefficient between the linear density of the yarns (Table 2) and the heat resistance of the different knitted fabrics (Table 5) is high for each knitted structure (Table 8).

Table 8. The correlation coefficients of heat resistance and selected yarn and fabric properties.

Knitted Sample Structure	Correlation Coefficients						
	Linear Density	Yarn Hairiness (Length of Protruding Fibres)				Thickness	Mass per Unit Area
		2 mm	4 mm	6 mm	8 mm		
Single plain	0.89241	0.98688	0.81619	0.63217	0.48558	0.63803	0.81236
Single miss	0.89081	0.96882	0.83147	0.66089	0.53297	0.79345	0.92944
Single tuck	0.99334	0.95358	0.95406	0.82584	0.70488	0.98268	0.82534
Double plain	0.95409	0.98832	0.88101	0.70842	0.56797	0.78978	0.90537
Double Milano rib	0.95421	0.99198	0.87418	0.69593	0.55038	0.70619	0.90318

In studies, the little-observed factor of the hairiness of the yarn (Table 3) shows a significant impact on heat resistance. Interestingly, in addition to the number of protruding fibers, the length of the protruding fibers also has a statistically decisive influence (Table 8). The number of the shortest protruding fibers shows the strongest correlation with heat resistance. To increase the heat resistance, it is important to increase the number of protruding fibers.

The influence of thickness on the air permeability of knitted fabrics is medium to high and inversely proportional. As the thickness of the knitted fabric increases, the air permeability decreases, especially in double structures. The exception is a single tuck fabric, where the greatest porosity is influenced more by the shape of tuck loops than the thickness (Table 9).

Table 9. The correlation coefficients between air permeability and selected yarn and fabric properties.

Knitted Sample Structure	Correlation Coefficients						
	Linear Density	Yarn Hairiness (Length of Protruding Fibres)				Thickness	Mass per Unit Area
		2 mm	4 mm	6 mm	8 mm		
Single plain	−0.80487	−0.58298	−0.89139	−0.93040	−0.91987	−0.50103	−0.44484
Single miss	−0.81007	−0.60602	−0.90414	−0.94283	−0.93351	−0.71800	−0.50944
Single tuck	−0.06720	0.30399	−0.27054	−0.52998	−0.66410	0.14010	0.50699
Double plain	−0.94761	−0.84736	−0.98875	−0.94508	−0.87497	−0.97376	−0.83646
Double Milano rib	−0.81155	−0.67561	−0.90661	−0.92479	−0.90892	−0.93086	−0.76062

It is similar when a correlation of air permeability with mass per unit area is observed. Mass per unit area also has a medium effect on air permeability, which is higher for double-knitted structures (Table 9).

The correlation coefficient between the linear density of the yarn (Table 2) and the air permeability of the different knitted fabrics is high and inversely proportional for each structure, except for the single tuck, where the structure and the shape of the loop are again more important (Table 9).

The hairiness of the yarn (Table 3) also shows a significant influence on air permeability. In addition to the number of protruding fibers, the length of the protruding fibers also has a statistically decisive influence (Table 9), although the influence is inversely proportional. To increase air permeability, it is important to reduce the number of protruding fibers. The hairiness of the yarn is again least important for single tuck fabric where the shape of the tuck loop has the greater influence.

The correlation coefficient between the linear density of the yarn (Table 2) and the water vapour resistance is relatively weak to moderate. Similar correlation coefficients were obtained for yarn hairiness. A positive correlation is observed for both properties of the yarn, i.e., the finer the yarn, the lower will be the water vapor resistance. Lower yarn hairiness, especially of the shortest and most numerous protruding fibers (2 mm) will result in lower water vapour resistance. Moderate to high positive correlation coefficients between the thickness and mass per unit area and water vapour resistance were obtained for all knitted structures. Higher thickness or mass per unit area will result in higher water vapour resistance. Again, the shape of the tuck loop is more important than the yarn parameters or the thickness of knitted material (Table 10).

Table 10. The correlation coefficients between water vapour resistance and selected yarn and fabric properties.

Knitted Sample Structure	Correlation Coefficients						
	Linear Density	Yarn Hairiness (Length of Protruding Fibres)				Thickness	Mass per Unit Area
		2 mm	4 mm	6 mm	8 mm		
Single plain	0.74784	0.90664	0.67931	0.49148	0.36833	0.79575	0.95664
Single miss	0.56481	0.72480	0.55035	0.43234	0.33876	0.67858	0.73407
Single tuck	0.37182	0.53672	0.43834	0.39050	0.37734	0.34807	0.70002
Double plain	0.47280	0.45394	0.55580	0.57826	0.55864	0.6553	0.54081
Double Milano rib	0.68288	0.761314	0.70517	0.62598	0.54825	0.7408	0.77297

At the same time, air permeability and heat resistance are inversely proportional and moderately strongly correlated, except for the single tuck fabric, where the structure and shape of the tuck loop is more important (Table 11).

Table 11. Correlation coefficient of air permeability and heat resistance for different structures.

Knitted Sample Structure	Correlation Coefficient
Single plain	−0.51528
Single miss	−0.69783
Single tuck	0.02073
Double plain	−0.87624
Double Milano rib	−0.71627

From the correlation analysis, it can be concluded that yarn linear density, yarn short fibers hairiness, and mass per unit area of knitted fabric have the greatest impact on heat resistance (Table 8). The relationship between heat resistance and yarn hairiness and mass per unit area can be expressed by the multiple linear regression equation (Table 12). The share of variance of the dependent variable (heat resistance) interpreted by the model is very high (R^2) for all knitted structures. The model is therefore representative. The statistical significance of the model is confirmed by *p*-value, significance $F < 0.05$ for all regression models.

Table 12. Relationship between heat resistance (R), yarn hairiness (X_1) and mass per unit area (X_2) for different knitted structures.

Knitted Sample Structure	Multiple Linear Regression Equations	R^2	Significance F
Single plain	$R = 0.0120697 + 3.24855 \times 10^{-5} X_1 + 9.05167 \times 10^{-7} X_2$	$R^2 = 0.97393$	Signif. F = 0.02606
Single miss	$R = 0.0018770 + 2.00342 \times 10^{-5} X_1 + 1.09454 \times 10^{-4} X_2$	$R^2 = 0.96437$	Signif. F = 0.03562
Single tuck	$R = 0.0699356 + 7.40042 \times 10^{-5} X_1 - 3.70694 \times 10^{-4} X_2$	$R^2 = 0.97055$	Signif. F = 0.02944
Double plain	$R = 0.0201838 + 4.54164 \times 10^{-5} X_1 - 3.27122 \times 10^{-5} X_2$	$R^2 = 0.97837$	Signif. F = 0.02162
Double Milano rib	$R = 0.0228920 + 3.71459 \times 10^{-5} X_1 - 3.19884 \times 10^{-5} X_2$	$R^2 = 0.98672$	Signif. F = 0.01327

The yarn linear density, the yarn hairiness of the longer protruding fibers, and the thickness of the knitted fabric have the greatest impact on air permeability, except in the case of tuck structures, where the shape of the tuck loop allows higher air permeability and all other parameters become less significant (Table 9). However, a statistically significant model can be constructed only for the double plain structure: $Ap = 95.55241537 - 0.280261395 X_1 - 0.052690916 X_2$, $R^2 = 0.96335$, significance F= 0.03664, where air permeability (Ap) is the dependent variable, X_1 is the yarn linear density, and X_2 is the yarn hairiness.

A different result was obtained for water vapour resistance, on which the mass and thickness of the knitted fabric have the strongest influence. A statistically significant model can be constructed only for the single plain structure: $Ret = -0.534779 - 3.13649 X_1 + 0.032617 X_2$,

$R^2 = 0.96147$, significance F= 0.03852, where water vapor resistance (Ret) is the dependent variable, X_1 is the thickness, and X_2 is the mass per unit area.

6. Conclusions

It can be stated that the single tuck structure, i.e., single piqué, made of wool, PAN or wool/ PAN blend, has the highest heat resistance, medium water vapour resistance, and at the same time the highest air permeability, which favours it for use for sportswear when body heat should be preserved, as well as après-sports and casual wear because of the comfort it could provide. Along with the highest hairiness of these yarns, a comfortable feel can be expected, but at the same time, pilling can occur. Single piqué made from 100% cotton yarn has a lower heat resistance, which qualifies it for summer clothing. The disadvantage of using this yarn could be the high twist and the resulting skewness of the structure.

The double plain knitted structure has the second highest heat resistance for all materials and the highest water vapour resistance. At the same time, its air permeability is low compared to other fabrics analysed, which, combined with the highest thickness of all the structures and materials studied, could make it the most suitable knitted fabric for outdoor use. The double plain structure is also very stretchy, which ensures a good fit to the body.

The single plain structure has the lowest heat resistance and the lowest water vapour resistance, which qualifies this structure for rapid loss of body heat, i.e., for clothing for active training. This structure has moderate air permeability, and the combination of a single plain and a single tuck structure would increase air permeability and give an ideal structure of knitted fabric for active training and rapid body heat loss.

Statistical analysis shows that when designing knitwear, it is important to consider not only the structure itself but also the raw material composition of the yarn, which is strongly correlated with thermal resistance and air permeability. A less considered factor, the hairiness of the yarn, correlates strongly with thermal resistance, water vapour resistance, and air permeability. To increase thermal resistance, it is important to increase the number of protruding fibers. Conversely, to increase air permeability, it is important to decrease the number of protruding fibers.

All the knitted structures studied were least air permeable when made from 100% PAN, which could affect their comfort. All structures made from PA filament showed significantly lower heat resistance, the lowest water vapour resistance, and high air permeability. The studied PA filament yarn is very stretchy and has high tenacity. All the studied structures are thinnest when made from the PA filament, which contributes to their lightness. It can be concluded that the PA filament is best suited for the use of sportswear in hot indoor or outdoor climates.

Author Contributions: Conceptualization, I.S.Č. and A.P.Č.; methodology, I.S.Č., V.M.P.M. and A.P.Č.; validation, I.S.Č., V.M.P.M. and A.P.Č.; formal analysis, I.S.Č. and V.M.P.M.; investigation, I.S.Č., V.M.P.M., Ž.P. and A.P.Č.; resources, A.P.Č.; data curation, Ž.P.; writing—original draft preparation, I.S.Č., V.M.P.M. and A.P.Č.; writing—review and editing, I.S.Č., V.M.P.M. and A.P.Č.; project administration, I.S.Č.; funding acquisition, I.S.Č. All authors have read and agreed to the published version of the manuscript.

Funding: This work has been fully supported by Croatian Science Foundation under the project IP-2020-02-5041 Textile Materials for Enhanced Comfort in Sports—TEMPO.

Institutional Review Board Statement: Not applicable.

Informed Consent Statement: Not applicable.

Data Availability Statement: Data is contained within the article.

Acknowledgments: . This work has been fully supported by Croatian Science Foundation under the project IP-2020-02-5041 Textile Materials for Enhanced Comfort in Sports.

Conflicts of Interest: The authors declare no conflict of interest.

References

1. Black, S. *Knitting: Fashion, Industry, Craft*; V&A Publishing: London, UK, 2012; p. 83.
2. Sportswear, Cambridge Dictionary. Available online: https://dictionary.cambridge.org/dictionary/english/sportswear (accessed on 1 June 2021).
3. Sportswear, Merriam-Webster. Available online: https://www.merriam-webster.com/dictionary/sportswear (accessed on 1 June 2021).
4. Athleisure, Merriam-Webster. Available online: https://www.merriam-webster.com/dictionary/athleisure (accessed on 1 June 2021).
5. Athleisure, Cambridge Dictionary. Available online: https://dictionary.cambridge.org/dictionary/english/athleisure (accessed on 1 June 2021).
6. Shishoo, R. (Ed.) *Textiles in Sports*; Woodhead Publishing: Cambridge, UK, 2005.
7. Market for Sports Apparel Is Expected to Grow. Available online: https://textilefocus.com/market-sports-apparel-expected-grow (accessed on 3 June 2021).
8. How Sportswear Became High Fashion, Origin. Available online: https://www.originoutside.com/insights/how-sportswear-became-high-fashion (accessed on 4 June 2021).
9. Fourt, L.; Hollies, N.R.S. *Clothing: Comfort and Functions*; Marcel Decker Inc.: New York, NY, USA, 1970.
10. Slater, K. *Human Comfort*; Charles, C., Ed.; Thomas Publisher: Springfield, IL, USA, 1985; p. 4.
11. *Ansi/Ashrae 55 Thermal Environmental Conditions for Human Occupancy*; ASHRAE Standards Committee, The ASHRAE Board of Directors, and the American National Standards Institute: Atlanta, GA, USA, 2017; Available online: https://www.ashrae.org (accessed on 1 June 2021).
12. Peng, Y.; Cui, Y. Advanced Textiles for Personal Thermal Management and Energy. *Joule* **2020**, *4*, 724–742. [CrossRef]
13. Fan, J.; Hunter, L. *Engineering Apparel Fabrics and Garments*; Woodhead Publishing: Cambridge, UK, 2009; p. 212.
14. Shishoo, R. (Ed.) *Textiles for Sportswear*; Woodhead Publishing: Cambridge, UK, 2015; p. 199.
15. Oğlakcioğlu, N.; Marmarali, A. Thermal comfort properties of some knitted structures. *Fibres Text. East. Eur.* **2007**, *15*, 94–96.
16. van Amber, R.R.; Wilson, C.A.; Laing, R.; Lowe, B.M. Thermal and moisture transfer properties of sock fabrics differing in fiber type, yarn, and fabric structure. *Text. Res. J.* **2015**, *85*, 1269–1280. [CrossRef]
17. Uçar, N.; Yilmaz, T. Thermal properties of 1 × 1, 2 × 2, 3 × 3 rib knit fabrics. *Fibres Text. East. Eur.* **2004**, *12*, 34–38.
18. Erdumlu, N.; Saricam, C. Investigating the effect of some fabric parameters on the thermal comfort properties of flat knitted acrylic fabrics for winter wear. *Text. Res. J.* **2017**, *87*, 1349–1359. [CrossRef]
19. Senthilkumar, P.; Suganthi, T. Influence of tuck stitch in course direction on thermal comfort characteristics of layered knitted fabrics. *Indian J. Fibre Text. Res. (IJFTR)* **2019**, *44*, 163–172.
20. Yang, Y.; Yu, X.; Chen, L.; Zhang, P. Effect of knitting structure and yarn composition on thermal comfort properties of bi-layer knitted fabrics. *Text. Res. J.* **2021**, *91*, 3–17. [CrossRef]
21. Özdil, N.; Marmarali, A.; Kretzschmar, S.D. Effect of yarn properties on thermal comfort of knitted fabrics. *Int. J. Therm. Sci.* **2007**, *46*, 1318–1322. [CrossRef]
22. Prakash, C.; Ramakrishnan, G.; Koushik, C.V. The effect of loop length and yarn linear density on the thermal properties of bamboo knitted fabric. *Autex Res. J.* **2011**, *11*, 102–105.
23. Mishra, R.; Jamshaid, H.; Siddique Yosfani, S.H.; Hussain, S.; Nadeem, M.; Petru, M.; Tichy, M.; Muller, M. Thermo physiological comfort of single jersey knitted fabric derivatives. *Fash. Text.* **2021**, *8*, 1–22. [CrossRef]
24. Oğlakcioğlu, N.; Çay, A.; Marmarali, A.; Mert, E. Characteristics of knitted structures produced by engineered polyester yarns and their blends in terms of thermal comfort. *J. Eng. Fibers Fabr.* **2015**, *10*, 32–41. [CrossRef]
25. Sampath, M.B.; Arputharaj, A.; Senthilkumar, M.; Nalankilli, G. Analysis of thermal comfort characteristics of moisture management finished knitted fabrics made from different yarns. *J. Ind. Text.* **2011**, *42*, 19–33. [CrossRef]
26. Gericke, A.; van der Pol, J. A comparative study of regenerated bamboo. cotton and viscose rayon fabrics. Part 1: Selected comfort properties. *J. Fam. Ecol. Consum. Sci.* **2010**, *38*, 63–73. [CrossRef]
27. Gun, A.D. Dimensional. physical and thermal comfort properties of plain knitted fabrics made from modal viscose yarns having microfibers and conventional fibers. *Fibers Polym.* **2011**, *12*, 258–267. [CrossRef]
28. Lizák, P.; Legerska, J.; Militký, J.; Mojumdar, S.C. Thermal transport characteristics of polypropylene fiber-based knitted fabrics. *J. Therm. Anal. Calorim.* **2012**, *108*, 837–841. [CrossRef]
29. Majumdar, A.; Mukhopadhyay, S.; Yadav, R. Thermal properties of knitted fabrics made from cotton and regenerated bamboo cellulosic fibres. *Int. J. Therm. Sci.* **2010**, *49*, 2042–2048. [CrossRef]

30. Bivainytė, A.; Mikučionienė, D.; Kerpauskas, P. Investigation on thermal properties of double-layered weft knitted fabrics. *Mater. Sci.* **2012**, *18*, 167–171. [CrossRef]
31. Jhanji, Y.; Gupta, D.; Kothari, V.K. Effect of fibre, yarn and fabric variables on heat and moisture transport properties of plated knit. *Indian J. Fibre Text. Res.* **2017**, *42*, 255–256.
32. Kaplan, S.; Yilmaz, B. Thermal Comfort Performances of Double-face Knitted Insulation Fabrics. *Fibers Polym.* **2022**, *23*, 537–545. [CrossRef]
33. Kumar, S.; Boominathan, S.H.; Kirubakar Raj, D.V. Comparative Analysis on Thermo-physiological Behavior of Eri Silk, Wool and Bamboo Knitted Fabrics Toward Sportswear. *J. Nat. Fibers* **2021**, *18*, 1–12. [CrossRef]
34. *ISO 2060*; Textiles—Yarn from Packages—Determination of Linear Density (Mass per Unit Length) by the Skein Method. International Organization for Standardization: Genève, Switzerland, 1994; p. 13.
35. *ISO 17202*; Textiles—Determination of Twist in Single Spun Yarns—Untwist/Retwist Method. International Organization for Standardization: Genève, Switzerland, 2002; p. 10.
36. *ISO 2062*; Textiles—Yarns from Packages—Determination of Single-End Breaking Force and Elongation at Break Using Constant Rate of Extension (CRE) Tester. International Organization for Standardization: Genève, Switzerland, 2009; p. 10.
37. *ISO 5084*; Textiles—Determination of Thickness of Textiles and Textile Products. International Organization for Standardization: Genève, Switzerland, 1996; p. 5.
38. Salopek Čubrić, I.; Skenderi, Z. Effect of finishing treatments on heat resistance of one- and two-layered fabrics. *Fibers Polym.* **2014**, *15*, 1635–1640. [CrossRef]
39. Potočić Matković, V.M.; Salopek Čubrić, I.; Skenderi, Z. Thermal resistance of polyurethane-coated knitted fabrics before and after weathering. *Text. Res. J.* **2015**, *84*, 2015–2025. [CrossRef]
40. Salopek Čubrić, I.; Skenderi, Z. Impact of Cellulose Materials Finishing on Heat and Water Vapor Resistance. *Fibres Text. East. Eur.* **2013**, *21*, 61–66.
41. *ISO 11092*; Textiles—Physiological Effects—Measurement of Thermal and Water-Vapour Resistance under Steady-State Conditions (Sweating Guarded-Hotplate Test). International Organization for Standardization: Genève, Switzerland, 2014; p. 15.
42. *ISO 9237*; Textiles—Determination of the Permeability of Fabrics to Air. International Organization for Standardization: Genève, Switzerland, 1995; p. 5.
43. Engel, F.; Sperlich, B. (Eds.) *Compression Garments in Sports: Athletic Performance and Recovery*; Springer: Berlin/Heidelberg, Germany, 2016; p. 21.
44. Spencer, D.J. *Knitting Technology. A Comprehensive Handbook and Practical Guide*, 3rd ed.; Woodhead Publishing: Cambridge, UK, 2001.
45. Tellier-Loumagne, F. *The Art of Knitting. Inspirational Stitches, Textures and Surfaces*; Thames & Hudson: London, UK, 2005.
46. *ISO 8388*; Knitted Fabrics—Types—Vocabulary. International Organization for Standardization: Genève, Switzerland, 1994; p. 85.
47. Fabric Structures—Single Jersey and Its Derivatives—Part 1, Textile Guide. Available online: https://www.textileguide.info/2019/12/fabric-structures-single-jersey-and-its.html (accessed on 26 June 2021).

materials

MDPI

Article

Analysis of Thermomechanical Properties of Polyethylene with Cement Addition

Adam Gnatowski [1,*], Agnieszka Kijo-Kleczkowska [1], Łukasz Suchecki [1], Paweł Palutkiewicz [1] and Jarosław Krzywański [2]

1 Faculty of Mechanical Engineering and Computer Science, Czestochowa University of Technology, 42-200 Czestochowa, Poland; a.kijo-kleczkowska@pcz.pl (A.K.-K.); lukasz.suchecki@pcz.pl (Ł.S.); pawel.palutkiewicz@pcz.pl (P.P.)
2 Faculty of Science and Technology, Jan Dlugosz University in Czestochowa, 42-200 Czestochowa, Poland; j.krzywanski@ujd.edu.pl
* Correspondence: adam.gnatowski@pcz.pl

Abstract: The paper undertakes preliminary research towards the identification of the use of plastic waste, taking into account the possibility of increasing their mechanical strength and reducing flammability, as well as reducing the emission of harmful compounds to the atmosphere through the addition of cement. This is extremely important not only from the point of view of the wide use of plastic products in the industry, but also their thermal utilization. The present study deals with the aspect of the utilization of waste polyethylene (HDPE) as a matrix in composites with filler in the form of cement at 5 and 10%. The composite samples were prepared by injection molding after the prior proper mixing of the components. Comparative thermomechanical (DSC, tensile strength, DMTA), microstructure and flammability results are presented for HDPE samples and their composites with cement. It was found that the addition of cement as a filler to polyethylene made it possible to obtain composites with good thermomechanical properties.

Keywords: cement; polymer composites; DSC testing; microstructure; combustion; DMTA testing; tensile strength

Citation: Gnatowski, A.; Kijo-Kleczkowska, A.; Suchecki, Ł.; Palutkiewicz, P.; Krzywański, J. Analysis of Thermomechanical Properties of Polyethylene with Cement Addition. *Materials* **2022**, *15*, 1587. https://doi.org/10.3390/ma15041587

Academic Editor: Igor Cretescu

Received: 12 January 2022
Accepted: 18 February 2022
Published: 20 February 2022

1. Introduction

The properties and structure of polymers depend on the conditions of their use, and the addition of fillers modifies the structure of polymeric materials. The development of civilization, on the one hand, requires continuous improvement of industrial technologies for the production of various types of polymer products, but on the other hand, it is associated with an increasing volume of multiple types of waste. Most plastic waste is still landfilled, and only a small amount is recycled or thermally disposed of. Due to the growing problem with this type of waste material all over the world, a way is currently being sought for their effective and possibly the most ecological way of recycling and disposal. That is why it is essential to conduct experimental research to determine the possibilities of rational management of the said waste. The literature has addressed the use of plastic waste, such as in the production of construction materials, conversion into fuel, use in the re-manufacturing of household goods and clothing, and use as a matrix in polymer composites.

Composites are a combination of at least two components (phases) with different properties, formed using different technologies, and the obtained new materials express new properties. In polymer composites, plastics are the matrix, while the other components are the filler. For example, polymers used as a matrix in ceramic-polymer composites have the following functions: allow loads to be transferred to the fibers, give the product the desired shape, determine the thermal and chemical properties and the flammability of composites, have a significant influence on the composite manufacturing methods [1,2].

Manufacturing composites with plastics not only reduces the amount of polymer waste, which is extremely important in terms of environmental protection, but also reduces the need to consume mined fossil resources.

In the literature, researchers undertake studies on the possibility of using various types of waste to produce construction materials. These include demolition materials [3], ceramic hollow bricks [4], sanitary ceramics [5], glass fibers and waste [6], rubber waste [7], cork [8], ashes from biomass co-firing [9], and polymeric materials in concretes and cement mortars [10,11].

Noteworthy are polymer-cement composites (PCC), which are obtained by adding polymer to the concrete mixture in an amount not exceeding 5% by the weight of cement. Smaller amounts of cement are classified as admixtures [12]. Due to chemical reactivity, we distinguish post-mix polymer modifiers, i.e., chemically cured synthetic (epoxy) resins that crosslink simultaneously with cement hydration and pre-mix polymer modifiers, i.e., essentially chemically unreactive polymers (e.g., polyacrylic esters) who acts mainly physiochemically in coalescence form of polymer particles to generating a continuous polymer film. The polymer may be introduced into the concrete mixture in various forms, such as an aqueous dispersion, emulsion, re-dispersible powder (after mixing with water, it again forms a dispersion), aqueous solution, or liquid synthetic resin (binary system resin-hardener) [13]. The addition of polymer positively influences the properties of cement composite. It particularly improves flexural and tensile strength, tightness against liquids and gases (thus material durability), and adhesion to various substrates, including concrete base. These are advantages for polymer-cement concretes and mortars widely used to repair and protect concrete structures, making pavements, e.g., industrial floors and precast elements [14]. Mineral additives with pozzolanic properties, especially fly ash, are now widely used as concrete additives and as the main ingredients of common types of cement [15]. The polymer added to cement causes changes, both in the technological and functional properties of the composite [12]. Most of these properties are improved (tensile and flexural strength, adhesion to the substrate, liquid- and gas-tightness). The beneficial properties of materials with polymer-cement binders ensure that their use is constantly increasing. The main areas of application of this type of material are repair and protection of structures from corrosion, road and bridge surfaces, industrial floors, and prefabricated elements. In paper [16], the authors also presented the results of research on agrocement, showing its good strength and durability.

Paper [17] extensively deals with the types of plastics and methods used to obtain plastic aggregate, methods of evaluating different properties of aggregate and concrete, properties of plastic aggregates, fresh and hardened concrete of cement mortar and concrete in the presence of plastic aggregate, and the use of plastic waste in concrete production.

Wang and Meyer [18] investigated the possibility of using recycled high-impact polystyrene (HIPS) as a substitute for sand in cement mortar. It was shown that the compressive and tensile strength of mortar decreases by replacing sand with HIPS, but the decrease in tensile strength is much less. HIPS makes the mortar more ductile. HIPS reduces dry density, dynamic modulus of elasticity, thermal conductivity, and water vapor permeability, but does not affect freeze-thaw resistance. The use of mortar made with different HIPS contents is promising, mainly due to better thermal insulation.

The authors of the paper [19] presented a study on PET, PP and other similar plastics used in reinforced concrete, mortars, and concrete mixes instead of natural materials such as sand. It was shown that the addition of plastics at 0.1% to 0.5% increases the durability of the mix by up to 14.28%. The tensile strength can be increased by 1.63% up to 23.6% by adding plastic to concrete at 0.5% to 1.5% by volume of the mix. However, some research results indicate that replacing 5% of the sand with plastic can cause a deterioration of up to 8% in flexural strength while replacing 15–50% causes up to an 18% decrease in these parameters. Researchers using SEM (scanning electron microscopy) methodology agree that 60% replacement of sand with plastic results in improved thermal conductivity up to

58%, while subsequent addition of even 1% plastic changes the properties of the mixture, reducing its thermal conductivity up to 18%.

According to [20], it is possible to manufacture plastic cement from polyethylene waste and Portland cement, in proportions of 60% and 40%. It was found to reduce its density, increase plasticity, and improve workability, i.e., produce lightweight materials. The density increases as the proportion of waste increases to 30% and then gradually decreases. The maximum density is 1.972 gm/cm^3, being lower than the density of cement mortar from sand and Portland cement. The moisture content of the plastic cement produced in this study ranged from 10.5% to 23.4% for the product immersed in water for seven days. However, for products immersed for 28 days, the moisture content decreased to levels in the range of 3.6% to 11.6%. The best moisture percentage was equal to 3.6 and 3.79% for products with 25% and 30% waste polyethylene, respectively. The best compressive strength for the product was found in the mixture of 25%, 30%, and 35% polyethylene. Their yield strengths are 971, 915, and 945 N for seven days of immersion in water, and 2352 for the 25% blend and 1271 N 30% blend, respectively, after 28 days of immersion. Products containing 25% to 30% waste polyethylene have good workability, allowing for trouble-free hole drilling. Based on the research, it has been shown that the optimum percentage of waste polyethylene is between 25 and 35%, which gives good mix properties.

It is important to more carefully identify cement as an additive to plastics, mainly in terms of economy (for example low cement price compared to other flame retardants) and ecological (cement is a mineral binder).

In the articles dealing with the addition of mineral and inorganic substances to plastics [21–29], it was shown, inter alia, that they can act as reinforcing fillers, processing aids, impact modifiers, and they are used to increase the polymeric material's modulus of elasticity and stiffness. In addition, flame retardants (including cement) used as additives to plastics should slow down the combustion process and smoke emission.

For example, in paper [24] the authors showed that talc enhances the composites' thermal conductivity and thermo-physical properties. No significant effect on the composite tensile yield and fracture strength was observed. The thermal composites' conductivity, thermal diffusivity, and specific density values increased almost linearly.

The authors of paper [25] showed that incorporation of carbon black in HDPE emphasizes the crystallinity and crystallized size more than that of inorganic filler-talc. The tensile strength of the composite decreases with the increase of both types of fillers, and this decrease is explained on the basis of the Nielson model. Flexural strength increased with the increase of carbon black content but decreased with the increase of talc content.

In the paper [26], the authors stated that, poor thermal properties are one of the main limitations in the application of polymers. Thus, HDPE/10% $CaCO_3$ nanocomposite was manufactured. The properties of HDPE and its nanocomposite were examined through differential scanning calorimetery (DSC) and thermo mechanical analyzer (TMA) tests. DSC tests showed that the addition of nano-sized calcium carbonate to HDPE caused an increase in heat capacity, sensible heat, and crystallinity index. The TMA results illustrated an increase in dimensional stability of HDPE as nano-sized calcium carbonate was added to it.

Some authors [27] showed that the presence of $CaCO_3$ does not have a considerable effect on the melting properties of the composites. TGA analyses showed that the thermal stability of the composites is better than the neat HDPE resin.

In paper [28], authors tested high-density polyethylene, which was filled with chalk in various concentrations ranging from 10% to 60% by weight. Ethylene oxide oligomer Mw = 300 was used as a liquid modifier for chalk in the filler amount of 0 ÷ 20 wt %. The mechanical properties of these composites showed that high-density polyethylene filled with chalk have quite high ultimate elongation and impact strength while their elastic modulus and tensile strength are very near to those values for pure high-density polyethylene. On the basis of mechanical properties and microscopical observations, the crack and microcrack damping is attributed to the presence of an ethylene oxide oligomer.

The authors of paper [29] showed that kaolin, mica, talc and calcium carbonate are the most often used as fillers to reduce the production cost and to improve the properties of the thermoplastics, including electrical and thermal conductivity hardness, strength, and rigidity, flexural modulus, dimensional stability, and crystallinity.

In this study, the effect of the cement filler, in the amounts of 5% and 10% (by weight), on the thermomechanical properties of HDPE moldings was analyzed. Thermal tests by DSC and flammability tests by UL 94 V, UL 94 HB methods, microstructure tests, as well as tensile strength tests and DMTA measurements, were performed.

2. Materials and Methods

To carry out the tests, post-production waste in the form of granules of plastic produced by Lotte Chemical with trade name Hivorex 2210J was used, processed in accordance with the manufacturer's data (Melt Flow Index-7.0 g/10 min, flow temperature: 190 °C, density: 0.959 g/cm^3; Vicat softening temperature: 122 °C [30]), to which Heidelberg Cement's SOLID 32.5 Portland cement [31] was added at 5 and 10% (by weight). The polymer material used for the tests was post-production waste obtained from a local manufacturer. In order to prepare the composites, the material in the form of granules was mechanically mixed with cement and introduced into the plasticizing system of the extruder, Rolbach SJ45, by means of which new material was produced using the parameters: cylinder heating temperature of zone I: 145 °C, zone II: 160 °C, zone III: 175 °C; the temperature of the extrusion head: 190 °C; auger rotation: 160 min^{-1}. This material was then milled in a Shini SG-2417-CE slow speed mill. The obtained recyclate was used to produce test samples using a Kraus Maffei KM 65-160C1 injection molding machine. The injection moldings had the form according to the PN-EN ISO 527-1 standard [32].

Optimal properties of the injected samples (both HDPE and composites) were obtained under the following processing conditions: pressure limit in the plasticizing unit: 110 MPa, holding pressure: 60 MPa, holding time: 18 s, cooling time: 10 s, melt temperature: 190 °C, mould temperature: 80 °C.

Thermal DSC and degree of crystallinity tests of the samples were performed using a NETZSCH PC 200 scanning microcalorimeter, according to PN-EN ISO 11357-3:2018-06 [33]. DSC curves were recorded while heating the samples, in nitrogen atmosphere, at a rate of 10 °C/min in the temperature range 40–200 °C. he formulations for DSC studies were cut perpendicular to the flow direction from samples obtained by injection molding to minimize the skin-core effect.

The Proteus software of the device was used to determine the degree of crystallinity; Polyma 214 NETZSCH software was used specifically. This program allowed studying the course of sample melting in the given temperature range and determining the area between the thermographic curve and baseline in the range of the occurrence of endothermic reflex. The mass of the samples was 11 mg. Samples were weighed with a SARTORIUS balance with an accuracy of 0.01 mg, with internal calibration and closed measuring space.

Investigation of the material structure was conducted in transmitted light using an optical microscope, Nikon Eclipse E 200. The samples were microtomed slices with a thickness of 11 ÷ 14 μm that were cut from the core of the injection moulded parts. A rotary microtome Thermo Shandon Finesse Me+ was used for this purpose. Determination of the flammability of the polymeric material and comparison of the flammability of the composites were carried out on a test stand (Figure 1.) using UL 94 V, UL 94 HB methods. In the UL 94HB flammability test, the melt time is measured from the first to the second measurement point (the distance between the measurement points is 75 mm). If the molded part does not ignite within 30 s, the test is aborted. The time it takes for the material to burn within the measuring point is the result of the flammability of the test specimen. The specimens to be tested were made by injection molding. The specimen length was 127 mm, and the width was 12.7 mm. The thickness of the test specimen must not be more than 12.7 mm. Preparation of test specimens: the test material was conditioned for 48 h at 50% humidity and 23 °C.

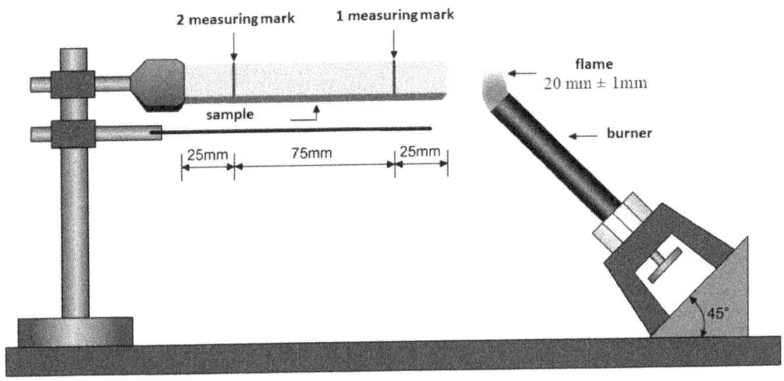

Figure 1. Schematic diagram of the apparatus for determining flammability by burner flame method UL 94HB [34].

In the UL-94-V method (Figure 2) for determining the flammability of the molded piece, a torch with a flame height of 20 ± 1 mm was used. The tested shape was subjected to the flame of the burner twice, each time for 10 s. The method consisted of moving the moving torch under the molding for 10 s. Then the flame was set aside for 7 s, after which it was moved under the molding again for another 10 s. The result of the measurement was the time it would take for the melting-lit shaper to ignite and burn the 0.08 g cotton just below it due to falling polymer droplets. If the tested shaped piece fails to ignite under the influence of the burner flame within a given time, the test is aborted. Table 1 shows the flammability classes defined by the UL 94-V method. The preconditioning of the fittings for the test consists of holding the fittings for 48 h at 50% humidity and 23 °C.

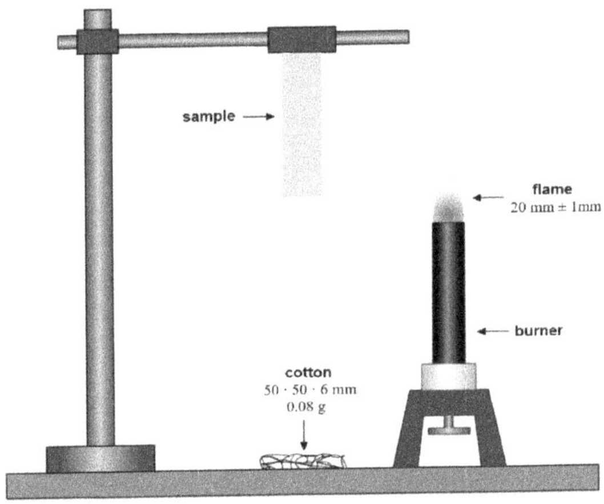

Figure 2. Schematic diagram of the apparatus for determining flammability with a burner by UL 94-V method [35].

The test specimens were made by injection molding. The length of the specimen was 127 mm, the width was 12.7 mm. The thickness of the test specimen must not be more than 12.7 mm.

A Hegewald & Peschke Meß- und Prüftechnik GmbH Inspekt 20 tensile testing machine was used to carry out the static tensile test, in which self-clamping jaws were set up for the static tensile test. The tests were carried out on standardized specimens according to EN ISO 527-2 [36], with dimensions of 150 × 10 × 4 mm. The test parameters were gauge length L_0 = 110 mm, tensile speed v = 50 mm/s (according to PN-EN ISO 527-1 [36]. All samples were conditioned equally at room temperature.

Table 1. Flammability classes as determined by UL 94-V [35].

Flammability Class	V-0	V-1	V-2
Burning time of specimen after burner application (s)	≤10	≤30	≤30
Total burning time (10 applications of flame) (s)	≤50	≤250	≤250
Burning time and glow after the second flame application (s)	≤30	≤60	≤60
Burning droplets appearance	yes	no	yes
Total combustion of specimen	no	no	no

NETZSCH's DMA 242 device was applied for dynamic mechanical properties tests, according to PN-EN ISO 6721-1:2019-07 [37], with a three-point free bending specimen holder in the form of a 50 × 10 × 4 mm beam. On the specimen placed in the holder, through the mandrel were introduced the actions of sinusoidally varying force with a frequency of 1 Hz and 10 Hz with constant amplitude of 120μm, while heating the specimen at a rate of 2 °C/min from a temperature of −150 °C to 100 °C. From the values of force and strain (read by measuring sensors), considering the dimensions of the specimen, the values of the storage modulus E' and the loss modulus E' and the loss tangent tanδ were calculated. The results are presented as a plot of the variation of the conservative modulus E' and the loss tangent tanδ as a function of temperature.

3. Results and Discussion

The results of DSC experiments are shown in Figure 3 and Table 2. Changes in the character of DSC curves and changes in the degree of crystallinity of the tested materials were found.

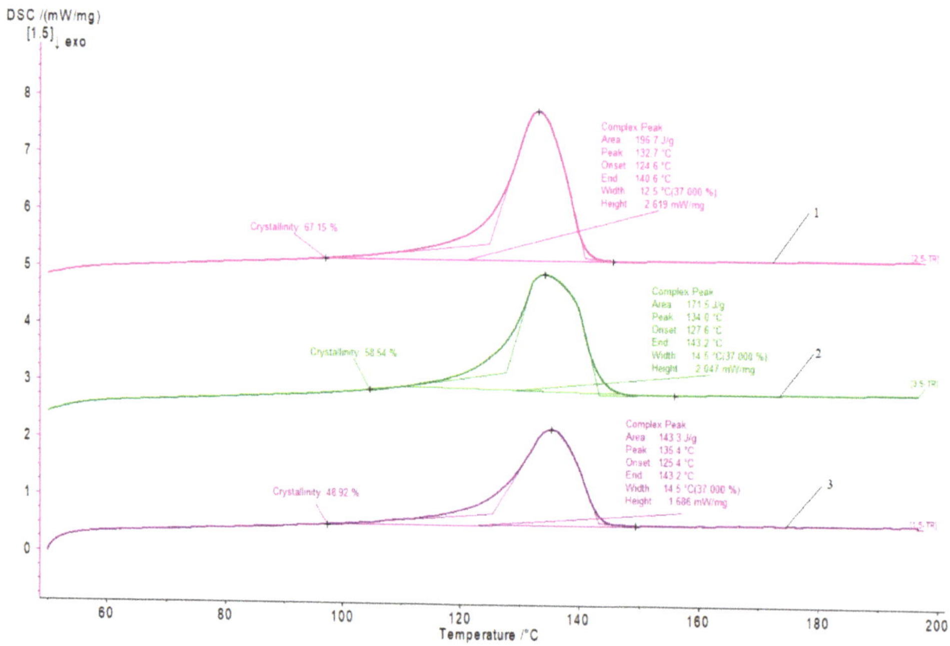

Figure 3. Example of DSC thermograms obtained in the first heating cycle (Netzsh Proteus program): (1) HDPE, (2) HDPE + 5% cement, (3) HDPE + 10% cement.

Table 2. Results of DSC investigations obtained from calculations by Netzch program (averaged values from three measurements).

Material	Enthalpy [J/g]	Melt Temperature Range [°C]	Melt Temperature–Peak Max. [°C]	Crystallinity Degree [%]
HDPE	195.8	124.7 ÷ 140.6	132.7	66.82
HDPE + 5% of cement	170.1	125.4 ÷ 142.8	134.7	58.10
HDPE + 10% of cement	141.7	125.5 ÷ 143.3	135.3	48.36

The DSC investigations prove the decrease in the crystallinity degree of the HDPE/cement composite after the cement addiction. However, as expected, the shape of the thermograms was insignificantly changed. Since HDPE is a semicrystalline polymer, a decrease in the crystalline phase after cement addiction was expected. The reduction in the value of the degree of crystallinity is probably due to the inhibition of the possibility of growth of spherulites after adding cement, which is confirmed by the microstructure tests.

For the composite samples, the temperature of the highest melting rate was not reduced, but the melting temperature range was changed significantly. With a higher cement content, a shift in the melting temperature range of the crystalline phase towards higher values was recorded.

In the case of composite samples, the amount of energy absorbed by the polymer is decreased. The melting enthalpy reached lowest values for the samples of HDPE with 10% of cement addiction.

Other conditions for cooling the polymer by adding cement may cause too small a chain mobility. Therefore they can create areas with maximum ordering, leading to a reduction in the degree of crystallinity. However, slight undercooling favors the formation of more structured composite structures, as evidenced by microstructure studies.

Figure 4 presents pictures of the microstructures for the polyethylene and composites.

Crystallization of the HDPE/cement composites proceeds through nucleation, the thermodynamically stable formation of embryos, and through a process of growth of the crystalline phase of polymer. The crystals grow faster in the pre-embryos created than evenly distributed throughout the amorphous phase. The emergence of any crystal growth process is initiated by the earlier formation of the embryo having a large surface area to its mass. The results of microscopic examination indicate that the nucleation is a heterogeneous composite produced by the presence of cement. The size of the crystal structures in the composite samples is smaller than samples of HDPE. In HDPE recyclate, spherulites have a size from 0.02 to 0.01 μm, while in samples with cement, they have a smaller size, from 0.01 to 0.006 μm. Adding cement blocks the development of crystalline structures. For the structure of a composite, characteristically, the size of the crystal structures decreases while their compaction increases in the case of higher cement content.

In this work, the flammability of polymeric material was determined, and the flammability of composites was compared using UL 94 V, UL 94 HB methods.

Thermal energy introduced to the polymer causes its heating and then melting. Further introduction of thermal energy leads to thermal decomposition processes: depolymerization, degradation, and destruction. High temperature evaporates the products of thermal decomposition (pyrolysis) and ignites the plastic. Sustaining the burning process depends on the amount of heat transferred from the flame to the plastic and the amount of flammable gases and oxygen in the environment. It follows that the combustion process will not be sustained if the decomposition of the plastic requires more heat than is supplied by the flame or if a solid, non-flammable residue covers the surface and isolates the flammable residue from the heat source.

(**a**)

(**b**)

(**c**)

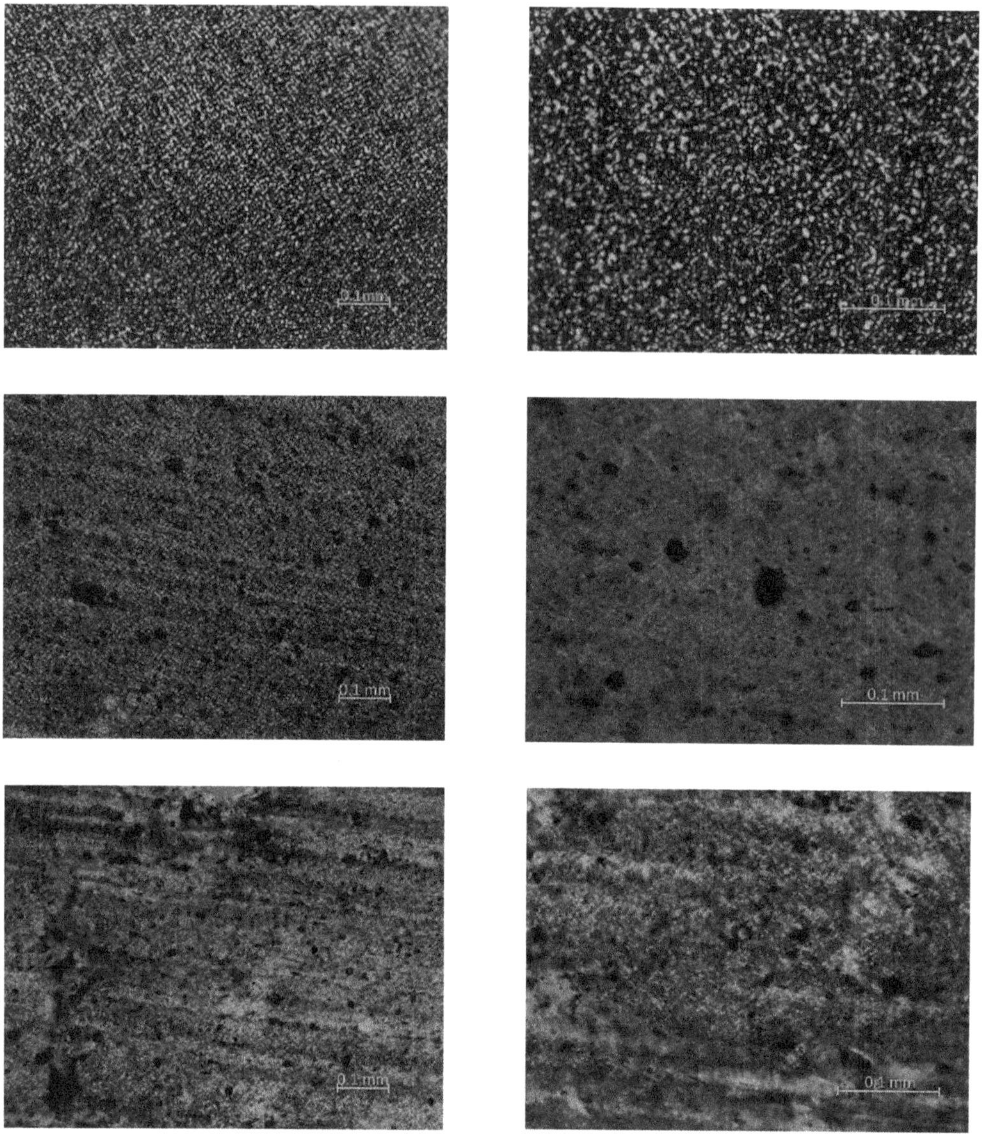

Figure 4. Structures from an optical microscope: (**a**) HDPE, (**b**) HDPE composite with 5% of cement, (**c**) HDPE composite with 10% of cement.

The combustion of gases resulting from the thermal decomposition of a polymeric material is an exothermic reaction. The materials tested burn with a surface diffusion flame. The factor that determines the rate of flame combustion is the heat of combustion. The ignitability of the tested material is influenced by the content of hydrogen atoms in the polymer molecule. The lower the number of hydrogen atoms, the fewer flammable gases are produced during the thermal process. Due to its low molecular weight and high heat of combustion, hydrogen has the highest energy-to-mass ratio, and the explosion force of hydrogen is 2.5 times greater than conventional hydrocarbon fuels. Hydrogen is characterized by a low ignition energy and a high combustion rate. The mechanism of combustion of polymers depends on their structure. The combustion of thermoplastics is diffusion combustion (the speed of mixing the ingredients is much slower than the rate of chemical reactions), with the characteristics of homogeneous (gases) and heterogeneous (solid material and gases) combustion. The combustion process taking place in the zone separating the combustible gas from the air is controlled by the mutual diffusion of both components. The flammability of polymers is reduced by changing the energy balance of their combustion process, which is achieved by introducing additional components into the plastic, the so-called flame retardants. They are divided into two groups:

- additive, which are substances bound to the polymer because of physical type interactions—such as the cement used in the study;
- reactive substances, which join with polymer by chemical bonds.

Boron and molybdenum compounds, silicates, carbonates, and hydrated aluminum trioxide are used as additive flame retardants. On the other hand, reactive flame retardants are some organophosphorus, halogen and nitrogen-containing, organohalogen compounds, and antimony trioxide.

The filler used in the study acted as a flame retardant. In flammability tests, the HDPE was classified into flammability class V-2, while the composite with higher cement content approaches the V-0 flammability class, which proves the effect of the filler as a fire retardant of the composite produced based on recycled materials.

The flammability characteristics recorded in the UL 94 V flammability tests are shown in Table 3.

Table 3. The UL 94 V flammability tests.

Material			
HDPE	burning time: 32 s	occurrence of burning drops: yes	complete combustion of the sample: yes
HDPE + 5% of cement	burning time 23 s	occurrence of burning drops: yes	complete burning of specimen: no
HDPE + 10% of cement	burning time less than 10 s	occurrence of burning drops: yes (sometimes)	complete burning of the sample no

The following flammability characteristics were recorded in the UL 94 HB flammability tests:

(1) for HDPE:
  - burning time: 6 min 05 s (the sample burned with a high flame with a yellow glow and a blue core during the flaming test droplets dripped);

(2) for HDPE + 5% cement:
  - burning time 3 min 05 s (the sample burned with a low flame with a yellow glow and a blue core during the flaming test droplets dripped);

(3) for HDPE + 10% cement:
  - burning time 2 min 20 s (the sample burned with a low flame with a yellow glow and a blue core. As the sample burned, a large piece of material was left pointing downward, which burned before the droplet fell).

Figure 5 illustrates the results of the tensile strength tests. It was found that the addition of cement to the plastic caused a decrease in the elongation of the composites concerning HDPE. The above graphs (Figure 5) show that as the content of cement increases, the tensile strength of the molded parts decreases slightly. Parts from HDPE with a 5% cement content had a 6% lower tensile strength than unfilled parts, and successively, parts with 10% cement were 8.5% lower than the unfilled samples. The samples from HDPE without filler did not break, and those samples had much higher elongation at break than the samples with even fewer amounts of filler (higher than 500%). Higher quantities of cement additions reduced the extension of molded parts from polyethylene. Pieces with a 10% cement content exhibit the smallest elongation value, amounting to 173%.

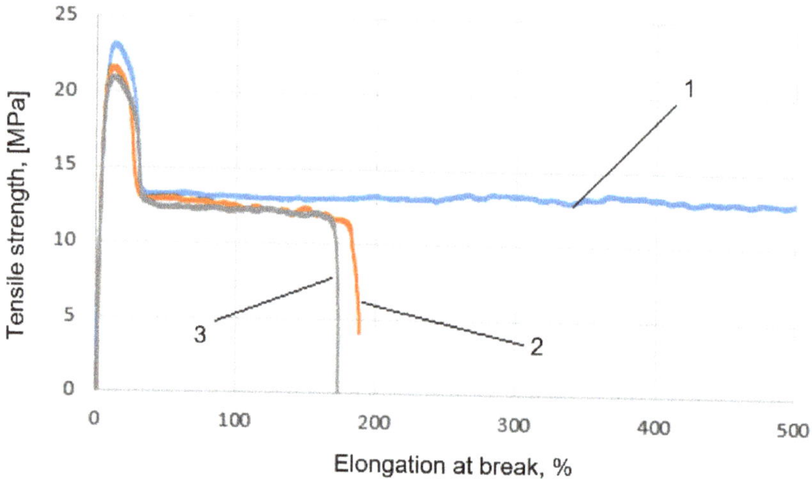

Figure 5. Example of diagram of the relation between tensile strength and elongation for: (1) HDPE, (2) HDPE + 5% cement (3) HDPE + 10% cement.

The results of thermomechanical properties are shown in Figure 6. The study shows that the addition of cement to polyethylene increases the e storage modulus values. The analysis of the recorded values of storage modulus and loss tangent tanδ shows the differences for the materials studied, but not significantly depending on the cement content in the composite. In the region of temperatures lower than the glass transition temperature, polyethylene is in a glassy state, and it is hard and brittle. In the glassy area, the thermal energy is insufficient to overcome the potential barrier rotational movement and displacement of the particle segments. The structure is in a state of thermodynamic imbalance. As the temperature increases, there is a decrease in the modulus value for HDPE. The material is in the glass transition region where the loss tangent reaches a maximum value at the glass transition temperature at a given strain frequency of 1 and 10 Hz. In the glass transition region, Brownian motion in the molecular chain is initiated. The thermal energy becomes comparable to the potential energy barrier for chain rotation. Near the glass transition temperature, the viscoelastic properties change very rapidly both with time and with changing temperature. By analyzing the changes in the values of the tangent of the loss tangent and the effect of cement addition on the material stiffness, differences in the values were found in the temperature range from −150 °C to −20 °C, both at strain frequencies of 1 and 10 Hz and different filler additions. As the temperature increases, the polyethylene changes to a highly elastic state. In the temperature range from 20 °C to 100 °C, the curves of HDPE and composites are similar to each other but differ significantly in values. In each case, the first reduction in E′ value corresponds to relaxation. The increase in the temperature leads to the second, substantial reduction in storage modulus that corresponds to the operation of relaxation of amorphous regions of HDPE. A flat profile zone can be observed for the curves within the range of temperature above vitrification temperature Tg, corresponding to the transition from the vitreous into the highly-plastic state. The increase in the temperature leads to a reduction in the value of module for composites.

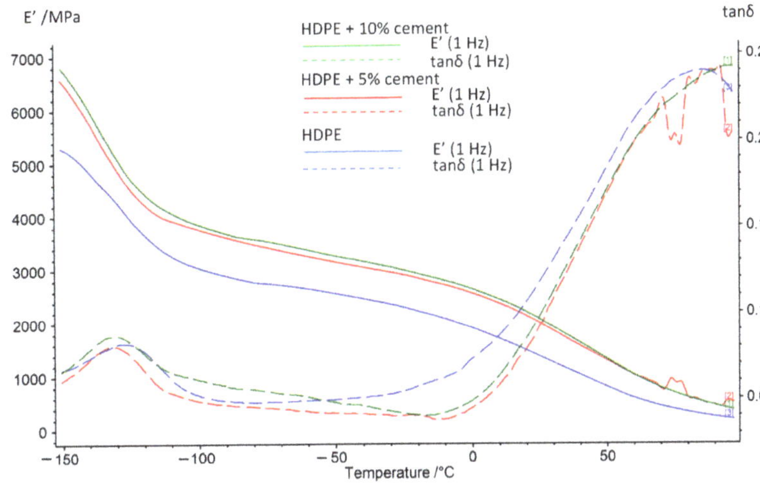

(**a**)

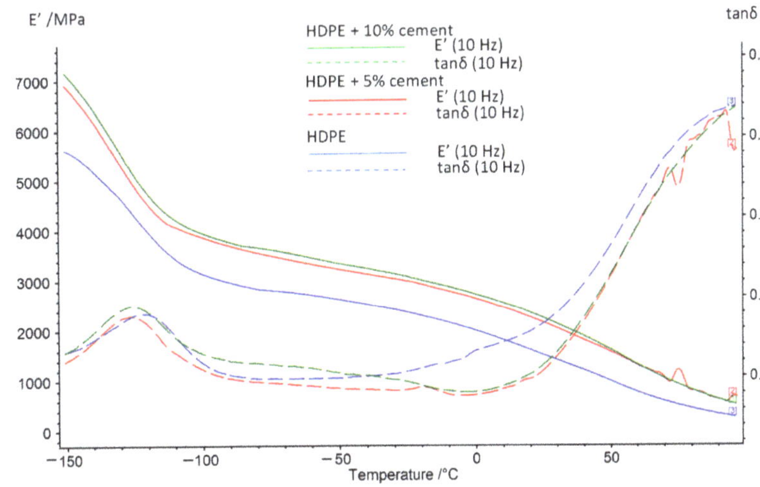

(**b**)

Figure 6. Example of DMTA plot of loss tangent tanδ and storage modulus E′ for (HDPE, HDPE + 5% cement, HDPE + 10% cement) for frequencies: (**a**) 1 Hz, (**b**) 10 Hz.

In the studied temperature range (from −150 to 100 °C), values E′ for the composites containing cement were considerably higher than those of E′ for HDPE without filler. The profile of changes in storage modulus versus temperature is similar for both frequencies studied (1 and 10 Hz). Storage modulus for the sample containing 10 wt% of filler at a temperature of 25 °C is 2070 MPa and is by 35% higher than E′ for HDPE, with a test frequency of 1 Hz. Therefore, the amorphous phase of the composite with 10 wt% of filler shows the strongest reinforcement with the particles of cement, which demonstrates a higher degree of dispersion of filler in the specimen compared other composites. It should be emphasized that the composites containing 5 and 10 wt% of filler are described by the similar and greater value of E′ over the whole temperature range compared to the non-filled polymer. Therefore, filler particles have a substantial reinforcing effect on the material for all compositions studied. The addition of cement causes a shift in a maximum of loss tangent tanδ towards higher temperatures. It reduces with the shift towards lower temperatures with the increase in cement content. These shifts in peaks result from the increasing stabilization of the composites resulting from the immobilization of the fragments of HDPE crystallites by the filler particles [38,39].

4. Conclusions

The addition of cement as a filler to polyethylene made it possible to obtain composites with good thermomechanical properties. The analyzed composites showed a decrease in the values of the degree of crystallinity, differences in the values of the melting point of the crystalline phase, a decrease in the elongation of the samples, and an increase in the values of the storage modulus of the samples compared to HDPE. The microstructure analysis of the composites confirmed the changes in the crystalline phase recorded in thermal studies by differential scanning calorimetry. The size of crystalline structures characteristically decreases for the composite, while their density increases with higher cement content. A significant effect of the filler as a flame retardant of the produced composite based on recycled materials was found in flammability tests.

The research results presented in the paper constitute an experimental cognitive element of the issue and require continuation and development in the aspect of testing composites with broader cement content.

Author Contributions: Conceptualization, A.G., A.K.-K., P.P. and J.K.; methodology, A.G. and A.K.-K.; formal analysis, A.G. and A.K.-K.; investigation, A.G., A.K.-K., Ł.S. and P.P.; resources, A.G., Ł.S. and P.P.; writing—original draft preparation, A.G., A.K.-K., Ł.S. and P.P.; visualization, A.G., A.K.-K. and J.K.; supervision, A.G., A.K.-K. and J.K. All authors have read and agreed to the published version of the manuscript.

Funding: This research received no external funding.

Institutional Review Board Statement: Not applicable.

Informed Consent Statement: Not applicable.

Data Availability Statement: Data is contained within the article.

Conflicts of Interest: The authors declare no conflict of interest.

References

1. Boczkowska, A.; Kapuściński, J.; Puciłowski, K.; Wojciechowski, S. *Composites*; Warsaw University of Technology Publishing House: Warsaw, Poland, 2000.
2. Ceramics—Polymer Composites. Available online: Ztnic.ch.pw.edu.pl/pdf/Kompozyty_ceramika-polimer.pdf (accessed on 16 December 2021).
3. Yuan, F.; Shen, L.; Li, Q. Emergy analysis of the recycling options for construction and demolition waste. *Waste Manag.* **2011**, *31*, 2503–2511. [CrossRef] [PubMed]
4. Binici, H. Effect of crushed ceramic and basaltic pumice as fine aggregates on concrete mortars properties. *Constr. Build. Mater.* **2007**, *21*, 1191–1197. [CrossRef]
5. Jura, J.; Halbiniak, J.; Ulewicz, M. Utilization of waste and sanitary ceramics in cement mortars. *Ceram. Mater.* **2015**, *4*, 438–442.
6. Park, S.; Lee, B. Studies on expansion properties in mortar containing waste glass and fibers. *Cem. Concr. Res.* **2004**, *34*, 1145–1152. [CrossRef]
7. Lia, G.; Garricka, G.; Eggersb, J.; Abadieb, C.; Stubblefieldc, M.; Pang, S. Waste tire fiber modified concrete. *Compos. Part B* **2004**, *35*, 305–312. [CrossRef]
8. Panesar, D.K.; Shindman, B. The mechanical, transport and thermal properties of mortar and concrete containing waste cork. *Cem. Concr. Compos.* **2012**, *34*, 982–992. [CrossRef]
9. Wang, S.; Miller, A.; Llamazos, E.; Fonseca, F.; Baxter, L. Biomass fly ash in concrete: Mixture proportioning and mechanical properties. *Fuel* **2008**, *87*, 365–371. [CrossRef]
10. Ismail, Z.Z.; Al-Hashmi, E.A. Use of waste plastic in concrete mixture as aggregate replacement. *Waste Manag.* **2008**, *28*, 2041. [CrossRef]
11. Jura, J.; Ulewicz, M. Use of the waste materials in cement mortars. *J. Civ. Eng. Environ. Archit.* **2016**, *63*, 247–254.

12. Łukowski, P.; Jaworska, B.; Sokołowska, J.; Kępniak, M. The influence of mineral addition on the properties of polymer-cement composites. In *Concrete and Concrete Structures—Tests*; Warsaw University of Technology Publishing House: Warsaw, Poland, 2015; pp. 19–27.
13. Łukowski, P. *The Role of Polymers in Shaping the Properties of Binders and Polymer-Cement Composites*; Publishing House of the Warsaw University of Technology: Warsaw, Poland, 2008.
14. Czarnecki, L.; Łukowski, P. Polymer-cement concretes. *Cem. Lime Concr.* **2010**, *5*, 243–258.
15. Giergiczny, Z. *Fly Ash in the Composition of Cement and Concreto*; Publishing House of the Silesian University of Technology: Gliwice, Poland, 2013.
16. He, J.; Kawasaki, S.; Achal, V. The Utilization of Agricultural Waste as Agro-Cement in Concrete: A Review. *Sustainability* **2020**, *12*, 6971. [CrossRef]
17. Saikia, N.; Brito, J. Use of plastic waste as aggregate in cement mortar and concrete preparation: A review. *Constr. Build. Mater.* **2012**, *34*, 385–401. [CrossRef]
18. Wang, R.; Meyer, C. Performance of cement mortar made with recycled high impact polystyrene. *Cem. Concr. Compos.* **2012**, *34*, 975–981. [CrossRef]
19. Ibrahim, H.; Alfahdawi, S.A.; Osman, R.; Al-Hadithi, H.A. Utilizing waste plastic polypropylene and polyethylene terephthalate as alternative aggregates to produce lightweight concrete: A review. *J. Eng. Sci. Technol.* **2016**, *11*, 1165–1173.
20. Jassim, A.K. Recycling of Polyethylene Waste to Produce Plastic Cement. *Procedia Manuf.* **2017**, *8*, 635–642. [CrossRef]
21. SpecialChem. The Material Selection Platform. Available online: https://polymer-additives.specialchem.com/selection-guide/flame-retardants-for-fire-proof-plastics (accessed on 23 June 2021).
22. Gnatowski, A.; Jelonkiewicz, J.; Laskowski, L.; Laskowska, M. Influence of the copper-containing SBA-15 silica fillers on the mechanical properties of high density polyethylene. *J. Nanomater.* **2016**, *2016*, 3291719. [CrossRef]
23. Gnatowski, A.; Kijo-Kleczkowska, A.; Golebski, R.; Mirek, K. Analysis of polymeric materials properties changes after addition of reinforcing fibers. *Int. J. Numer. Methods Heat Fluid Flow* **2020**, *6*, 2833–2843. [CrossRef]
24. Mehrjerdi, A.K.; Adl-Zarrabi, B.; Cho, S.W.; Skrifvars, M. Mechanical and thermo-physical properties of high-density polyethylene modified with talc. *J. Appl. Polym. Sci.* **2013**, *4*, 2128–2138. [CrossRef]
25. Parvin, N.; Ullah, S.; Mina, F.; Gafur, A. Structures and mechanical properties of talc and carbon black reinforced high density polyethylene composites: Effects of organic and inorganic fillers. *J. Bangladesh Acad. Sci.* **2013**, *1*, 11–20. [CrossRef]
26. Sahebian, S.; Zebarjad, S.M.; Vahdati Khaki, J.; Sajjadi, S.A. The effect of nano-sized calcium carbonate on thermodynamic parameters of HDPE. *J. Mater. Processing Technol.* **2009**, *3*, 1310–1317. [CrossRef]
27. Elleithy, R.H.; Ali, I.; Ali, M.A.; Al-Zahrani, S.M. High density polyethylene/micro calcium carbonate composites: A study of the morphological, thermal, and viscoelastic properties. *J. Appl. Polym. Sci.* **2010**, *4*, 2413–2421. [CrossRef]
28. Badran, B.M.; Gałeski, A.; Kryszewski, M. High-density polyethylene filled with modified chalk. *J. Appl. Polym. Sci.* **1982**, *10*, 3669–3681. [CrossRef]
29. Tasdemir, M.; Ersoy, S. Mechanical, morphological and thermal properties of HDPE polymer composites filled with talc, calcium carbonate and glass spheres. *Rom. J. Mater.* **2015**, *2*, 147–154.
30. Lotte Chemnical. Available online: www.b2bpolymers.com/TDS/Honam_Hivorex_2210J.pdf (accessed on 2 January 2013).
31. Heidelberg Cement Group. Available online: https://www.gorazdze.pl/pl/Cement-SOLID-32-5 (accessed on 29 March 2021).
32. *Differential Scanning Calorimetry (DSC)—Part 3: Determination of Temperature and Enthalpy of Melting and Crystallization*; PN-EN ISO 11357-3:2018-06; ISO: Warsaw, Poland, 2018.
33. *BS EN 60695-11-10/IEC 60695-11-10*; Method A Horizontal Test Rate of Burning and/or Extent and Time of Burning of Plastics in a Horizontal Position. Impact Tech, Inc.: London, UK, 2014.
34. *BS EN 60695-11-10/IEC 60695-11-10*; Method B Vertical Test Measuring the Comparative Burning Characteristics of Solid Plastics in a Vertical. Impact Tech, Inc.: London, UK, 2014.
35. *PN-EN ISO 527-2*; Determination of Mechanical Properties under Static Stretching. ISO: Warsaw, Poland, 2013.
36. *PN-EN ISO 527-1*; Determination of Mechanical Properties under Static Stretching. ISO: Warsaw, Poland, 2020.
37. *PN-EN ISO 6721-1:2019-07*; Determination of Dynamic Mechanical Properties. ISO: Warsaw, Poland, 2019.
38. Popli, R.; Glotin, M.; Mandelkern, L.; Benson, R.S. Dynamic mechanical studies of α and β relaxations of polyethylenes. *J. Polym. Sci.* **1984**, *3*, 407–448. [CrossRef]
39. Cerrada, M.L.; Benavente, R.; Perez, E. Effect of short glass fiber on structure and mechanical behavior of an ethylene–1-octene copolymer. *Macromol. Chem. Phys.* **2001**, *13*, 2686–2695. [CrossRef]

MDPI
St. Alban-Anlage 66
4052 Basel
Switzerland
www.mdpi.com

Materials Editorial Office
E-mail: materials@mdpi.com
www.mdpi.com/journal/materials

MDPI

www.ingramcontent.com/pod-product-compliance
Lightning Source LLC
Chambersburg PA
CBHW061456210926
51850CB00008B/1730
9783725802883